IET CONTROL, ROBOTICS AND SENSORS SERIES 126

Fault Diagnosis and Fault-Tolerant Control of Robotic and Autonomous Systems

Other volumes in this series:

Volume 8	**A History of Control Engineering, 1800–1930** S. Bennett
Volume 9	**Embedded Mechatronics System Design for Uncertain Environments: Linux-based, MATLAB® xPC Target, PIC, Arduino and Raspberry Pi Approaches** C.S. Chin
Volume 18	**Applied Control Theory, 2nd Edition** J.R. Leigh
Volume 20	**Design of Modern Control Systems** D.J. Bell, P.A. Cook and N. Munro (Editors)
Volume 28	**Robots and Automated Manufacture** J. Billingsley (Editor)
Volume 33	**Temperature Measurement and Control** J.R. Leigh
Volume 34	**Singular Perturbation Methodology in Control Systems** D.S. Naidu
Volume 35	**Implementation of Self-tuning Controllers** K. Warwick (Editor)
Volume 37	**Industrial Digital Control Systems, 2nd Edition** K. Warwick and D. Rees (Editors)
Volume 39	**Continuous Time Controller Design** R. Balasubramanian
Volume 40	**Deterministic Control of Uncertain Systems** A.S.I. Zinober (Editor)
Volume 41	**Computer Control of Real-time Processes** S. Bennett and G.S. Virk (Editors)
Volume 42	**Digital Signal Processing: Principles, devices and applications** N.B. Jones and J.D.McK. Watson (Editors)
Volume 44	**Knowledge-based Systems for Industrial Control** J. McGhee, M.J. Grimble and A. Mowforth (Editors)
Volume 47	**A History of Control Engineering, 1930–1956** S. Bennett
Volume 49	**Polynomial Methods in Optimal Control and Filtering** K.J. Hunt (Editor)
Volume 50	**Programming Industrial Control Systems Using IEC 1131-3** R.W. Lewis
Volume 51	**Advanced Robotics and Intelligent Machines** J.O. Gray and D.G. Caldwell (Editors)
Volume 52	**Adaptive Prediction and Predictive Control** P.P. Kanjilal
Volume 53	**Neural Network Applications in Control** G.W. Irwin, K. Warwick and K.J. Hunt (Editors)
Volume 54	**Control Engineering Solutions: A practical approach** P. Albertos, R. Strietzel and N. Mort (Editors)
Volume 55	**Genetic Algorithms in Engineering Systems** A.M.S. Zalzala and P.J. Fleming (Editors)
Volume 56	**Symbolic Methods in Control System Analysis and Design** N. Munro (Editor)
Volume 57	**Flight Control Systems** R.W. Pratt (Editor)
Volume 58	**Power-plant Control and Instrumentation: The control of boilers and HRSG systems** D. Lindsley
Volume 59	**Modelling Control Systems Using IEC 61499** R. Lewis
Volume 60	**People in Control: Human factors in control room design** J. Noyes and M. Bransby (Editors)
Volume 61	**Nonlinear Predictive Control: Theory and practice** B. Kouvaritakis and M. Cannon (Editors)
Volume 62	**Active Sound and Vibration Control** M.O. Tokhi and S.M. Veres
Volume 63	**Stepping Motors, 4th Edition** P.P. Acarnley
Volume 64	**Control Theory, 2nd Edition** J.R. Leigh
Volume 65	**Modelling and Parameter Estimation of Dynamic Systems** J.R. Raol, G. Girija and J. Singh
Volume 66	**Variable Structure Systems: From principles to implementation** A. Sabanovic, L. Fridman and S. Spurgeon (Editors)
Volume 67	**Motion Vision: Design of compact motion sensing solution for autonomous systems** J. Kolodko and L. Vlacic
Volume 68	**Flexible Robot Manipulators: Modelling, simulation and control** M.O. Tokhi and A.K.M. Azad (Editors)
Volume 69	**Advances in Unmanned Marine Vehicles** G. Roberts and R. Sutton (Editors)
Volume 70	**Intelligent Control Systems Using Computational Intelligence Techniques** A. Ruano (Editor)

Volume 71 **Advances in Cognitive Systems** S. Nefti and J. Gray (Editors)
Volume 72 **Control Theory: A guided tour, 3rd Edition** J.R. Leigh
Volume 73 **Adaptive Sampling with Mobile WSN** K. Sreenath, M.F. Mysorewala, D.O. Popa and F.L. Lewis
Volume 74 **Eigenstructure Control Algorithms: Applications to aircraft/rotorcraft handling qualities design** S. Srinathkumar
Volume 75 **Advanced Control for Constrained Processes and Systems** F. Garelli, R.J. Mantz and H. De Battista
Volume 76 **Developments in Control Theory towards Glocal Control** L. Qiu, J. Chen, T. Iwasaki and H. Fujioka (Editors)
Volume 77 **Further Advances in Unmanned Marine Vehicles** G.N. Roberts and R. Sutton (Editors)
Volume 78 **Frequency-Domain Control Design for High-Performance Systems** J. O'Brien
Volume 80 **Control-oriented Modelling and Identification: Theory and practice** M. Lovera (Editor)
Volume 81 **Optimal Adaptive Control and Differential Games by Reinforcement Learning Principles** D. Vrabie, K. Vamvoudakis and F. Lewis
Volume 83 **Robust and Adaptive Model Predictive Control of Nonlinear Systems** M. Guay, V. Adetola and D. DeHaan
Volume 84 **Nonlinear and Adaptive Control Systems** Z. Ding
Volume 86 **Modeling and Control of Flexible Robot Manipulators, 2nd edition** M.O. Tokhi and A.K.M. Azad
Volume 88 **Distributed Control and Filtering for Industrial Systems** M. Mahmoud
Volume 89 **Control-based Operating System Design** A. Leva *et al.*
Volume 90 **Application of Dimensional Analysis in Systems Modelling and Control Design** P. Balaguer
Volume 91 **An Introduction to Fractional Control** D. Valério and J. Costa
Volume 92 **Handbook of Vehicle Suspension Control Systems** H. Liu, H. Gao and P. Li
Volume 93 **Design and Development of Multi-Lane Smart Electromechanical Actuators** F.Y. Annaz
Volume 94 **Analysis and Design of Reset Control Systems** Y. Guo, L. Xie and Y. Wang
Volume 95 **Modelling Control Systems Using IEC 61499, 2nd Edition** R. Lewis and A. Zoitl
Volume 96 **Cyber-Physical System Design with Sensor Networking Technologies** S. Zeadally and N. Jabeur (Editors)
Volume 99 **Practical Robotics and Mechatronics: Marine, Space and Medical Applications** I. Yamamoto
Volume 100 **Organic Sensors: Materials and Applications** E. Garcia-Breijo and P. Cosseddu (Editors)
Volume 102 **Recent Trends in Sliding Mode Control** L. Fridman J.P. Barbot and F. Plestan (Editors)
Volume 104 **Control of Mechatronic Systems** L. Guvenc, B.A. Guvenc, B. Demirel, M.T. Emirler
Volume 105 **Mechatronic Hands: Prosthetic and Robotic Design** P.H. Chappell
Volume 107 **Solved Problems in Dynamical Systems and Control** D. Valério, J.T. Machado, A.M. Lopes and A.M. Galhano
Volume 108 **Wearable Exoskeleton Systems: Design, Control and Applications** S. Bai, G.S. Virk and T.G. Sugar
Volume 111 **The Inverted Pendulum in Control Theory and Robotics: From theory to new innovations** O. Boubaker and R. Iriarte (Editors)
Volume 112 **RFID Protocol Design, Optimization, and Security for the Internet of Things** A.X. Liu, M. Shahzad, X. Liu and K. Li

Volume 113 **Design of Embedded Robust Control Systems Using MATLAB®/Simulink®** P.H. Petkov, T.N. Slavov and J.K. Kralev

Volume 114 **Signal Processing and Machine Learning for Brain–Machine Interfaces** T. Tanaka and M. Arvaneh (Editor)

Volume 116 **Imaging Sensor Technologies and Applications** W Yang (Editor)

Volume 117 **Data Fusion in Wireless Sensor Networks** D. Ciuonzo and P.S. Rossi (Editors)

Volume 118 **Modeling, Simulation and Control of Electrical Drives** M.F. Rahman and S.K. Dwivedi (Editors)

Volume 119 **Swarm Intelligence Volumes 1–3** Y. Tan (Editor)

Volume 121 **Integrated Fault Diagnosis and Control Design of Linear Complex Systems** M. Davoodi, N. Meskin and K. Khorasani

Volume 123 **Data-Driven Modeling, Filtering and Control: Methods and applications** C. Novara and S. Formentin (Editors)

Volume 127 **Sensors, Actuators, and Their Interfaces: A multidisciplinary Introduction, 2nd Edition** N. Ida

Volume 129 **Signal Processing to Drive Human–Computer Interaction: EEG and eye-controlled interfaces** I. Kompatsiaris, C. Kumar and S. Nikolopoulos (Editors)

Fault Diagnosis and Fault-Tolerant Control of Robotic and Autonomous Systems

Edited by
Andrea Monteriù, Alessandro Freddi and Sauro Longhi

The Institution of Engineering and Technology

Published by The Institution of Engineering and Technology, London, United Kingdom

The Institution of Engineering and Technology is registered as a Charity in England & Wales (no. 211014) and Scotland (no. SC038698).

First published 2020

The Institution of Engineering and Technology
Michael Faraday House
Six Hills Way, Stevenage
Herts, SG1 2AY, United Kingdom

www.theiet.org

British Library Cataloguing in Publication Data
A catalogue record for this product is available from the British Library

ISBN 978-1-78561-830-7 (hardback)
ISBN 978-1-78561-831-4 (PDF)

Typeset in India by MPS Limited

Contents

List of contributors

Masoud Abbaszadeh
GE Global Research
General Electric
Niskayuna 12309, New York
USA
[Chapter 8]

Guillem Alenyà
Institut de Robòtica i Informàtica Industrial (CSIC-UPC)
Llorens i Artigas, 4-6, 08028 Barcelona
Spain
[Chapter 6]

Filippo Arrichiello
Department of Electrical and Information Engineering (DIEI)
University of Cassino and Southern Lazio
Urban area of Folcara, 03043 Cassino
Italy
[Chapter 7]

Alessandro Baldini
Department of Information Engineering
Università Politecnica delle Marche
Via Brecce Bianche, 60131 Ancona
Italy
[Chapter 5]

Hakan Basargan
Department of Control for Transportation and Vehicle Systems
Budapest University of Technology and Economics
Stoczek street 2, 1111 Budapest
Hungary
[Chapter 10]

József Bokor
Systems and Control Laboratory
Institute for Computer Science and Control
Kende u. 13-17, 1111 Budapest
Hungary
[Chapter 11]

Paolo Castaldi
Department of Electrical, Electronic and Information Engineering (DEI)
University of Bologna
Via dell'Università 50, 47522 Cesena
Italy
[Chapter 9]

Andrea Cristofaro
Department of Computer, Control and Management Engineering (DIAG)
Sapienza University of Rome
Via Ariosto 25, 00185 Rome
Italy
[Chapter 4]

Riccardo Felicetti
Department of Information Engineering
Università Politecnica delle Marche
Via Brecce Bianche, 60131 Ancona
Italy
[Chapter 5]

Alessandro Freddi
Department of Information Engineering
Università Politecnica delle Marche
Via Brecce Bianche, 60131 Ancona
Italy
[Chapter 5]

Péter Gáspár
Systems and Control Laboratory
Institute for Computer Science and Control
Kende u. 13-17, 1111 Budapest
Hungary
[Chapter 10]

Giuseppe Gillini
Department of Electrical and Information Engineering (DIEI)
University of Cassino and Southern Lazio
Urban area of Folcara, 03043 Cassino
Italy
[Chapter 7]

Jianda Han
State Key Laboratory of Robotics
Shenyang Institute of Automation
Chinese Academy of Sciences
110016, Shenyang, Liaoning Province
P.R. China
Institutes for Robotics and Intelligent Manufacturing
Chinese Academy of Sciences
110016, Shenyang, Liaoning Province
P.R. China
Institute of Robotics and Automatic Information Systems
College of Artificial Intelligence, and Tianjin Key Laboratory of Intelligent Robotics
Nankai University
300350, Tianjin
P.R. China
[Chapter 3]

Kazuhiko Hasegawa
Department of Naval Architecture and Ocean Engineering
Osaka University
2-1 Yamadaoka, Suita, Osaka
Japan
[Chapter 5]

Yuqing He
State Key Laboratory of Robotics
Shenyang Institute of Automation
Chinese Academy of Sciences
110016, Shenyang, Liaoning Province
P.R. China
Institutes for Robotics and Intelligent Manufacturing
Chinese Academy of Sciences
110016, Shenyang, Liaoning Province
P.R. China
[Chapter 3]

Yuchen Jiang
Department of Control Science and Engineering
Harbin Institute of Technology
92 West Dazhi Street, Harbin
China
Department of Electrical and Computer Engineering
Technical University of Munich
Theresienstr. 90, 80333 Munich
Germany
[Chapter 12]

Tor Arne Johansen
Center for Autonomous Marine Operations and Systems
Department of Engineering Cybernetics
NTNU
O.S. Bragstads plass 2D, Trondheim
Norway
[Chapter 4]

Martina Lippi
Department of Information and Electrical Engineering and Applied Mathematics (DIEM)
University of Salerno
Via Giovanni Paolo II 132, 84084 Fisciano
Italy
[Chapter 7]

Vincenzo Lippiello
PRISMA Lab
Department of Electrical Engineering and Information Technology
Università degli Studi di Napoli Federico II
Via Claudio 21, 80125, Naples
Italy
[Chapter 2]

György Lipták
Systems and Control Laboratory
Institute for Computer Science and Control
Kende u. 13-17, 1111 Budapest
Hungary
[Chapter 11]

Zhong Liu
State Key Laboratory of Robotics
Shenyang Institute of Automation
Chinese Academy of Sciences

110016, Shenyang, Liaoning Province
P.R. China

Institutes for Robotics and Intelligent Manufacturing
Chinese Academy of Sciences
110016, Shenyang, Liaoning Province
P.R. China
University of Chinese Academy of Sciences
100049, Beijing
P.R. China
[Chapter 3]

Alessandro Marino
Department of Electrical and Information Engineering (DIEI)
University of Cassino and Southern Lazio
Urban area of Folcara, 03043 Cassino
Italy
[Chapter 7]

András Mihály
Systems and Control Laboratory
Institute for Computer Science and Control
Kende u. 13-17, 1111 Budapest
Hungary
[Chapter 10]

Nicola Mimmo
Department of Electrical, Electronic and Information Engineering (DEI)
University of Bologna
Viale del Risorgimento 2, 40136 Bologna
Italy
[Chapter 9]

Andrea Monteriù
Department of Information Engineering
Università Politecnica delle Marche
Via Brecce Bianche, 60131 Ancona
Italy
[Chapter 5]

Jyotsna Pandey
Department of Naval Architecture and Ocean Engineering
Osaka University
2-1 Yamadaoka, Suita, Osaka
Japan
[Chapter 5]

Tamás Péni
Systems and Control Laboratory
Institute for Computer Science and Control
Kende u. 13-17, 1111 Budapest
Hungary
[Chapter 11]

Francesco Pierri
School of Engineering
University of Basilicata
Via Nazario Sauro 85, 85100 Potenza
Italy
[Chapter 7]

Vicenç Puig
Institut de Robòtica i Informàtica Industrial (CSIC-UPC)
Llorens i Artigas, 4-6, 08028 Barcelona
Spain
[Chapter 6]

Gerasimos Rigatos
Unit of Industrial Automation
Industrial Systems Institute
26504, Rion Patras
Greece
[Chapter 8]

Damiano Rotondo
Department of Electrical and Computer Engineering (IDE)
University of Stavanger (UiS)
Kristine Bonnevies vei 22, 4021 Stavanger
Norway
[Chapter 4]

Fabio Ruggiero
PRISMA Lab
Department of Electrical Engineering and
Information Technology
Università degli Studi di Napoli Federico II
Via Claudio 21, 80125, Naples
Italy
[Chapter 2]

Alberto San Miguel
Institut de Robòtica i Informàtica Industrial (CSIC-UPC)
Llorens i Artigas, 4-6, 08028 Barcelona
Spain
[Chapter 6]

Diana Serra
PRISMA Lab
Department of Electrical Engineering and Information Technology
Università degli Studi di Napoli Federico II
Via Claudio 21, 80125, Naples
Italy
[Chapter 2]

Bruno Siciliano
PRISMA Lab
Department of Electrical Engineering and Information Technology
Università degli Studi di Napoli Federico II
Via Claudio 21, 80125, Naples
Italy
[Chapter 2]

Silvio Simani
Department of Engineering
University of Ferrara
Via Saragat 1, 44122 Ferrara
Italy
[Chapter 9]

Zoltán Szabó
Systems and Control Laboratory
Institute for Computer Science and Control
Kende u. 13-17, 1111 Budapest
Hungary
[Chapter 11]

Didier Thelliol
Faculte des Sciences et Techniques
CRAN UMR 7039
CNRS, University of Lorraine
B.P. 70239, 54506 Vandœuvre-lés-Nancy
France
[Chapter 3]

Bálint Vanek
Systems and Control Laboratory
Institute for Computer Science and Control
Kende u. 13-17, 1111 Budapest
Hungary
[Chapter 11]

Ban Wang
Department of Integrated Technology and Control Engineering
School of Aeronautics
Northwestern Polytechnical University
Xi'an 710072, Shaanxi
China
[Chapter 1]

Patrice Wira
IRIMAS
Université d'Haute Alsace
68039, Mulhouse
France
[Chapter 8]

Liying Yang
State Key Laboratory of Robotics
Shenyang Institute of Automation
Chinese Academy of Sciences
110016, Shenyang, Liaoning Province
P.R. China
Institutes for Robotics and Intelligent Manufacturing
Chinese Academy of Sciences
110016, Shenyang, Liaoning Province
P.R. China
[Chapter 3]

Shen Yin
Department of Control Science and Engineering
Harbin Institute of Technology
92 West Dazhi Street, Harbin
China
[Chapter 12]

Youmin Zhang
Department of Mechanical, Industrial and Aerospace Engineering
Concordia University
Montreal H3G 1M8, Quebec
Canada
[Chapter 1]

About the editors

Andrea Monteriù is an associate professor at Università Politecnica delle Marche (Ancona, Italy). His main research interests include fault diagnosis, fault-tolerant control, nonlinear, dynamics and control, periodic and stochastic system control, applied in different fields including aerospace, marine and robotic systems. He published more than 130 papers in international journals and conferences and is involved both in national and international research projects. He is the author of the book *Fault Detection and Isolation for Multi-Sensor Navigation Systems: Model-Based Methods and Applications*; and a co-editor or author of four books on Ambient Assisted Living including the IET book entitled *Human Monitoring, Smart Health and Assisted Living: Techniques and Technologies*

Alessandro Freddi is an assistant professor at Università Politecnica delle Marche (Ancona, Italy), where he teaches "Preventive Maintenance for Robotics and Smart Automation" and is a founder member of "Syncode", a startup operating in the field of industrial automation. His main research activities cover fault diagnosis and fault-tolerant control with applications to robotics, and development and application of assistive technologies. He published more than 70 papers in international journals and conferences and is involved both in national and international research projects.

Sauro Longhi is a full professor at Università Politecnica delle Marche (Ancona, Italy), where he acted as a Rector from 2013 to 2019. Since May 2014 he has been the president of the GARR Consortium, the Italian national computer network for universities and research. Since September 2019 he has also been the president of SIDRA, the Italian Society of Researchers and Professors of Automation. His research interests include modelling, identification and control of linear and nonlinear systems, control of mobile robots, underwater vehicles, vessels and unmanned aerial vehicle, cooperative control of autonomous agents, service robots for assistive applications supporting mobility and cognitive actions, home and building automation, web technology in process control and remote control laboratories, decentralized control over networks, sensors networks, power management in hybrid cars, electric motor control, embedded control system, management and control of renewable energy resources, efficient management of energy systems, automatic fault detection and isolation. He has published more than 400 papers on international journals and conferences.

Foreword

Research and development of robotic systems has been attracting the interest of both the academic and professional world since the early 1960s, when robots were employed for the very first time in industrial applications. Since then, they have experienced an exponential growth, which led to modern robotic and autonomous systems, ranging from industrial to medical, and from exploration to service applications. This incredible adaptability to different application fields was made possible thanks to the development of robots of different types (e.g. stationary, legged, wheeled, flying, swimming) and sizes (macro, micro and nano), together with the advances in sensor technology and computer learning. Despite the type, the size and the field of application, modern robotic and autonomous systems require an increasing level of reliability and safety, i.e. they must be able to perform a task and not cause harm even in the presence of faults and failures. For instance, an autonomous aerial vehicle experiencing a fault should be at least able to land in order to avoid a crash, or a faulty industrial robot should be at least able to stop before causing harm to a human operator.

The challenge of increasing the reliability and safety of such systems is usually addressed both during the design and the operational phase. During the design phase, the system is studied in terms of reliability, availability, maintainability, safety and dependability. During the operational phase, the occurrence of one or more faults is addressed by means of fault detection and diagnosis (FDD) together with fault-tolerant control (FTC) and condition-based maintenance, which can be model-based, signal-based or data-driven. The growth in the knowledge of both design and diagnostic capabilities is now evolving into the future trend of fault prognosis.

The aim of this book is thus to gather contributions, both from the academic and professional world, to address the challenge of increasing the reliability and safety of robotic and autonomous systems. Beyond its survey aspects, the book also aims to present relevant theoretical findings as well as challenging applications in various fields. The book is composed of 13 chapters that are organized as follows.

Chapters 1–5 cover the topics of **fault diagnosis and FTC applied to unmanned vehicles**, with a main focus on aerial and marine applications, which represent the most challenging ones from a diagnosis and control point of view. In detail, **Chapter 1 – "Fault diagnosis and fault-tolerant control of unmanned aerial vehicles"** by *Ban Wang and Youmin Zhang* aims to design and develop an active FTC scheme for unmanned aerial vehicles (UAVs) to ensure their safety and reliability. The chapter first details the main concepts of FDD and FTC, thus providing an introduction to academics and practitioners outside these research fields. Then, the chapter focuses on the specific field of multirotor UAVs and proposes a fault estimation approach via recurrent neural networks, where adaptive sliding mode control guarantees system tracking performance in the presence of model uncertainties. In this way, the aerial system has the capability of tolerating certain faults without imperiling itself and its surroundings, posing the base to the so-called self-repairing "smart" flight control systems whose aim is to improve reliability and survivability of UAVs.

Chapter 2 – "Control techniques to deal with the damage of a quadrotor propeller" by *Fabio Ruggiero, Diana Serra, Vincenzo Lippiello and Bruno Siciliano* can be seen as a tutorial to guide the reader toward the implementation of an active FTC system (AFCTS) dealing with the damage of a UAV propeller. In particular, the quadrotor with fixed propellers is the aerial device taken into account and, differently from the previous chapter, this one focuses on the problem of keeping the UAV from crashing in the case of complete loss of a propeller. The adopted solution is to completely turn off the damaged motor, which requires in turn that the motor aligned to the same quadrotor axis of the broken one is turned off as well, thus resulting in the so-called birotor configuration. Control designed is based on PID and backstepping approaches.

Chapter 3 – "Observer-based LPV control design of quad-TRUAV under rotor-tilt axle stuck fault" by *Zhong Liu, Didier Theilliol, Liying Yang, Yuqing He and Jianda Han* deals with a very promising class of UAVs, namely the tilt-rotor UAV (TRUAV). The TRUAV is a transformable aircraft that possesses the advantages of both multirotor and fixed-wing vehicles, since they are capable of both hovering and high-speed cruise, thus representing a promising candidate for building aerial robotic platforms in the near future. In detail, the transition control of a quad-TRUAV is faced via linear parameter varying (LPV) method, and the FTC strategy is considered further for rotor-tilt axle stuck fault.

As Chapters 1–3 were mainly concerned on how to handle faults of electrical and/or mechanical nature, **Chapter 4 – "An unknown input observer-based framework for fault and icing detection and accommodation in overactuated unmanned aerial vehicles"** by *Andrea Cristofaro, Damiano Rotondo and Tor Arne Johansen* pays attention to the problem of faults due to environmental factors instead. Indeed, the use of UAVs as support to operations in remote areas and harsh environments, such as marine operations in the Arctic, is becoming more and more crucial. These systems are expected to operate in very critical weather conditions, and for this reason they are subject to the occurrence of icing. The formation of ice layers on airfoils decreases the lift, while it increases the drag and the mass of the vehicle: for this reason, icing can be regarded as a structural fault. The chapter adopts tools from the control allocation framework and presents an unknown input observer (UIO) approach for fault and icing diagnosis and accommodation in UAVs equipped with a redundant suite of effectors and actuators.

Among unmanned vehicles, UAVs are those which can benefit the most of fault diagnosis and fault-tolerant techniques, since they are inherently unstable, and a crash may result both in economical and safety problems. Another class of unmanned vehicles that requires high autonomy and reliability, however, is that of unmanned surface vehicles (USVs), marine vehicles that have many applications in scientific, industrial and military operations. **Chapter 5 – "Actuator fault tolerance for a WAM-V catamaran with azimuth thrusters"** by *Alessandro Baldini, Riccardo Felicetti, Alessandro Freddi, Kazuhiko Hasegawa, Andrea Monteriù and Jyotsna Pandey* presents an FTC scheme for an over-actuated USV, namely a wave adaptive modular vessel catamaran equipped with two azimuth thrusters. The scheme manages the most common actuator faults, i.e., loss of efficiency of the thruster and lock-in-place of the azimuth angle. The scheme is based on a three-layer architecture: a heuristic-based control policy for proper reference generation, a control law for the vehicle dynamics to achieve speed tracking of the generated reference, and a control allocation level for optimally redistributing the control efforts among the thrusters, even in the presence of actuator faults and failures. The control allocation and control policy layers permit to perform fault/failure tolerance, without varying the control law.

Chapters 6–8 move the attention from unmanned vehicles to full-fledged **mobile robots with manipulation capabilities**. In detail, **Chapter 6 – "Fault-tolerant control of a service robot"** by *Alberto San Miguel, Vicenç Puig and Guillem Alenyà* proposes an FTC scheme applied to a mobile service robot named TIAGo (PAL robotics), a robot designed to successfully perform tasks on highly dynamic and unpredictable anthropic domains. First, a quick introduction on service robots is provided, together with a description of the FDD algorithms to cope with several faults (from low-level actuators and sensors to the high-level decision layers). Then, the fault-tolerant solution is presented, based on a robust UIO (RUIO) that estimates the fault as well as the robot state. These estimations are integrated within the control algorithm based on an observer-based state feedback control. After the fault occurrence, from the fault estimation, a feedforward control action is added to the feedback control action to compensate the fault effect. To cope with the robot nonlinearity, its nonlinear model is transformed into a Takagi–Sugeno model. State-feedback and RUIO are designed using a linear matrix inequalities approach considering a gain-scheduling scheme.

Chapter 7 – "Distributed fault detection and isolation strategy for a team of cooperative mobile manipulators" by *Giuseppe Gillini, Martina Lippi, Filippo Arrichiello, Alessandro Marino and Francesco Pierri* extends the problem of fault detection and isolation (FDI) to a team of cooperative robots. Teams of robots result especially useful when complex and/or heavy tasks in a high-performance manner must be accomplished, such as exploration, navigation and surveillance or cooperative manipulation and load transportation tasks. The adoption of multiple cooperative robots improves both efficiency and robustness to faults of the overall system; however FDI strategies are much more challenging in the case of distributed control architecture. This chapter investigates a general distributed framework for multiple mobile-manipulator systems that allows each robot to detect and isolate possible faults of any other robot in the team, without the need of direct communication with it. By using an observer–controller scheme, each robot estimates the overall system state by means of a local observer and, then, it exploits this estimate to compute the local control input to achieve a desired cooperative task. Based on the same local observer, residual signals are also defined, which enable one to detect and identify possible faults in the network.

Chapter 8 – "Nonlinear optimal control for aerial robotic manipulators" by *Gerasimos Rigatos, Masoud Abbaszadeh and Patrice Wira* faces the critical problem of controlling an aerial robotic manipulator, namely a quadrotor equipped with a robotic arm with flexible joints. The state-space model of the aerial manipulator is linearized around a temporary operating point. The feedback gains of the H-infinity controller are selected by iteratively solving an algebraic Riccati equation. The proposed approach can cope with model uncertainties and external perturbations, including faults, and can be extended so that it does not require full measure of the state vector.

As highlighted in these chapters, modern robots require an increasing level of autonomy and the capability of interacting with humans. As automation and intelligence is getting more common in transportation systems, safety and autonomy of such systems is becoming more important as well. As a consequence, modern transportation systems are slowly evolving into autonomous intelligent machines, resembling robotic systems in the need of fault diagnosis control and FTC techniques. **Chapters 9 and 10** described two typical scenarios of how FDI and FTC can cope with **safety and autonomy of modern transportation systems**, thus giving an idea on the future of autonomous transportation systems. In detail **Chapter 9 – "Fault diagnosis and fault-tolerant control techniques for aircraft systems"** by *Paolo Castaldi, Nicola Mimmo and Silvio Simani* analyses and discusses an AFCTS for avionic applications. The approach applies to an aircraft longitudinal autopilot in the presence of faults affecting the system actuators. The key point of the developed FTC scheme consists of its active feature, since the fault diagnosis module provides a robust and reliable estimation of the fault signals, which are thus compensated. The design technique relies on a nonlinear geometric approach.

As for autonomous cars, instead, the focus of current research is mainly on electric vehicles and in-wheel motor solutions are among the most investigated. **Chapter 10 – "Fault-tolerant trajectory tracking control of in-wheel motor vehicles with energy-efficient steering and torque distribution"** by *Péter Gáspár, András Mihály and Hakan Basargan* has the objective to design a novel fault-tolerant and energy-efficient torque distribution method for an autonomous four-wheel independently-actuated electric vehicle. In normal operating conditions, the presented high-level reconfigurable controller and the low-level wheel torque optimization method grant better energy efficiency. In the case of actuator failure or performance degradation, the high-level controller performs a reconfiguration to apply more steering or more yawing torque according to the actual vehicle state and the corresponding priority regarding safety. The main goal of the design is to maximize the battery state-of-charge to enhance the range of the autonomous electric vehicle, while preserving safe motion in the case of a fault event.

Despite the type, the size and the field of application of modern robots, some FDD and FTC schemes are "general", in the sense that they can be adapted to work with different types of systems. The final chapters of the book, namely **Chapters 11 and 12**, propose two common approaches to **fault diagnosis and fault-tolerance in**

modern robotic and autonomous systems. As for **Chapter 11 – "Nullspace-based input reconfiguration architecture for over-actuated aerial vehicles"** by *Tamás Péni, Bálint Vanek, György Lipták, Zoltán Szabó and József Bokor*, it wants to solve the problem of actuator failures via control input reallocation, where the aim is to compensate the actuator failures by reconfiguring the remaining control inputs such that the performance degradation caused by the failure is as small as possible. One possible approach to solve this problem is based on the nullspace (or kernel). These algorithms can be applied only if control input redundancy is available in the system, with the advantage that there is no restriction for the synthesis method used to design the controller. The approach is applied on an aircraft model as example. Finally, **Chapter 12 – "Data-driven approaches to fault-tolerant control of industrial robotic systems"** by *Yuchen Jiang and Shen Yin* faces the problem of FTC on a different perspective, namely that of data-driven approaches. The chapter proposes to design intelligent learning-aided controllers under the framework of Youla parameterization. The goal is to design optimal controllers that have fault-tolerance capability and increased robustness. It can be implemented in a modular and plug-and-play manner, where Reinforcement Learning acts as a supervising module that calculates optimized Youla matrix design parameters in the design phase. The approach is applied to a wheeled robot as example.
Chapter 13 – "Conclusions" by *Andrea Monteriù, Alessandro Freddi and Sauro Longhi* provides a summary of the main results described in each chapter, together with some final thoughts by the editors.

Ancona, Italy
April 2020

Andrea Monteriù
Alessandro Freddi
Sauro Longhi

Chapter 1

Fault diagnosis and fault-tolerant control of unmanned aerial vehicles

Ban Wang[1] *and Youmin Zhang*[2]

With the increasing demand for unmanned aerial vehicles (UAVs) in both military and civilian applications, critical safety issues need to be specially considered in order to make better and wider use of them. UAVs are usually employed to work in hazardous and complex environments, which may seriously threaten the safety and reliability of UAVs. Therefore, the safety and reliability of UAVs are becoming imperative for development of advanced intelligent control systems. The key challenge now is the lack of fully autonomous and reliable control techniques in face of different operation conditions and sophisticated environments. Further development of UAV control systems is required to be reliable in the presence of system component faults and to be insensitive to model uncertainties and external environmental disturbances.

This chapter aims to design and develop an active fault-tolerant control (FTC) scheme for UAVs to ensure their safety and reliability. First, a cost-effective fault estimation scheme with a parallel bank of recurrent neural networks (RNNs) is proposed to accurately estimate actuator fault magnitude. Then, an adaptive sliding mode control (SMC) is proposed to guarantee system-tracking performance in the presence of model uncertainties without stimulating control chattering. Finally, a reconfigurable active FTC framework is established for a closed-loop unmanned quadrotor helicopter system by combing the developed fault estimation scheme with the proposed adaptive SMC.

1.1 Introduction

1.1.1 Unmanned aerial vehicle

Over the last decades, UAVs have been widely used by commercial industries, research institutes, and military sectors with applications to payload transportation [1,2], forest fire detection and fighting [3], surveillance [4,5], environmental monitoring [6],

[1]Department of Integrated Technology and Control Engineering, School of Aeronautics, Northwestern Polytechnical University, Xi'an, China

[2]Department of Mechanical, Industrial and Aerospace Engineering, Concordia University, Montreal, Canada

remote sensing [7], aerial mapping [8], etc. More recently, with the development of automation technologies, more and more small-scale UAVs are coming out, which further extends the applications of UAVs. New generations of UAVs need to be designed to achieve their missions with not only increased efficiency but also more safety and security. Future UAVs will be operated with algorithms capable of monitoring the aircraft's health status and of taking actions if needed [9].

For many applications, in order to accomplish the specific task, different sensors and instrument systems need to be incorporated into the assigned UAV to make it fully functional. In this sense, a UAV can be regarded as an aerial sensor/payload carrier, and usually the cost of those onboard instruments can easily exceed the cost of the UAV itself, especially for small-scale UAVs. Moreover, UAVs are usually employed to work in complex and hazardous environments, such as urban surveillance, firefighting, and special military actions. These situations may seriously threaten the safety and reliability of the UAVs and the expensive onboard instruments/payloads. Especially, for those applications carried out in urban area, in-flight faults exposed on UAVs may endanger human life and property in addition to the loss of the UAV itself. Therefore, the reliability and survivability of UAVs are becoming imperative, and critical safety issues need to be specially considered. As argued in [10–12], the increasing demands for safety, reliability, and high system performance have stimulated research in the area of FTC with the development in control theory and computer technology. Fault-tolerant capability is an important feature for safety-critical systems [10], such as aircraft [13,14] and spacecraft [15,16]. It turns out that it will be beneficial to have a UAV system with the capability of tolerating certain faults without imperiling itself and its surroundings. From the control point of view, it is to design a so-called self-repairing "smart" flight control system to improve reliability and survivability of UAVs, which is also the requirement for the next generation flight control system according to unmanned aircraft systems roadmap released in 2005 by the US Department of Defense.

1.1.2 Fault detection and diagnosis

Fault detection and diagnosis (FDD) is used to detect faults and diagnose their location and magnitude in a system. It has the following tasks: fault detection, fault isolation, and fault identification [17]. Fault detection indicates that something is wrong in the system, for example, the occurrence of a fault and the time of the fault occurrence. Fault isolation determines the location and the type of a fault. Fault identification determines the magnitude of a fault. FDD addresses the challenge of real-time monitoring the occurrence of fault in a system. Due to real-time requirements and the dynamic nature of the mechanical system, there is usually only a very limited amount of time available to carry out the post-fault model construction and control reconfiguration actions.

Fault diagnosis algorithms are generally classified into two types known as model-based and data-driven methods. Model-based approaches are mostly based on analytical redundancy and require an analytical mathematical model of the system. The presence of faults is detected by means of the so-called residuals, which can be generated in different ways: parity equations, state estimation-based methods, and parameter estimation-based methods. The performance of model-based approaches depends strongly on the usefulness of the constructed model [18]. The constructed model must include all situations of the system under study. It must be able to handle changes at the operating point. If the constructed model fails, the whole diagnostic system will also fail. However, in practice, it is usually quite challenging and difficult to meet all the requirements of model-based approaches due to the inevitable unmodeled dynamics, uncertainties, model mismatches, noises, disturbances, and inherent nonlinearities [19]. The sensitivity to modeling errors has become the key problem in the application of model-based approaches. In contrast, data-driven diagnosis approaches, such as neural network-based intelligent methods, mostly rely on historical and current data from the sensors and do not require a detailed mathematical model of the system but need representative training data. The idea is that the operation of the system is classified according to measurement data. Formally, this is a mapping from measurement space into decision space [18]. Therefore, data play a very important role in this kind of method. With application to UAVs, model-free neural network-based approaches are preferred due to the cost limitation and short period of development compared to conventional physical modeling of an aircraft. The capabilities of neural networks for function approximation, classification, and their ability to deal with uncertainties and parameter variations make them a viable and an alternative choice for using in fault diagnosis problems [20].

1.1.3 Fault-tolerant control

As modern technological systems become complex, their corresponding control systems are designed more and more sophisticated, including the urgent need to increase reliability of the systems. This stimulates the study of reconfigurable control systems. Most research works in reconfigurable control systems focus on FDD that can serve as a monitoring system by detecting, localizing, and identifying faults in a system. FDD is a very important procedure, but it is not sufficient to ensure safe operation of the system. For some safety-critical systems such as aircraft and spacecraft, the continuation of operation is a key feature and the closed-loop system should be capable of maintaining its prespecified performance in terms of quality, safety, and stability despite the presence of faults. This calls for the appearance of FTC system (FTCS) [10]. More precisely, FTCS is a control system that can accommodate system component faults and is able to maintain system stability and performance in both fault-free and faulty conditions [21]. Generally speaking, FTCS can be classified into two types known as passive FTCS (PFTCS) and active FTCS (AFTCS) [10]. A PFTCS is designed to be robust against a class of presumed faults without requiring online detection of faults [22,23], while an AFTCS is based on controller reconfiguration or selection of predesigned controllers with the help of a FDD unit [24,25]. A comparative study between AFTCS and PFTCS is presented by Jiang and Yu in [26]. From performance perspective, a PFTCS focuses more on the robustness of the control system to accommodate multiple system faults without striving for optimal performance for any specific faulty conditions. Since stability is the prior consideration in a passive approach, the designed controller turns to be more conservative from the performance point of view. An AFTCS typically consists of a FDD unit, a reconfigurable controller, and a controller reconfiguration mechanism. These three units have to work in harmony to complete successful control tasks, and an optimal solution with certain preset performance criteria can be found. However, there are other issues that can impair an AFTCS to achieve its mission. Usually, system component faults can render the system unstable, and therefore there is only a limited amount of time available for an AFTCS to react to the faults and to make corrective control actions [10].

Generally speaking, a fault is an unpermitted deviation of at least one characteristic property of the system from the acceptable, usual, and standard condition, whereas a failure is a permanent interruption of a system's ability to perform a required function under specified operating conditions [27]. Based on this definition, the impact of a fault can be a small reduction in efficiency that may degrade the overall functioning of the system. Resulting from one or more faults, a failure is therefore an event that terminates the functioning of a unit in the system. Faults can take place in different parts of the controlled system. According to their occurrence location, faults can be classified into actuator faults, sensor faults, and other system component faults [10,28]. Since actuators play an important role of connecting control signals to physical movements of the system to accomplish specific objectives, there will be an emphasis on actuator faults in this chapter. Indeed, a sensor fault/failure does not directly affect the flight performance of the aircraft. The sensor fault/failure can be handled either by using a redundant sensor if available, or by reconstructing the missing measurement data with the knowledge of the system and the measurement data furnished by the remaining sensors [29]. However, as soon as there is an actuator fault, the aircraft flight performance will be inevitably degraded, and immediate action must be taken to preserve its original performance. Partial loss of control effectiveness of an actuator is a common fault occurring in aircraft systems that can result from hydraulic or pneumatic leakage, increased resistance or a fall in the supply voltage [30]. Moreover, duplicating the actuators in the system in order to achieve increased fault-tolerance is often not an option due to their high price and large size and mass compared to sensors in the system.

From the control point of view, SMC is an approach to design robust control systems for handling large uncertainties with discontinuous control strategy, which demonstrates invariance to so-called matched uncertainties while on a reduced order sliding surface [31,32]. It was introduced for the first time in the context of the variable structure system [33]. It was developed for the control of electric drives by Utkin [34]. During the last three decades, it has shown to be a very effective approach in robust control area. To be more precise, SMC utilizes a high-speed switching control law to achieve two objectives. First, it drives the nonlinear system's state trajectory onto a specified surface in the state space that is called sliding surface. Second, it maintains the system's state trajectory on this surface for all subsequent times. During this process, the structure of the control system varies from one to another and thereby it is called as SMC to emphasize the importance of the sliding mode [35]. Intuitively, SMC uses practically infinite gain to force the trajectory of a dynamic system to slide

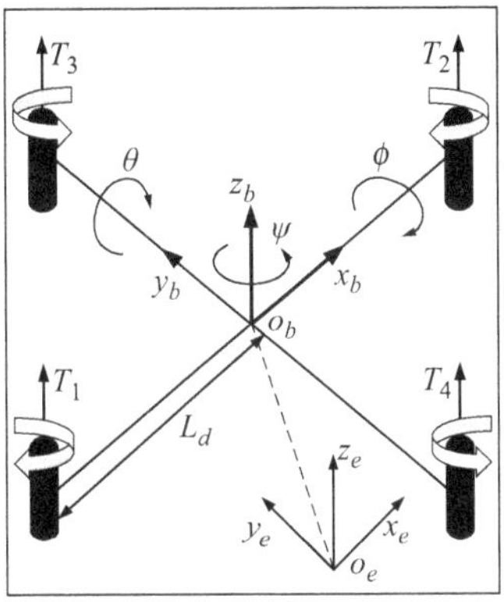

Figure 1.1 Configuration of a quadrotor helicopter

along the restricted sliding mode subspace, which leads to significantly improved control performance. The main advantage of SMC over the other nonlinear control approaches is its robustness to external disturbances, model uncertainties, and system parametric variations. One design parameter is synthesized in the discontinuous control part and the control can be as simple as a switching between two states; therefore, it does not need to be precise and will not be sensitive to parameter variations that enter into the control channel [36]. The property of insensitivity to parametric uncertainties and external disturbances makes SMC as one of the most promising control approaches.

1.2 Modeling of an unmanned quadrotor helicopter

This research work focuses on the study of UAVs. As an example of UAVs, a quadrotor helicopter is a relatively simple and easy-to-fly system [13,37]. Indeed, the quadrotor helicopter, as shown in Figure 1.1, is a small-scale UAV with four propellers placed around a main body. The main body includes power source, sensors, and control hardware. The vehicle is controlled by the four propellers, and their rotational speeds are independent, which makes it possible to control the roll, pitch, and yaw angle of the vehicle. And, its displacement is produced by the total thrust of the four propellers.

1.2.1 Kinematic equations

In order to model the quadrotor helicopter, we first discuss the coordinate systems used in this section. Two coordinate systems are employed: the local navigation frame and the body-fixed frame. The axes of the body-fixed frame are denoted as (o_b, x_b, y_b, z_b), and the axes of the local navigation frame are denoted as (o_e, x_e, y_e, z_e). The position $X^I = [x_e, y_e, z_e]^T$ and attitude $\Theta^I = [\phi, \theta, \psi]^T$ of the quadrotor helicopter are defined in the local navigation frame that is regarded as the inertial reference frame.

The translational velocity $V^B = [u, v, w]^T$ and rotational velocity $\omega^B = [p, q, r]^T$ are defined in the body-fixed frame.

To derive the equations of motion, the Newton–Euler formulation is employed in this work. However, Newton's equations of motion are usually derived relative to a fixed inertial reference frame, whereas the motion of object is most easily described in a body-fixed frame and onboard sensors also collect data with respect to the body-fixed frame. Therefore, a transformation matrix from the body-fixed frame to the inertial reference frame is required to help model the quadrotor helicopter. In this work, the transformation matrix is obtained by premultiplying the three basic rotation matrices in the order of roll-pitch-yaw (i.e., in the sequence of x, y, and z axis) as follows:

$$
\begin{aligned}
R_B^I &= R(\psi, z)R(\theta, y)R(\phi, x) \\
&= \begin{bmatrix} c\psi & -s\psi & 0 \\ s\psi & c\psi & 0 \\ 0 & 0 & 1 \end{bmatrix} \begin{bmatrix} c\theta & 0 & s\theta \\ 0 & 1 & 0 \\ -s\theta & 0 & c\theta \end{bmatrix} \begin{bmatrix} 1 & 0 & 0 \\ 0 & c\phi & -s\phi \\ 0 & s\phi & c\phi \end{bmatrix} \\
&= \begin{bmatrix} c\theta c\psi & s\phi s\theta c\psi - c\phi s\psi & c\phi s\theta c\psi + s\phi s\psi \\ c\theta s\psi & s\psi s\theta s\psi + c\phi c\psi & c\phi s\theta s\psi - s\phi c\psi \\ -s\theta & s\phi c\theta & c\phi c\theta \end{bmatrix}
\end{aligned}
\tag{1.1}
$$

where $c\phi = \cos\phi$ and $s\phi = \sin\phi$, which is the same for θ and ψ.

Then, another transformation matrix T_B^I can be determined by resolving the Euler angular rates into rotational velocities defined in the body-fixed frame as follows:

$$
\begin{aligned}
\begin{bmatrix} p \\ q \\ r \end{bmatrix} &= \begin{bmatrix} \dot{\phi} \\ 0 \\ 0 \end{bmatrix} + R(\phi, x) \begin{bmatrix} 0 \\ \dot{\theta} \\ 0 \end{bmatrix} + R(\phi, x)R(\theta, y) \begin{bmatrix} 0 \\ 0 \\ \dot{\psi} \end{bmatrix} \\
&= (T_B^I)^{-1} \begin{bmatrix} \dot{\phi} \\ \dot{\theta} \\ \dot{\psi} \end{bmatrix}
\end{aligned}
\tag{1.2}
$$

$$
T_B^I = \begin{bmatrix} 1 & \sin\phi\tan\theta & \cos\phi\tan\theta \\ 0 & \cos\phi & -\sin\phi \\ 0 & \sin\phi\sec\theta & \cos\phi\sec\theta \end{bmatrix}. \tag{1.3}
$$

According to the abovementioned transformation matrices, it is possible to describe the kinematic equations in the following matrix manner:

$$
\dot{\xi}^I = \begin{bmatrix} \dot{X}^I \\ \dot{\Theta}^I \end{bmatrix} = \begin{bmatrix} R_B^I & 0_{3\times 3} \\ 0_{3\times 3} & T_B^I \end{bmatrix} \begin{bmatrix} V^B \\ \omega^B \end{bmatrix} = J_B^I \nu^B. \tag{1.4}
$$

1.2.2 Dynamic equations

In order to derive the dynamic equations of the quadrotor helicopter, two assumptions need to be addressed first [38]:

1. The origin of the body-fixed frame coincides with the center of mass (COM) of the quadrotor helicopter.
2. The axes of the body-fixed frame are coincident with the principal axes of inertia of the quadrotor helicopter.

With the abovementioned assumptions, the inertial matrix becomes diagonal, and there is no need to take another point, COM, into account for deriving the dynamic equations.

By employing the Newton–Euler formulation, the forces and moments equations can be expressed as follows [37]:

$$\begin{bmatrix} F^I \\ \tau^B \end{bmatrix} = \begin{bmatrix} m & 0_{3\times3} \\ 0_{3\times3} & I \end{bmatrix} \begin{bmatrix} \ddot{X}^I \\ \dot{\omega}^B \end{bmatrix} + \begin{bmatrix} 0 \\ \omega^B \times I\,\omega^B \end{bmatrix} \tag{1.5}$$

where $F^I = [F_x, F_y, F_z]^T$ and $\tau^B = [\tau_x, \tau_y, \tau_z]^T$ are the force and moment vectors with respect to the inertial reference frame and the body-fixed frame, respectively. m is the total mass of the quadrotor helicopter, and I is the diagonal inertial matrix defined as

$$I = \begin{bmatrix} I_{xx} & 0 & 0 \\ 0 & I_{yy} & 0 \\ 0 & 0 & I_{zz} \end{bmatrix}. \tag{1.6}$$

The forces on the quadrotor helicopter are composed of three parts: gravitation (G), thrust (T), and the translational motion induced drag force (D), given by

$$\begin{aligned} F^I &= G + R_B^I T + D \\ &= \begin{bmatrix} 0 \\ 0 \\ -mg \end{bmatrix} + R_B^I \begin{bmatrix} 0 \\ 0 \\ U_z \end{bmatrix} + \begin{bmatrix} -K_1\dot{x}_e \\ -K_2\dot{y}_e \\ -K_3\dot{z}_e \end{bmatrix} \end{aligned} \tag{1.7}$$

where K_1, K_2, and K_3 are the drag coefficients, g is the acceleration of gravity, and U_z is the total thrust from the four propellers.

By substituting (1.7) into (1.5), we can obtain

$$\begin{bmatrix} \ddot{x}_e \\ \ddot{y}_e \\ \ddot{z}_e \end{bmatrix} = \begin{bmatrix} 0 \\ 0 \\ -g \end{bmatrix} + \frac{1}{m} \begin{bmatrix} (\cos\phi \sin\theta \cos\psi + \sin\phi \sin\psi)U_z - K_1\dot{x}_e \\ (\cos\phi \sin\theta \sin\psi - \sin\phi \cos\psi)U_z - K_2\dot{y}_e \\ (\cos\phi \cos\theta)U_z - K_3\dot{z}_e \end{bmatrix}. \tag{1.8}$$

Similarly, the torques on the quadrotor helicopter are composed of gyroscopic torque (M_g), the torque generated by the propellers (M_T), and the rotational motion-induced torque (M_f), described as

$$\begin{aligned}\tau^B &= M_g + M_T + M_f \\ &= -\sum_{i=1}^{4} I_r\left(\omega^B \times \begin{bmatrix} 0 \\ 0 \\ 1 \end{bmatrix}\right)(-1)^{i+1}\Omega_i + \begin{bmatrix} U_\phi \\ U_\theta \\ U_\psi \end{bmatrix} + \begin{bmatrix} -K_4 L_d \dot{\phi} \\ -K_5 L_d \dot{\theta} \\ -K_6 \dot{\psi} \end{bmatrix}\end{aligned} \tag{1.9}$$

where I_r is the inertial moment of the propeller and Ω_i is the ith propeller's rotational speed. K_4, K_5, and K_6 are the drag coefficients, L_d is the distance between each motor and the COM of the quadrotor helicopter, and U_ϕ, U_θ, and U_ψ are the generated torques with respect to the body-fixed frame.

Then, by substituting (1.9) into (1.5), the following equation can be obtained:

$$\begin{aligned}\begin{bmatrix} \dot{p} \\ \dot{q} \\ \dot{r} \end{bmatrix} = I^{-1}\Bigg(&-\begin{bmatrix} 0 & I_{zz}r & -I_{yy}q \\ -I_{zz}r & 0 & I_{xx}p \\ I_{yy}q & -I_{xx}p & 0 \end{bmatrix}\begin{bmatrix} p \\ q \\ r \end{bmatrix} + I_r\begin{bmatrix} -q \\ p \\ 0 \end{bmatrix}\Omega \\ &+ \begin{bmatrix} U_\phi \\ U_\theta \\ U_\psi \end{bmatrix} + \begin{bmatrix} -K_4 L_d \dot{\phi} \\ -K_5 L_d \dot{\theta} \\ -K_6 \dot{\psi} \end{bmatrix}\Bigg)\end{aligned} \tag{1.10}$$

where $\Omega = \Omega_1 + \Omega_2 - \Omega_3 - \Omega_4$ is the residual of the overall propellers' speed.

1.2.3 Control mixing

Due to the configuration of the quadrotor helicopter, the attitude (ϕ, θ) is coupled with the position (x_e, y_e), and a pitch or roll angle is required in order to move the quadrotor helicopter along the x_e- or y_e-direction. The virtual control inputs (U_z, U_ϕ, U_θ, U_ψ) as shown in (1.8) and (1.10) for moving and stabilizing the quadrotor helicopter are mapped from the thrusts generated by the four independent motors. The relationship between the motor inputs and the intermediate virtual control inputs can be given in the following matrix fashion:

$$\begin{bmatrix} U_z \\ U_\phi \\ U_\theta \\ U_\psi \end{bmatrix} = \begin{bmatrix} K_u & K_u & K_u & K_u \\ 0 & 0 & K_u L_d & -K_u L_d \\ K_u L_d & -K_u L_d & 0 & 0 \\ K_y & K_y & -K_y & -K_y \end{bmatrix} \cdot \begin{bmatrix} u_1 \\ u_2 \\ u_3 \\ u_4 \end{bmatrix} \tag{1.11}$$

where K_u is a positive gain related to propeller generated thrust, K_y is a positive gain related to propeller generated torque, and u_i $(i = 1, 2, 3, 4)$ is the pulse-width modulation (PWM) input of the ith motor.

In order to facilitate the controller design, assume that the changes of roll and pitch angles are very small, such that the transformation matrix T_B^I as shown in (1.3) is very close to an identity matrix. Therefore, the rotational velocities can be replaced directly by Euler angular rates. Finally, the nonlinear dynamics of the quadrotor helicopter can be represented as

$$
\begin{aligned}
\ddot{x}_e &= \frac{(\cos\phi\sin\theta\cos\psi + \sin\phi\sin\psi)(u_1+u_2+u_3+u_4)K_u}{m} - \frac{K_1\dot{x}_e}{m} \\
\ddot{y}_e &= \frac{(\cos\phi\sin\theta\sin\psi - \sin\phi\cos\psi)(u_1+u_2+u_3+u_4)K_u}{m} - \frac{K_2\dot{y}_e}{m} \\
\ddot{z}_e &= \frac{(\cos\phi\cos\theta)(u_1+u_2+u_3+u_4)K_u}{m} - \frac{K_3\dot{z}_e}{m} - g \\
\ddot{\phi} &= \frac{(I_{yy}-I_{zz})\dot{\theta}\dot{\psi}}{I_{xx}} + \frac{K_uL_d(u_3-u_4)}{I_{xx}} - \frac{I_r\Omega\dot{\theta}}{I_{xx}} - \frac{K_4L_d\dot{\phi}}{I_{xx}} \\
\ddot{\theta} &= \frac{(I_{zz}-I_{xx})\dot{\phi}\dot{\psi}}{I_{yy}} + \frac{K_uL_d(u_1-u_2)}{I_{yy}} + \frac{I_r\Omega\dot{\phi}}{I_{yy}} - \frac{K_5L_d\dot{\theta}}{I_{yy}} \\
\ddot{\psi} &= \frac{(I_{xx}-I_{yy})\dot{\phi}\dot{\theta}}{I_{zz}} + \frac{K_y(u_1+u_2-u_3-u_4)}{I_{zz}} - \frac{K_6\dot{\psi}}{I_{zz}}.
\end{aligned}
\tag{1.12}
$$

1.2.4 Actuator fault formulation

Actuator faults represent malfunctioning of the actuators in a system, which may be due to hydraulic leakages, broken wires, or stuck control surfaces in an aircraft. Such faults can be modeled as multiplicative faults as follows:

$$u_f(t) = (I_m - L_f(t))u(t) \tag{1.13}$$

where $u_f(t) = [u_{f1}(t), u_{f2}(t), \dots, u_{fm}(t)]^T$ is the faulty control input vector, $I_m \in \mathbb{R}^{m\times m}$ is an identity matrix, and $L_f(t) = \text{diag}([l_{f1}(t), l_{f2}(t), \dots, l_{fm}(t)])$ represents the loss of control effectiveness level of the actuators, where $l_{fi}(t)$ $(i = 1, 2, \dots, m)$ is a scalar satisfying $0 \le l_{fi}(t) \le 1$. If $l_{fi}(t) = 0$, the ith actuator works perfectly; otherwise, the ith actuator suffers certain level of fault with a special case $l_{fi}(t) = 1$ denoting the complete failure of the ith actuator.

Note that although such multiplicative actuator faults do not directly affect the dynamics of the controlled system itself, they can significantly affect the dynamics of the closed-loop system and may even affect the controllability of the system [28].

1.3 Active fault-tolerant control

In this section, an active FTC strategy is proposed for an unmanned quadrotor helicopter against actuator faults and model uncertainties based on adaptive SMC and RNNs. The schematic of the proposed control strategy is illustrated in Figure 1.2. There are two problems to be addressed in the following sections. First, in virtue of the proposed adaptive SMC, the system tracking performance can be guaranteed in the presence of model uncertainties and disturbances without stimulating control

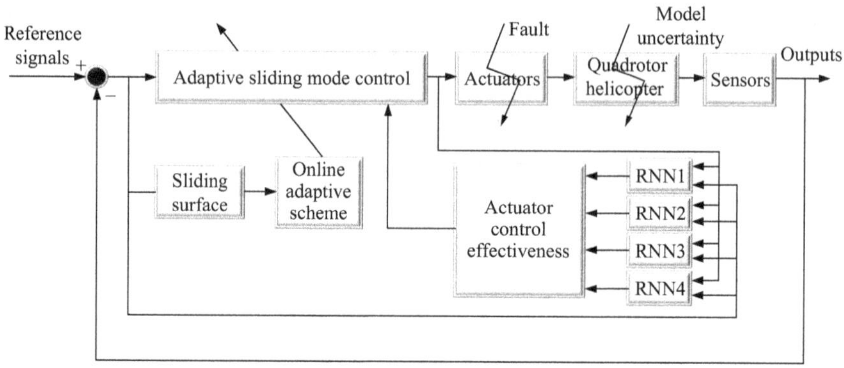

Figure 1.2 The schematic of the proposed active FTC strategy

chattering. Then, due to the fact that model-based fault estimation schemes may fail to correctly estimate faults in the presence of model uncertainties, a fault estimation scheme is proposed by designing a parallel bank of RNNs. For constructing the RNNs, two hidden layers are employed, and the outputs of the first hidden layer are fed back to form a context layer with purposes of capturing more information and achieving better learning capability. The weighting parameters of the RNNs are updated by using a back-propagation-through-time-based extended Kalman filter. With the trained RNNs, the severity of actuator faults can be precisely and reliably estimated. Finally, by synthesizing the fault estimation scheme with adaptive SMC, an active FTC mechanism is established. A number of numerical simulations are carried out to validate and demonstrate the effectiveness and advantages of the proposed control strategy.

1.3.1 Problem statement

Consider an integral-chain nonlinear affine system with model uncertainties and disturbances subject to actuator faults as

$$\begin{aligned} \dot{x}_1(t) &= x_2(t) \\ \dot{x}_2(t) &= F(x(t)) + G_B B_u L_c(t) u(t) + d(t) \\ y(t) &= Cx(t) + w(t) \end{aligned} \tag{1.14}$$

where $x(t) = [x_1(t), x_2(t)] \in \mathbb{R}^n$ is the state vector, $y(t) \in \mathbb{R}^q$ is the system output vector, $u(t) \in \mathbb{R}^m$ is the control input vector, $B_u \in \mathbb{R}^{p\times m}$ is the control effectiveness matrix, $C \in \mathbb{R}^{q\times n}$, and $G_B \in \mathbb{R}^{p\times p}$. The vector $F(x(t)) \in \mathbb{R}^p$ is a nonlinear function containing model uncertainties, and the bounds of these uncertainties are unknown *a priori*. $d(t) \in \mathbb{R}^p$ represents disturbances and noises that are unknown but bounded, i.e., $\|d(t)\| \leq D_d$, and $w(t) \in \mathbb{R}^q$ represents sensor modeling uncertainties and noises. $L_c(t) = I_m - L_f(t)$ represents the remaining useful level of the actuators.

For simplicity of expression, the notation t is omitted in the following sections, e.g., $x(t)$ is expressed as x.

1.3.2 Adaptive sliding mode control

In order to facilitate the state feedback control design, define the system state as $x = [\phi, \dot{\phi}, \theta, \dot{\theta}, \psi, \dot{\psi}]^T$. Then, the nonlinear system described in (1.14) is rewritten as

$$\begin{aligned} \dot{x}_{2i-1} &= x_{2i} \\ \dot{x}_{2i} &= h_{1i} f_{1i}(x) + h_{2i} x_{2i-1} + g_i \nu_i + d_i \\ \nu_i &= B_{ui} L_c u \end{aligned} \tag{1.15}$$

where $i = 1, 2, 3$ represents each subsystem, $x_{2i-1} = [\phi, \theta, \psi]^T$, and $x_{2i} = [\dot{\phi}, \dot{\theta}, \dot{\psi}]^T$.

By denoting x_{2i-1}^d as the desired trajectory, the tracking error vector is defined as

$$\tilde{x}_{2i-1} = x_{2i-1} - x_{2i-1}^d = \begin{bmatrix} \phi - \phi_d \\ \theta - \theta_d \\ \psi - \psi_d \end{bmatrix}. \tag{1.16}$$

With this tracking error vector, the integral sliding surface for the system is designed as

$$\sigma_i = \dot{\tilde{x}}_{2i-1} + k_{c2i} \tilde{x}_{2i-1} + k_{c1i} \int_{t_0}^{t} \tilde{x}_{2i-1}(\tau) d\tau - k_{c2i} \tilde{x}_{2i-1}(t_0) - \dot{\tilde{x}}_{2i-1}(t_0) \tag{1.17}$$

where t_0 is the initial time instant, k_{c1i} and k_{c2i} are the design parameters.

After designing the sliding surface, the next step is to design an appropriate control law to drive the sliding variable for reaching the designed sliding surface and thereafter keeping it within the close neighborhood of the sliding surface. In this process, model uncertainties and actuator faults are not taken into account. The corresponding control law can be designed in the following form:

$$\nu_i = \nu_{i0} + \nu_{i1} \tag{1.18}$$

where ν_{i0} is the continuous nominal control part to stabilize the ideal system without uncertainties and disturbances, and ν_{i1} is the discontinuous control part for compensating the perturbations and disturbances to ensure the expected sliding motion.

The nominal control part ν_{i0} is designed by equalizing $\dot{\sigma}_i = 0$, and the disturbance d_i is omitted in this case as shown in the following:

$$\nu_{i0} = g_i^{-1} (\ddot{x}_{2i-1}^d - k_{c2i} \dot{\tilde{x}}_{2i-1} - k_{c1i} \tilde{x}_{2i-1} - h_{1i} f_{1i}(x) - h_{2i} x_{2i-1}). \tag{1.19}$$

The discontinuous control part, which is synthesized to reject the disturbance d_i, is accordingly designed as

$$\nu_{i1} = -k_{c3i} \text{sat}(\sigma_i) \tag{1.20}$$

where k_{c3i} is a positive high gain that makes the sliding surface attractive, and the sat function is defined as [39]

$$\text{sat}(\sigma_i) = \begin{cases} \text{sign}(\sigma_i) & \text{if } |\sigma_i| > \Phi_i \\ \sigma_i/\Phi_i & \text{if } |\sigma_i| \leq \Phi_i \end{cases} \tag{1.21}$$

with Φ_i denoting the boundary layer thickness with a positive value.

In this way, without considering actuator faults and model uncertainties, the control law can be developed as

$$\begin{aligned} u = {} & B_{ui}^{+} g_i^{-1}(\ddot{x}_{2i-1}^{d} - k_{c2i}\dot{\tilde{x}}_{2i-1} - k_{c1i}\tilde{x}_{2i-1} - h_{1i}f_{1i}(x) - h_{2i}x_{2i-1}) \\ & - B_{ui}^{+}k_{c3i}\text{sat}(\sigma_i) \end{aligned} \tag{1.22}$$

where $B_{ui}^{+} = B_{ui}^{T}(B_{ui}B_{ui}^{T})^{-1}$.

With the consideration of model uncertainties, the estimated parameters $\hat{h}_{1i}$, $\hat{h}_{2i}$, and $\hat{g}_i$ need to be employed to derive the corresponding control law. In order to fully use the discontinuous control strategy of SMC, instead of adaptively changing the value of k_{c3i}, the estimated parameter $\hat{g}_i$ is synthesized with k_{c3i} to make change of the discontinuous control gain. Then, by denoting $\hat{\Gamma}_{1i} = \hat{g}_i^{-1}\hat{h}_{1i}$, $\hat{\Gamma}_{2i} = \hat{g}_i^{-1}\hat{h}_{2i}$, and $\hat{\Psi}_i = \hat{g}_i^{-1}$, the developed control law in (1.22) can be rearranged as

$$\begin{aligned} u = {} & B_{ui}^{+}\hat{\Psi}_i(\ddot{x}_{2i-1}^{d} - k_{c2i}\dot{\tilde{x}}_{2i-1} - k_{c1i}\tilde{x}_{2i-1} - k_{c3i}\text{sat}(\sigma_i)) \\ & - \hat{\Gamma}_{1i}f_{1i}(x) - \hat{\Gamma}_{2i}x_{2i-1}. \end{aligned} \tag{1.23}$$

The corresponding online adaptive schemes for estimating the uncertain parameters are designed as

$$\begin{aligned} \dot{\hat{\Gamma}}_{1i} &= f_{1i}(x)\sigma_{\Delta i} \\ \dot{\hat{\Gamma}}_{2i} &= x_{2i-1}\sigma_{\Delta i} \\ \dot{\hat{\Psi}}_i &= (-\ddot{x}_{2i-1}^{d} + k_{c2i}\dot{\tilde{x}}_{2i-1} + k_{c1i}\tilde{x}_{2i-1} + k_{c3i}\text{sat}(\sigma_i))\sigma_{\Delta i} \end{aligned} \tag{1.24}$$

where $\sigma_{\Delta i}$ is the measurement of the algebraic distance between the current sliding variable and the boundary layer defined as

$$\sigma_{\Delta i} = \sigma_i - \Phi_i\text{sat}(\sigma_i). \tag{1.25}$$

It features $\dot{\sigma}_{\Delta i} = \dot{\sigma}_i$ if the sliding variable is outside the boundary layer and $\sigma_{\Delta i} = 0$ if the sliding variable is inside the boundary layer.

1.3.3 Construction of reconfigurable mechanism

After taking into account model uncertainties in controller design, the next problem is to carry out actuator fault estimation and construction of reconfigurable mechanism. In this section, the detailed procedures of designing an active FTC for an unmanned quadrotor helicopter are explained. In order to effectively estimate the fault severity for each actuator, a parallel bank of RNNs are employed as fault identifiers corresponding to the various but limited number of faulty modes that are of most interest or possible in an unmanned quadrotor helicopter. Since there are four actuators in the studied quadrotor helicopter, four RNNs are designed for each individual actuator in order to easily capture fault symptoms with extracted features. Therefore, the fault severity assessment and fault isolation tasks can be performed simultaneously. This proposed scheme makes the problem of fault severity assessment more reliable and yields significant improvement with comparison to a single neural network-based approach that may not remain valid given the occurrence of multiple faults in different actuators.

Considering the closed-loop quadrotor helicopter, after fault occurrence, the onboard control system will increase the corresponding control input to the actuator, and the system tracking error will be changed as well. Therefore, in this section, the tracking errors between the reference command inputs and the quadrotor helicopter outputs, and the control inputs to the actuators can be served as the proper features and used as the inputs of the proposed RNNs. The output of each neural network is the estimated fault severity of the corresponding actuator that is represented by the control effectiveness level of the actuator.

The employed recurrent network has two hidden layers, and the outputs of the first hidden layer are fed back to form a context layer. Then, the general dynamic behavior of the RNN in response to an input vector $U(k) = [u_{n1}(k), u_{n2}(k), \ldots, u_{nr}(k)]^T$ can be described as

$$\begin{aligned} X_{h1}(k+1) &= \varphi_{h1}([X_{h1}(k), U(k)]^T, W_{h1}(k)) \\ X_{h2}(k+1) &= \varphi_{h2}(X_{h1}(k+1), W_{h2}(k)) \\ Y(k+1) &= W_O X_{h2}(k+1) \end{aligned} \tag{1.26}$$

where $U(k) \in \mathbb{R}^r$ is the input vector of the RNN, $X_{h1}(k+1) \in \mathbb{R}^{N1}$ is the output vector of the first hidden layer at time instant $k+1$, $X_{h2}(k+1) \in \mathbb{R}^{N2}$ is the output vector of the second hidden layer, and $Y(k+1) \in \mathbb{R}^q$ is the output vector of the RNN. $W_{h1}(k) \in \mathbb{R}^{N1\times(r+N1)}$, $W_{h2}(k) \in \mathbb{R}^{N2\times N1}$, and $W_O(k) \in \mathbb{R}^{q\times N2}$ represent the synaptic weighting matrix between the input layer and the first hidden layer, between the first hidden layer and the second hidden layer, and between the second hidden layer and the output layer, respectively. φ_{h1} and φ_{h2} represent the activation function for the first hidden layer and the second hidden layer, respectively.

Then, an extended Kalman filter based on back-propagation-through-time is employed to train the designed network. Consider the designed RNN in (1.26) with s synaptic weights and q output nodes. By denoting k as a time instant in the supervised training of the RNN, the entire set of synaptic weights in the RNN computed at time instant k is represented by the vector W_k. It is constructed by stacking the weights associated with the first neuron in the first hidden layer on top of each other, followed by those of the second neuron, and carrying on the same procedure until all the neurons in the first hidden layer are accounted for. Then, the same procedure will be applied to the second hidden layer and the output layer in the network to stack the weights into the vector W_k in the same orderly manner. Therefore, by choosing W_k as the state of the designed RNN, the state-space model of the network under training can be defined as

$$\begin{aligned} W_{k+1} &= W_k + \omega_k \\ D_k &= f(W_k, V_k, U_k) + \nu_k \end{aligned} \tag{1.27}$$

where U_k is the vector of input signals applied to the network, and V_k is the internal state inside the network representing the network activities. The dynamic noise ω_k and measurement noise ν_k are both uncorrelated Gaussian noises with zero mean and covariance matrices $Q_{\omega,k} \in \mathbb{R}^{s\times s}$ and $Q_{\nu,k} \in \mathbb{R}^{q\times q}$, respectively. f is the function denoting the overall nonlinearity of the RNN, and $D_k \in \mathbb{R}^q$ is the desired response.

The next problem is to carry out the supervised training process by using a sequential state estimator. Based on the designed RNN, a back-propagation-through-time-based extended Kalman filter is employed as the training algorithm with consideration of the nonlinear measurement model described in (1.27).

In order to implement the extended Kalman filter training algorithm, the measurement model should be linearized. By applying Taylor series expansion, (1.27) can be approximated as follows:

$$\begin{aligned} W_{k+1} &= W_k + \omega_k \\ \hat{D}_k &= F_k W_k + \nu_k \end{aligned} \tag{1.28}$$

where $F_k \in \mathbb{R}^{q\times s}$ is the measurement matrix of the linearized model that can be represented as

$$F_k = \begin{bmatrix} \dfrac{\partial y_1}{\partial w_1} & \dfrac{\partial y_1}{\partial w_2} & \cdots & \dfrac{\partial y_1}{\partial w_s} \\ \dfrac{\partial y_2}{\partial w_1} & \dfrac{\partial y_2}{\partial w_2} & \cdots & \dfrac{\partial y_2}{\partial w_s} \\ \vdots & \vdots & \vdots & \vdots \\ \dfrac{\partial y_q}{\partial w_1} & \dfrac{\partial y_q}{\partial w_2} & \cdots & \dfrac{\partial y_q}{\partial w_s} \end{bmatrix} \tag{1.29}$$

with q and s denoting the number of output neurons and synaptic weights, respectively, and $y_i (i = 1, 2, \ldots, q)$ denoting the ith output neuron.

Given the training sample $\Gamma = \{U_k, D_k\}_{k=1}^N$, for $k = 1, 2, \ldots, N$, the RNN weighting parameters are updated as follows:

$$\begin{aligned} G_k &= P_{k|k-1} F_k^T [F_k P_{k|k-1} F_k^T + Q_{\nu,k}] \\ \alpha_k &= d_k - Y_k(\hat{W}_{k|k-1}, V_k, U_k) \\ \hat{W}_{k|k} &= \hat{W}_{k|k-1} + \eta_k G_k \alpha_k \\ \hat{W}_{k+1|k} &= \hat{W}_{k|k} \\ P_{k|k} &= P_{k|k-1} - G_k F_k P_{k|k-1} \\ P_{k+1|k} &= P_{k|k} + Q_{\omega,k} \end{aligned} \tag{1.30}$$

which is initialized as

$$\begin{aligned} \hat{W}_{1|0} &= \mathbb{E}(W_1) \\ P_{1|0} &= \delta^{-1} I \end{aligned} \tag{1.31}$$

where $\hat{W}_{k|k-1}$ is the predicted estimation of the weight vector W_k, $\hat{W}_{k|k}$ is the filtered estimation of the weight vector W_k, and $G_k \in \mathbb{R}^{s\times q}$ is the Kalman gain matrix. $P_{k|k-1} \in \mathbb{R}^{s\times s}$ is the prediction error covariance matrix, and $P_{k|k} \in \mathbb{R}^{s\times s}$ is the filtering error covariance matrix. η_k is the learning rate, δ is a constant with small positive value, and $I \in \mathbb{R}^{s\times s}$ is an identity matrix.

By employing the estimated value of the remaining actuator useful level $\hat{L}_c$, the ultimate control law is developed as

$$\begin{aligned} u =& \hat{L}_c^{-1} B_{ui}^{+} \hat{\Psi}_i (\ddot{x}_{2i-1}^d - k_{c2i}\dot{\tilde{x}}_{2i-1} - k_{c1i}\tilde{x}_{2i-1} - k_{c3i}\text{sat}(\sigma_i)) \\ &- \hat{\Gamma}_{1i} f_{1i}(x) - \hat{\Gamma}_{2i} x_{2i-1}. \end{aligned} \tag{1.32}$$

Consider the actuator fault estimation error and denote $\tilde{L}_c^{-1} = L_c^{-1} - \hat{L}_c^{-1}$, the actual generated virtual control input can be formulated as

$$\nu_i = \nu_{id} - B_{ui} L_c \tilde{L}_c^{-1} B_{ui}^{+} \nu_{id} \tag{1.33}$$

where ν_{id} is the demanded virtual control input from the developed adaptive SMC.

Then, denoting $\tilde{\nu}_i = -B_{ui} L_c \tilde{L}_c^{-1} B_{ui}^{+} \nu_{id}$ and substituting (1.32) into (1.15), the following equation can be obtained:

$$\begin{aligned} \ddot{x}_{2i-1} &= h_{1i} f_{1i}(x) + h_{2i} x_{2i-1} + g_i \nu_{id} + g_i \tilde{\nu}_i + d_i \\ &= h_{1i} f_{1i}(x) + h_{2i} x_{2i-1} + (g_i + \tilde{g}_i)\nu_{id} + d_i \\ &= h_{1i} f_{1i}(x) + h_{2i} x_{2i-1} + \hat{g}_i \nu_{id} + d_i. \end{aligned} \tag{1.34}$$

With the designed adaptive control schemes in (1.24), the fault estimation error can be compensated by changing the parameter related to virtual control inputs without affecting system tracking performance.

Theorem 1.1. *Consider an integral-chain nonlinear affine system in (1.15) with actuator faults, model uncertainties, and bounded disturbances. Given the designed integral sliding surface (1.17), by employing the fault severity estimation scheme (1.28) and the state feedback control law (1.32) that is updated by (1.24), the desired tracking performance can be guaranteed with the discontinuous gain chosen as* $k_{c3i} \geq \eta_i + D_d$.

Proof of Theorem 1.1. Consider the following Lyapunov candidate function:

$$\begin{aligned} V =& \sum_{i=1}^{n} \frac{1}{2}\big[\sigma_{\Delta i}^2 + \Psi_i^{-1}(\hat{\Gamma}_{1i} - \Gamma_{1i})^2 + \Psi_i^{-1}(\hat{\Gamma}_{2i} - \Gamma_{2i})^2 \\ &+ \Psi_i^{-1}(\hat{\Psi}_i - \Psi_i)^2\big]. \end{aligned} \tag{1.35}$$

Then, the derivative of the selected Lyapunov candidate function can be calculated as

$$\begin{aligned}
\dot{V} &= \sum_{i=1}^{n} \big[\sigma_{\Delta i}\dot{\sigma}_{\Delta i} + \Psi_i^{-1}(\hat{\Gamma}_{1i} - \Gamma_{1i})\dot{\hat{\Gamma}}_{1i} + \Psi_i^{-1}(\hat{\Gamma}_{2i} - \Gamma_{2i})\dot{\hat{\Gamma}}_{2i} \\
&\quad + \Psi_i^{-1}(\hat{\Psi}_i - \Psi_i)\dot{\hat{\Psi}}_i\big] \\
&= \sum_{i=1}^{n} \big[\sigma_{\Delta i}(\Psi_i^{-1}\Gamma_{1i}f_1(x_2) + \Psi_i^{-1}\Gamma_{2i}f_2(x_2) + \Psi_i^{-1}\hat{\Psi}_i(\ddot{x}_{2i-1}^d \\
&\quad - k_{c2i}\dot{\tilde{x}}_{2i-1} - k_{c1i}\tilde{x}_{2i-1} - k_{c3i}\mathrm{sat}(\sigma_i)) - \Psi_i^{-1}\hat{\Gamma}_{1i}f_1(x_2) \\
&\quad - \Psi_i^{-1}\hat{\Gamma}_{2i}f_2(x_2) + d_i - \ddot{x}_{2i-1}^d + k_{c2i}\dot{\tilde{x}}_{2i-1} + k_{c1i}\tilde{x}_{2i-1}) \\
&\quad + \Psi_i^{-1}(\hat{\Gamma}_{1i} - \Gamma_{1i})\dot{\hat{\Gamma}}_{1i} + \Psi_i^{-1}(\hat{\Gamma}_{2i} - \Gamma_{2i})\dot{\hat{\Gamma}}_{2i} + \Psi_i^{-1}(\hat{\Psi}_i - \Psi_i)\dot{\hat{\Psi}}_i\big] \qquad (1.36) \\
&= \sum_{i=1}^{n} \big[\Psi_i^{-1}(\hat{\Gamma}_{1i} - \Gamma_{1i})(\dot{\hat{\Gamma}}_{1i} - f_1(x_2)\sigma_{\Delta i}) + \Psi_i^{-1}(\hat{\Gamma}_{2i} - \Gamma_{2i})(\dot{\hat{\Gamma}}_{2i} \\
&\quad - f_2(x_2)\sigma_{\Delta i}) + (\Psi_i^{-1}\hat{\Psi}_i - 1)\dot{\hat{\Psi}}_i + (\Psi_i^{-1}\hat{\Psi}_i - 1)(\ddot{x}_{2i-1}^d - \\
&\quad k_{c2i}\dot{\tilde{x}}_{2i-1} - k_{c1i}\tilde{x}_{2i-1})\sigma_{\Delta i} + (d_i - \Psi_i^{-1}\hat{\Psi}_i k_{c3i}\mathrm{sat}(\sigma_i))\sigma_{\Delta i}\big] \\
&= \sum_{i=1}^{n} \big[\Psi_i^{-1}(\hat{\Gamma}_{1i} - \Gamma_{1i})(\dot{\hat{\Gamma}}_{1i} - f_1(x_2)\sigma_{\Delta i}) + \Psi_i^{-1}(\hat{\Gamma}_{2i} - \Gamma_{2i})(\dot{\hat{\Gamma}}_{2i} \\
&\quad - f_2(x_2)\sigma_{\Delta i}) + (\Psi_i^{-1}\hat{\Psi}_i - 1)(\dot{\hat{\Psi}}_i + (\ddot{x}_{2i-1}^d - k_{c2i}\dot{\tilde{x}}_{2i-1} \\
&\quad - k_{c1i}\tilde{x}_{2i-1} - k_{c3i}\mathrm{sat}(\sigma_i))\sigma_{\Delta i}) + d_i\sigma_{\Delta i} - k_{c3i}\mathrm{sat}(\sigma_i)\sigma_{\Delta i}\big].
\end{aligned}$$

Substituting (1.24) into (1.36) leads to

$$\begin{aligned}
\dot{V} &= \sum_{i=1}^{n} \big[-k_{c3i}\mathrm{sat}(\sigma_i)\sigma_{\Delta i} + d_i\sigma_{\Delta i}\big] \\
&\leq \sum_{i=1}^{n} \big[-(\eta_i + D_d)\mathrm{sat}(\sigma_i)\sigma_{\Delta i} + d_i\sigma_{\Delta i}\big] \qquad (1.37) \\
&\leq \sum_{i=1}^{n} \big[-\eta_i|\sigma_{\Delta i}|\big].
\end{aligned}$$

Thus, the system satisfies the standard η-reachability condition, and the tracking performance of the considered system can be guaranteed with the proposed active FTC scheme in the presence of actuator faults, model uncertainties, and bounded disturbances. □

1.4 Simulation results

In order to verify the performance and capability of the proposed active FTC scheme, simulations with respect to different scenarios are performed in this section.

Denoting t_f as fault occurrence time, the element $l_{fi}(t)$ in the diagonal matrix $L_f(t)$ can be modeled as [40]

$$l_{fi}(t) = \begin{cases} 0 & \text{if } t < t_f \\ 1 - e^{-\alpha_i(t-t_f)} & \text{if } t \geq t_f \end{cases} \tag{1.38}$$

where $\alpha_i > 0$ is a scalar denoting the unknown fault evolution rate.

Large values of α_i characterize fast faults development, which are also known as abrupt faults. Due to the fast changes, the ability to correctly detect these abrupt changes is a great challenge for most of the fault diagnosis algorithms [41]. Small values of α_i represent slow faults development, namely, incipient faults. The difficulty in dealing with incipient faults is their small effects on system performance in a closed-loop system, which can be eliminated by the onboard control system.

Therefore, in order to make the proposed fault diagnosis strategy more general and practical, the aforementioned two types of fault modes are considered, i.e., abrupt and incipient faults. An example of the injected abrupt and incipient fault is shown in Figure 1.3. Beyond these faults, noises are added to actuators to make the diagnosis process more challenging and closer to practical cases for the proposed fault diagnosis scheme. In addition, the model uncertainties related to the moment of inertia of the quadrotor helicopter is set to be 20% of the nominal values, $I_{un} = 0.2I_n$ where $I_n = [I_{xx}, I_{yy}, I_{zz}]^T = [0.03, 0.03, 0.04]^T$. Therefore, the moment of inertia used to develop the active FTC scheme is chosen as $I = [0.024, 0.024, 0.032]^T$.

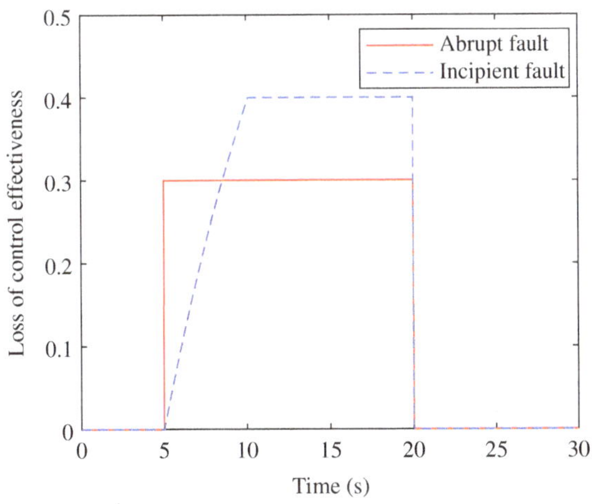

Figure 1.3 Two types of faults injected to actuators

Table 1.1 Fault scenarios considered for active FTC problem in simulation

Fault scenario	Type and magnitude	Position	Time period (s)
1	Abrupt fault	Actuator 1	5–20
2	Incipient fault	Actuator 1	5–20

The drag coefficients are chosen as $[K_4, K_5, K_6]^T = [0.01, 0.01, 0.01]^T$. The value of α_i to generate abrupt fault and incipient fault for the ith actuator is chosen as 10 and 0.05, respectively. Focusing on the fault pattern of the actuators, two different faulty scenarios are considered, which are detailed in Table 1.1.

According to Figure 1.2, four RNNs are employed for actuator faults diagnosis in the closed-loop quadrotor helicopter system. Each network has four layers where the input layer has two neurons, each of the two hidden layers has ten neurons with sigmoidal activation functions, and the output layer has one neuron with linear function. The network weights are initialized with small random values to prevent early saturation of the hidden layer neurons. The inputs of the four networks are $[u_1, \tilde{\theta}]^T$, $[u_2, \tilde{\theta}]^T$, $[u_3, \tilde{\phi}]^T$, and $[u_4, \tilde{\phi}]^T$, respectively. u_i $(i = 1, 2, 3, 4)$ is the PWM input of the ith motor of the quadrotor helicopter as the control input to the system. $\tilde{\theta}$ and $\tilde{\phi}$ represent pitch angle tracking error and roll angle tracking error, respectively. The output of each network is the remaining useful level of the corresponding actuator, which can be directly used by the control system to reconfigure the controller and accommodate actuator faults.

To train the designed neural networks, multiple faults with different types and magnitudes are considered and included in the training data. Due to physical limits for avoiding losing control of the quadrotor helicopter, complete actuator failure cases are not considered in this work. Thus, for generating the training samples, the loss of control effectiveness magnitude is set to be between 0 and 0.8. The neural networks are individually trained by using 8,000 data points with 200 epochs. The learning rate changes from 1 to 1×10^{-5}, the covariance matrix of the dynamic noise changes from 1×10^{-2} to 1×10^{-6}, and the covariance matrix of the measurement noise changes from 100 to 5 with the change of training epoch.

1.4.1 Fault estimation and accommodation results

Fault scenario 1: An abrupt fault with 25% loss of control effectiveness is injected into actuator 1 between 5 and 20 s. In this case, the control input u_1 and pitch angle tracking error $\tilde{\theta}$ will be increased. By employing these two inputs, the fault estimation performance is shown in Figure 1.4. After fault occurrence at 5 s, it can be precisely estimated by the proposed fault estimation strategy. Then, at 20 s when the actuator fault is recovered, the proposed fault estimation strategy can also give a correct result. By using the estimated fault magnitude, the original system tracking performance is maintained as shown in Figure 1.5.

Fault scenario 2: An incipient fault with 25% loss of control effectiveness is introduced to actuator 1 between 5 and 20 s. The difficulty for estimating such kind

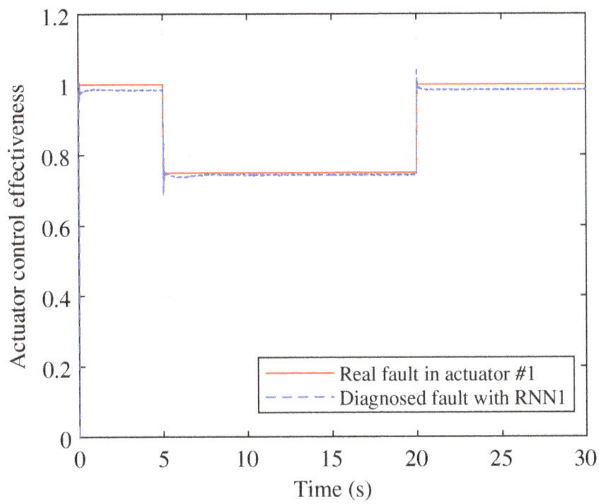

Figure 1.4 *Fault estimation performance of RNN1 for actuator 1 in the presence of abrupt fault*

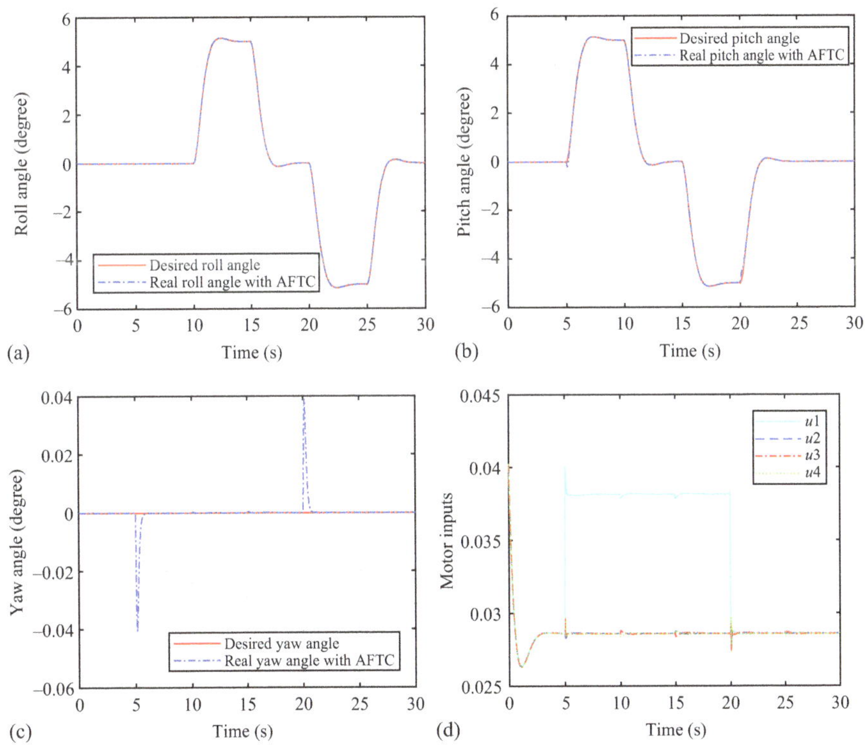

Figure 1.5 *System tracking performance in the presence of abrupt fault: (a) roll motion, (b) pitch motion, (c) yaw motion, and (d) motor inputs of the system*

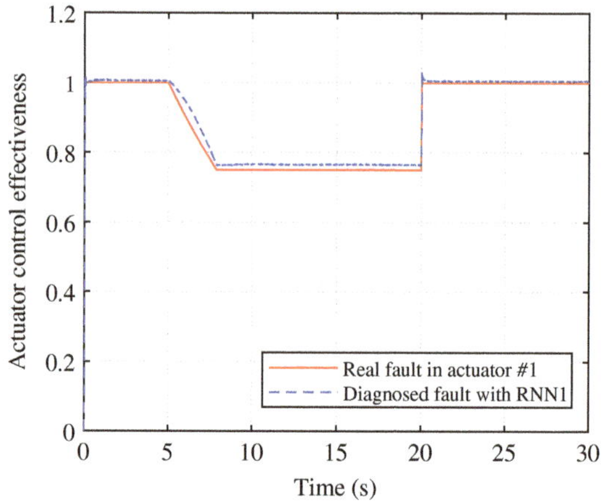

Figure 1.6 Fault estimation performance of RNN1 for actuator 1 in the presence of incipient fault

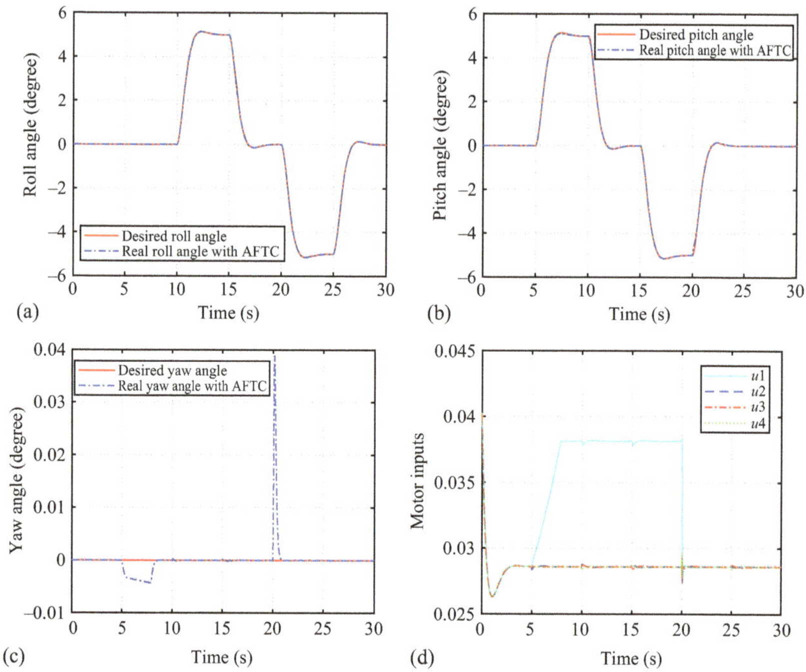

Figure 1.7 System tracking performance in the presence of incipient fault: (a) roll motion, (b) pitch motion, (c) yaw motion, and (d) motor inputs of the system

of faults in the closed-loop system lies in that the feedback control system will help one to maintain system tracking performance due to the low fault evolution rate. The fault severity assessment performance and corresponding system tracking performance with the proposed AFTC are illustrated in Figures 1.6 and 1.7. Similar to scenario 1, the proposed fault estimation scheme shows that an accurate fault severity assessment result and system tracking performance is ensured as well.

1.5 Conclusions

In this chapter, an active FTC strategy based on adaptive SMC and RNN is proposed to accommodate actuator faults and model uncertainties for an unmanned quadrotor helicopter. With the help of the developed online adaptive schemes, the proposed control strategy can adaptively generate appropriate control signals to compensate model uncertainties without the knowledge of uncertainty bounds to maintain tracking performance and stability of the quadrotor helicopter. Then, by designing a parallel bank of RNNs, the severity of the actuator faults can be precisely and reliably estimated, which is then synthesized with the proposed adaptive SMC to reconfigure the controller and accommodate actuator faults. The simulation results show that the proposed scheme can effectively estimate and accommodate various kinds of actuator faults in the closed-loop quadrotor helicopter system. Although the developed FDD and FTC scheme in this chapter has been based on a quadrotor UAV, it can be easily extended and applied to other UAVs and robotic systems as well.

References

[1] Lee H, Kim HJ. Estimation, control, and planning for autonomous aerial transportation. IEEE Transactions on Industrial Electronics. 2017;64(4): 3369–3379.

[2] Michael N, Fink J, Kumar V. Cooperative manipulation and transportation with aerial robots. Autonomous Robots. 2011;30(1):73–86.

[3] Yuan C, Zhang YM, Liu ZX. A survey on technologies for automatic forest fire monitoring, detection, and fighting using unmanned aerial vehicles and remote sensing techniques. Canadian Journal of Forest Research. 2015;45(7): 783–792.

[4] Capitán J, Merino L, Ollero A. Cooperative decision-making under uncertainties for multi-target surveillance with multiples UAVs. Journal of Intelligent & Robotic Systems. 2016;84(1–4):371–386.

[5] Li Z, Liu Y, Walker R, *et al.* Towards automatic power line detection for a UAV surveillance system using pulse coupled neural filter and an improved Hough transform. Machine Vision and Applications. 2010;21(5):677–686.

[6] Dunbabin M, Marques L. Robots for environmental monitoring: Significant advancements and applications. IEEE Robotics & Automation Magazine. 2012;19(1):24–39.

[7] Xiang H, Tian L. Development of a low-cost agricultural remote sensing system based on an autonomous unmanned aerial vehicle (UAV). Biosystems Engineering. 2011;108(2):174–190.

[8] Siebert S, Teizer J. Mobile 3D mapping for surveying earthwork projects using an unmanned aerial vehicle (UAV) system. Automation in Construction. 2014;41:1–14.

[9] Ducard GJ. Fault-Tolerant Flight Control and Guidance Systems: Practical Methods for Small Unmanned Aerial Vehicles. London: Springer Science & Business Media; 2009.

[10] Zhang YM, Jiang J. Bibliographical review on reconfigurable fault-tolerant control systems. Annual Reviews in Control. 2008;32(2):229–252.

[11] Yu X, Jiang J. A survey of fault-tolerant controllers based on safety-related issues. Annual Reviews in Control. 2015;39:46–57.

[12] Yin S, Xiao B, Ding SX, *et al.* A review on recent development of spacecraft attitude fault tolerant control system. IEEE Transactions on Industrial Electronics. 2016;63(5):3311–3320.

[13] Zhang YM, Chamseddine A, Rabbath C, *et al.* Development of advanced FDD and FTC techniques with application to an unmanned quadrotor helicopter testbed. Journal of the Franklin Institute. 2013;350(9):2396–2422.

[14] Wang B, Yu X, Mu LX, *et al.* Disturbance observer-based adaptive fault-tolerant control for a quadrotor helicopter subject to parametric uncertainties and external disturbances. Mechanical Systems and Signal Processing. 2019;120:727–743.

[15] Hu Q, Xiao B. Fault-tolerant sliding mode attitude control for flexible spacecraft under loss of actuator effectiveness. Nonlinear Dynamics. 2011; 64(1–2):13–23.

[16] Xiao B, Huo M, Yang X, *et al.* Fault-tolerant attitude stabilization for satellites without rate sensor. IEEE Transactions on Industrial Electronics. 2015;62(11):7191–7202.

[17] Shen Q, Jiang B, Shi P. Fault Diagnosis and Fault-Tolerant Control Based on Adaptive Control Approach. Cham: Springer Science & Business Media; 2017.

[18] Pouliezos A, Stavrakakis GS. Real Time Fault Monitoring of Industrial Processes. Dordrecht: Springer Science & Business Media; 2013.

[19] Mohammadi R, Naderi E, Khorasani K, *et al.* Fault diagnosis of gas turbine engines by using dynamic neural networks. In: Proceedings of the ASME Turbo Expo 2010: Power for Land, Sea, and Air; The American Society of Mechanical Engineers, Glasgow, UK, June 14–18, 2010. p. 365–376.

[20] Talebi HA, Khorasani K, Tafazoli S. A recurrent neural-network-based sensor and actuator fault detection and isolation for nonlinear systems with application to the satellite's attitude control subsystem. IEEE Transactions on Neural Networks. 2009;20(1):45–60.

[21] Mahmoud M, Jiang J, Zhang YM. Active Fault Tolerant Control Systems: Stochastic Analysis and Synthesis. Berlin, Heidelberg: Springer Science & Business Media; 2003.

[22] Jiang J, Zhao Q. Design of reliable control systems possessing actuator redundancies. Journal of Guidance, Control, and Dynamics. 2000;23(4):709–718.

[23] Liao F, Wang JL, Yang GH. Reliable robust flight tracking control: An LMI approach. IEEE Transactions on Control Systems Technology. 2002;10(1): 76–89.

[24] Zhang YM, Jiang J. Active fault-tolerant control system against partial actuator failures. IEE Proceedings-Control Theory and Applications. 2002;149(1): 95–104.

[25] Badihi H, Zhang YM, Hong H. Fuzzy gain-scheduled active fault-tolerant control of a wind turbine. Journal of the Franklin Institute. 2014;351(7): 3677–3706.

[26] Jiang J, Yu X. Fault-tolerant control systems: A comparative study between active and passive approaches. Annual Reviews in Control. 2012;36(1):60–72.

[27] Isermann R. Fault-Diagnosis Applications: Model-based Condition Monitoring: Actuators, Drives, Machinery, Plants, Sensors, and Fault-Tolerant Systems. Berlin, Heidelberg: Springer Science & Business Media; 2011.

[28] Edwards C, Lombaerts T, Smaili H. Fault Tolerant Flight Control: A Benchmark Challenge. Berlin, Heidelberg: Springer Science & Business Media; 2010.

[29] Chen RH, Speyer JL. Sensor and actuator fault reconstruction. Journal of Guidance Control and Dynamics. 2004;27(2):186–196.

[30] Wang B, Zhang YM. Adaptive sliding mode fault-tolerant control for an unmanned aerial vehicle. Unmanned Systems. 2017;5(04):209–221.

[31] Utkin VI. Sliding Modes in Control and Optimization. Berlin, Heidelberg: Springer Science & Business Media; 2013.

[32] Shtessel Y, Edwards C, Fridman L, *et al.* Sliding Mode Control and Observation. New York, NY: Birkhäuser; 2014.

[33] Utkin VI. Variable structure systems with sliding modes. IEEE Transactions on Automatic Control. 1977;22(2):212–222.

[34] Utkin VI. Sliding mode control design principles and applications to electric drives. IEEE Transactions on Industrial Electronics. 1993;40(1):23–36.

[35] Edwards C, Spurgeon S. Sliding Mode Control: Theory and Applications. Boca Raton, FL: CRC Press; 1998.

[36] Liu J, Wang X. Advanced Sliding Mode Control for Mechanical Systems: Design, Analysis and MATLAB Simulation. Beijing: Tsinghua University Press and Berlin Heidelberg: Springer-Verlag; 2011.

[37] Wang B, Zhang YM. An adaptive fault-tolerant sliding mode control allocation scheme for multirotor helicopter subject to simultaneous actuator faults. IEEE Transactions on Industrial Electronics. 2018;65(5):4227–4236.

[38] Bresciani T. Modelling, Identification and Control of a Quadrotor Helicopter. Lund University. Sweden; 2008. Available from: http://lup.lub.lu.se/student-papers/record/8847641.

[39] Slotine JJ, Sastry SS. Tracking control of non-linear systems using sliding surfaces, with application to robot manipulators. International Journal of Control. 1983;38(2):465–492.

[40] Zhang X, Polycarpou MM, Parisini T. A robust detection and isolation scheme for abrupt and incipient faults in nonlinear systems. IEEE Transactions on Automatic Control. 2002;47(4):576–593.

[41] Abbaspour A, Aboutalebi P, Yen KK, *et al.* Neural adaptive observer-based sensor and actuator fault detection in nonlinear systems: Application in UAV. ISA Transactions. 2016;67:317–329.

Chapter 2

Control techniques to deal with the damage of a quadrotor propeller

Fabio Ruggiero[1], Diana Serra[1], Vincenzo Lippiello[1], and Bruno Siciliano[1]

This chapter can be considered a tutorial to guide the readers toward the implementation of active fault-tolerant control systems dealing with the damage of a propeller of an unmanned aerial vehicle. The addressed aerial device is a quadrotor with fixed propellers. The presented methodology also supposes to turn off the motor, the opposite of the broken one. In this way, a birotor configuration with fixed propellers is achieved. State-of-the-art approaches, using a PID-based controller and a backstepping controller, are presented in a tutorial form, thus neglecting the stability proofs, to leave room to a fast and concise description of the implementation procedures.

2.1 Introduction

Civilians, servicemen, mass media, and researchers paid close attention to unmanned aerial vehicles (UAVs) in the last decade. The growth of applications in which vertical takeoff and landing UAVs are becoming present in everyday life is incredible. Professional photographers and film-makers are now always accompanied by certified UAV pilots to see the scene from different points of view. One of the most prominent electronic commerce websites around the world is planning to deliver packages to customers' hands in 30 min, or less, using hexacopters [1]. The White House and the National Science Foundation accelerated the use of UAVs for civilian uses in monitoring and inspection of physical infrastructures, quick response to disasters, agricultural and meteorological domains with a considerable amount of funding [2]. Some companies started to think about a sort of personal UAV equipped with a camera to record self-movies [3], particularly indicated for a sportsman. Oil and gas facilities see the nondestructive measurement tests as a prosperous application for the UAVs in refineries since they are currently performed by human operators climbing huge and costly scaffolding. Several robotic commercial solutions are currently

[1]PRISMA Lab, Department of Electrical Engineering and Information Technology, University of Naples Federico II, Naples, Italy

(or are ready-to-be) available like the APELLIX drone [www.apellix.com], the Texo Drone Survey and Inspection platform [www.texodroneservices.co.uk/blog/56], and the RoNik Inspectioneering UT device [www.inspectioneering.eu].

Researchers see in the UAVs an unstable dynamic system that perfectly fits the challenge of implementing safe and robust control and mechatronic designs to accomplish every application mentioned above and go further beyond. As a matter of fact, the UAVs can be both employed for passive tasks like inspection and surveillance, and active tasks like grasping and aerial manipulation. This scenario requires the introduction not only of rules and regulation but also of safe controllers. An overview about control applied to UAVs, and the related applications in aerial manipulation, can be found in [4]. The design of controllers for safety-critical systems is thus required: fault detection, diagnosis, and tolerance approaches become essential. Fault tolerance methods try to maintain the same functionalities of the system, allowing reduced performance when damage appears [5]. Passive fault-tolerant control systems (PFTCSs) do not alter the structure of the controller, while the active fault tolerant control systems (AFTCSs) reconfigure the control actions to guarantee stability and acceptable performance of the system [6].

This chapter wants to be a tutorial to guide the reader toward the implementation of an AFCTS dealing with the damage of a UAV propeller. In particular, the quadrotor with fixed propellers is the aerial device taken into account. The quadrotor is a versatile and agile device with four rotors lying on the same plane, with the adjacent propellers rotating in the opposite directions [https://www.youtube.com/watch?v=w2itwFJCgFQ]. This chapter resumes the results provided in [7,8] in a tutorial form, thus neglecting the stability proofs, to leave room to a fast and concise description of the implementation procedures.

The outline of the chapter is as follows. The problem statement and the literature review about it are provided in the next section. The notation and mathematical modeling are introduced in Section 2.3. The control methodologies are described in Section 2.4 with the related control schemes. The comparison between the proposed methodologies is carried out through numerical simulation within Section 2.5. Section 2.6 concludes the chapter.

2.2 Problem statement

In this chapter, the damage of a quadrotor propeller is addressed. To cope with such a situation, the adopted solution is to turn off the damaged motor completely. Besides, the motor aligned on the same quadrotor axis of the broken propeller is turned off as well. This configuration is a birotor with fixed propellers. Such a solution also includes the case of two broken propellers provided that they are aligned on the same axis. Other case studies are not taken into account.

It is assumed that the system already detected the failure: it is out the scope of this work the implementation of a fault-detection procedure. Besides, the control methodologies addressed in this chapter starts from a birotor configuration with a given initial condition of the state. The switch between the nominal quadrotor working condition into the birotor emergency condition is out of this scope of this chapter. Therefore, it is possible to provide the following problem formulation.

- **Problem**. *Control the position of the birotor from an initial position to the desired one in the Cartesian space, along a desired trajectory.*

The next sections solve the problem above by providing a tutorial extracting the solutions from [7,8].

In the literature, several alternative solutions can be found solving similar problems, or considering the parts that are neglected in this chapter.

Regarding the fault-detection, a method was proposed in [9]. A Luenberger observer was instead employed in [10] along with a sliding mode controller. When the loss in the efficiency is verified in each of the four propellers, the gain scheduling approach proposed in [11] may be employed.

A thorough literature review reveals that there exist many methods addressing the control of a UAV in the case of one or more motor partial failures. Most of them belong to the AFTCS class. Instead, a PFTCS for a coaxial counter-rotating octorotor based on the super-twisting algorithm was designed in [12]. The proposed second-order sliding mode technique ensured robustness with respect to uncertainties and disturbances, and it could also deal directly with faults and failures by compensating for the actuator loss in the system without prior knowledge on the fault, its location, and severity. About AFTCS, a method was tackled in [13] to estimate the aerial vehicle model after the failure detection (50% loss in the efficiency of a quadrotor propeller), guaranteeing the stability of the UAV. A backstepping approach was proposed in [14] but only 25% performance loss in the motors was considered. A comparison among different methodologies was compared in [15] for a 50% loss in propellers performance.

On the other hand, additional methods consider the complete failure of a quadrotor propeller. Despite the impossibility to control the yaw angle, a feedback linearization with a proportional-derivative (PD)-based controller was employed in [16] to control a quadrotor with a wholly broken motor. However, no stability analysis and coupling effects between inner and outer loops were considered as instead in [7]. A controller for an equidistant trirotor was designed in [17], but the formulation is available only for spiral motions. An H-infinity loop shaping technique was adopted in [18] for the safe landing of a quadrotor with a propeller failure. Periodic solutions were exploited in [19] together with a linear-quadratic regulator (LQR) to control the quadrotor in the case of single, two opposing, or three propellers failure. A reconfigurability analysis of UAVs with four and six rotors was investigated in [20]. The hovering control in the presence of the blockage or the complete loss of a rotor was pursued. The quadrotor is shown to be not reconfigurable, while the hexrotors can handle a blockage without difficulties. Furthermore, it is shown that whenever a UAV is stabilisable, it is also possible to recover the system's behavior accurately. A three-loop hybrid nonlinear controller was instead designed in [21] to achieve high-speed flight of a quadrotor with a complete loss of a single rotor. Robustness against complex aerodynamic effects brought by both fast translational and fast spinning motion of a damaged quadrotor was shown. As a result, the UAV can continue high-speed missions instead of having an emergency landing immediately. A morphing quadrotor was shown in [22] that can stay in hover after rotor failure. Similarly to [7,8], an emergency fault-tolerant controller was developed in [23] employing a birotor through a bounded control law. Instead of using a birotor, an emergency controller based on quadrotor to trirotor conversion maneuver was applied experimentally in [24] on the AR Drone 2 platform to recover from a total failure of one rotor. The proposed controller was based on control re-allocation, where the infected actuator was exempted from the control effect, and control efforts are redistributed among the remaining actuators. Hence, the emergency controller transformed the infected quadrotor into a trirotor, using a simple proportional–integral–derivative (PID)-based controller. To control the attitude of a quadrotor and to increase robustness against model uncertainties and actuator faults, a sliding mode controller was applied in [25]. Besides, an adaptive fuzzy system was employed to compensate for the estimation error of nonlinear functions and faulty parts. In order to avoid instability and to increase the robustness of the closed-loop system, a new parallel fuzzy system was proposed along with a main fuzzy system. The adaptation rules of the main and parallel fuzzy systems were extracted from Lyapunov's stability theory.

A complete AFCTS architecture, including error detection, fault isolation, and system recovery, was presented in [26]. The diagnosis system was based on the motor speeds and currents measurements. Once the motor failure or the rotor loss is diagnosed, a recovery algorithm was applied using the pseudo-inverse control allocation approach to redistribute the control efforts among the remaining actuators. An approach coping with not only fault detection and isolation but also fault-tolerant control was proposed in [27]. An incremental nonlinear dynamic inversion approach was introduced to design the fault-tolerant controller for the quadrotor in the presence of the fault. The complete AFTCS enabled the quadrotor to achieve any position even after the complete loss of one rotor.

Recently, multicopter (more than four or six) and tilting propellers can provide novel solutions and control approaches and also achieve an actuator redundancy in the system in the case of failure of one or more motors.

2.3 Modeling

In this section, the mathematical models of both the quadrotor and the birotor are introduced.

2.3.1 Quadrotor

Let $\Sigma_i - \{x_i, y_i, z_i\}$ and $\Sigma_b - \{x_b, y_b, z_b\}$ be the inertial world-fixed frame and the body frame placed at the center of the UAV, respectively. To clarify the notation, the main terms are resumed in the following:

$R_b \in SO(3)$	rotation matrix from Σ_b to Σ_i
$\eta_b = \begin{bmatrix}\phi & \theta & \psi\end{bmatrix}^T \in \mathbb{R}^3$	roll, pitch, and yaw angles, respectively, representing a minimal orientation representation

$\omega_b^b \in \mathbb{R}^3$	angular velocity of Σ_b with respect to Σ_i expressed in Σ_b
$m \in \mathbb{R}$	mass of the vehicle
$I_b = \text{diag}\{I_x, I_y, I_z\} \in \mathbb{R}^{3\times3}$	inertia matrix with respect to Σ_b
$g \in \mathbb{R}$	gravity acceleration
$p_b = \begin{bmatrix} x & y & z \end{bmatrix}^T \in \mathbb{R}^3$	position of Σ_b in Σ_i
$u \in \mathbb{R}^+$	the total thrust perpendicular to the propeller's rotation plane
$\tau_b^b = \begin{bmatrix} \tau_\phi & \tau_\theta & \tau_\psi \end{bmatrix}^T \in \mathbb{R}^3$	torque around the axes of Σ_b expressed in the same frame
$I_n \in \mathbb{R}^{n\times n}$	identity matrix of proper dimensions
$e_3 = \begin{bmatrix} 0 & 0 & 1 \end{bmatrix}^T \in \mathbb{R}^3$	definition of the vertical axis vector in Σ_i
$F_p = \text{diag}\{F_{px}, F_{py}, F_{pz}\} \in \mathbb{R}^{3\times3}$	linear velocity air drag matrix
$F_o = \text{diag}\{F_{ox}, F_{oy}, F_{oz}\} \in \mathbb{R}^{3\times3}$	angular velocity air drag matrix

Neglecting the gyroscopic torques, due to the combination of the UAV rotation and the propellers, the equations of motion for a UAV can be written as

$$m\ddot{p}_b = mg - uRe_3 - F_p\dot{p}_b, \tag{2.1a}$$

$$I_b\dot{\omega}_b^b = -\omega_b^b \times I_b\omega_b^b - F_o\omega_b + \tau_b^b, \tag{2.1b}$$

$$\dot{R}_b = R_bS(\omega_b^b), \tag{2.1c}$$

where $\times$ represents the cross-product operator, $S(\cdot) \in \mathbb{R}^{3\times3}$ is the skew-symmetric matrix. The configuration of the UAV is thus defined by the position p_b and the attitude expressed by the rotation matrix R_b, which lies within $SO(3) = \{R \in \mathbb{R}^{3\times3} : R^TR = I_3, \det(R) = 1\}$.

The dynamic model (2.1a)–(2.1c) has the linear configuration expressed with respect to Σ_i, while the angular configuration is expressed with respect to Σ_b. For control purposes, neglecting air drag, the equations of motion can be both expressed in the inertial frame as

$$m\ddot{p}_b = mg - uRe_3, \tag{2.2a}$$

$$M\ddot{\eta}_b = -C\dot{\eta}_b + Q^T\tau_b^b, \tag{2.2b}$$

with $M = Q^TI_bQ \in \mathbb{R}^{3\times3}$ the positive definite mass matrix, $C = Q^TS(Q\dot{\eta}_b)I_bQ + Q^TI_B\dot{Q} \in \mathbb{R}^{3\times3}$ the Coriolis matrix, and

$$Q = \begin{bmatrix} 1 & 0 & -s_\theta \\ 0 & c_\phi & c_\theta s_\phi \\ 0 & -s_\phi & c_\theta c_\phi \end{bmatrix} \in \mathbb{R}^{3\times3},$$

a suitable transformation matrix such that $\omega_b^b = Q\dot{\eta}_b$. Notice the use of notation s_α and c_α for the $\sin(\alpha)$ and $\cos(\alpha)$, respectively.

The dynamic models (2.1a)–(2.1c) and (2.2a) and (2.2b) can be referred to most of the available UAVs. The control inputs are generalized as the total thrust u and the torques τ_b^b in the body frame. In this chapter, a particular UAV like the quadrotor is addressed. Looking at Figure 2.1, this vehicle has four rotors and propellers located in the same plane in a cross or X configuration symmetrical to the center of mass. Neglecting the dynamics of the rotors and the propellers, it is assumed that the torque generated by each propeller is directly proportional to the thrust. Notice that the rotation clockwise of propellers 1 and 3, and counterclockwise of 2 and 4, generate a thrust always directed in the opposite direction of z_b-axis, but along it. These common assumptions yield the following equations between the control input and the propeller speed $\omega_i \in \mathbb{R}$, with $i = 1, \ldots, 4$,

$$u = \rho_u(\omega_1^2 + \omega_2^2 + \omega_3^2 + \omega_4^2), \tag{2.3a}$$

$$\tau_\phi = l\rho_u(\omega_2^2 - \omega_4^2), \tag{2.3b}$$

$$\tau_\theta = l\rho_u(\omega_3^2 - \omega_1^2), \tag{2.3c}$$

$$\tau_\psi = \rho_c(\omega_1^2 - \omega_2^2 + \omega_3^2 - \omega_4^2), \tag{2.3d}$$

with $l > 0$ the distance between each propeller and the quadrotor center of mass, $\rho_u > 0$ and $\rho_c > 0$ two aerodynamic parameters.

2.3.2 Birotor

This chapter proposes the following technique to deal with motor failure. Suppose that a motor is either completely broken or losing power. The choice carried out here is to completely turn off both such a damaged motor and the symmetric one to the center of mass, even though if this last is correctly working. Without loss of generality, with reference to Figure 2.1, suppose that the motor 2 is damaged. Then, it is decided to turn off also motor 4. Therefore, $\omega_2 = \omega_4 = 0$ yields $\tau_\phi = 0$ from (2.3b). The resulting configuration is a birotor with fixed propellers. The equations of motion are

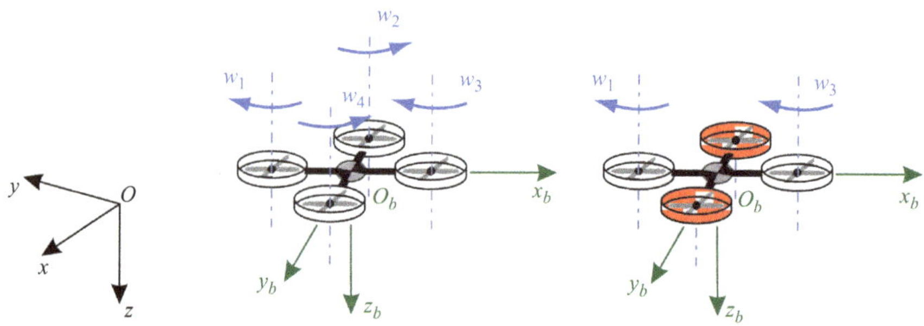

Figure 2.1 Left: The quadrotor and related frames: black; the inertial frame Σ_i: green; the body frame Σ_b: blue; the speed and label of each motor. Right: The birotor configuration with in red the turned-off propellers

again either (2.1a)–(2.1c) or (2.2a) and (2.2b), while the computation of the propellers speed in (2.3a)–(2.3d) changes into

$$u = \rho_u(\omega_1^2 + \omega_3^2), \tag{2.4a}$$
$$\tau_\phi = 0, \tag{2.4b}$$
$$\tau_\theta = l\rho_u(\omega_3^2 - \omega_1^2), \tag{2.4c}$$
$$\tau_\psi = \rho_c(\omega_1^2 + \omega_3^2). \tag{2.4d}$$

If motor 4 is damaged, the situation is identical by turning off motor 2. Instead, suppose that either motor 1 or 3 is damaged, then motor 3 or 1 is turned off, respectively. In this case, the computation of the propellers speed in (2.3a)–(2.3d) changes into

$$u = \rho_u(\omega_2^2 + \omega_4^2),$$
$$\tau_\phi = l\rho_u(\omega_2^2 - \omega_4^2),$$
$$\tau_\theta = 0,$$
$$\tau_\psi = -\rho_c(\omega_2^2 + \omega_4^2).$$

Without loss of generality, in the remainder of the chapter, only the case with motor 2 and 4 turned is considered. Notice that the case of two defective motors but not aligned on the same quadrotor axis is out of the scope of this work.

An analysis of (2.4a)–(2.4d) reveals that τ_ψ cannot be freely controlled since it is impossible to change its sign. On the other hand, it is possible to control the total thrust u and the pitch torque τ_θ independently. Folding (2.4a) into (2.4d) yields

$$\tau_\psi = \bar{\tau}_\psi = \frac{\rho_c}{\rho_u}u, \tag{2.5}$$

which is the spinning torque of the birotor around the z_b-axis of Σ_b and depending on the current total thrust scaled by some aerodynamic parameters. The effect is a continuous rotation of the birotor around its vertical axis.

2.4 Control design

This section addresses two control solutions to deal with birotor control. In both of them it is demonstrated that, apart from the yaw angle that is uncontrollable as seen above, the birotor can reach any position in the Cartesian space. Despite the uncontrollability of the yaw angle, the device can be safely landed in a predetermined location following a desired emergency landing trajectory.

The two control techniques are, namely, a PID-based approach and a backstepping procedure. Since this chapter is a tutorial to guide the reader step-by-step in the implementation of the control laws, the proofs of stability of each of the control schemes mentioned above are not here reported. The interested reader can find them in [7,8], respectively.

In order to implement the control schemes, the following assumptions must be introduced

- **Assumption 1.** The birotor configuration is always within the configuration space defined by $\mathcal{Q} = \{p_b \in \mathbb{R}^3, \eta_b \in \mathbb{R}^3 : \theta \neq \pi/2 + k\pi, \phi \neq \pi/2 + k\pi, k \in \mathbb{Z}\}$.
- **Assumption 2.** The planned linear acceleration $\ddot{p}_d$ is bounded.

While the second assumption can be easily guaranteed by a suitable planner, the first assumption usually depends on initial conditions of the birotor once the controller is activated. Notice that the desired quantities are denoted by the subscript *d*.

2.4.1 PID control scheme

The design of the PID-based control scheme starts from the following control law:

$$\tau_\theta = \left(\frac{I_y c_\phi + I_y s_\phi^2}{c_\phi}\right)\bar{\tau}_\theta + \alpha_1, \tag{2.6}$$

where $\bar{\tau}_\theta \in \mathbb{R}$ is a virtual control input, and

$$\begin{aligned}
\alpha_1 &= I_y s_\phi \frac{\bar{\tau}_\psi - \alpha_2}{I_z c_\phi} + \alpha_3,\\
\alpha_2 &= -I_x\dot{\phi}\dot{\theta}c_\phi + I_y\dot{\phi}\dot{\theta}c_\phi - I_z\dot{\phi}\dot{\theta}c_\phi + I_x\dot{\psi}^2 c_\theta s_\phi s_\theta\\
&\quad -I_y\dot{\psi}^2 c_\theta s_\phi s_\theta - I_x\dot{\phi}\dot{\psi}c_\theta s_\phi + I_y\dot{\phi}\dot{\psi}c_\theta s_\phi\\
&\quad -I_z\dot{\phi}\dot{\psi}c_\theta s_\phi + I_x\dot{\psi}\dot{\theta}c_\phi s_\theta - I_y\dot{\psi}\dot{\theta}c_\phi s_\theta - I_z\dot{\psi}\dot{\theta}c_\phi s_\theta,\\
\alpha_3 &= -I_x\dot{\phi}\dot{\theta}s_\phi - I_y\dot{\phi}\dot{\theta}s_\phi + I_z\dot{\phi}\dot{\theta}s_\phi - I_x\dot{\psi}^2 c_\phi c_\theta s_\theta\\
&\quad +I_z\dot{\psi}^2 c_\phi c_\theta s_\theta + I_x\dot{\phi}\dot{\psi}c_\phi c_\theta + I_y\dot{\phi}\dot{\psi}c_\phi c_\theta\\
&\quad -I_z\dot{\phi}\dot{\psi}c_\phi c_\theta + I_x\dot{\psi}\dot{\theta}s_\phi s_\theta - I_y\dot{\psi}\dot{\theta}s_\phi s_\theta - I_z\dot{\psi}\dot{\theta}s_\phi s_\theta.
\end{aligned}$$

Substituting (2.6) into (2.2b) and considering (2.4b) and (2.5) yield

$$\ddot{\theta} = \bar{\tau}_\theta, \tag{2.7}$$

meaning that it is possible to entirely control the pitch angle through $\bar{\tau}_\theta$. Let $\theta_d \in \mathbb{R}$ the desired pitch angle. A simple PD controller can be designed for the pitch angle as follows:

$$\bar{\tau}_\theta = \ddot{\theta}_d + k_{d,\theta}\dot{e}_\theta + k_{p,\theta}e_\theta, \tag{2.8}$$

with $e_\theta = \theta_d - \theta \in \mathbb{R}$, $\dot{e}_\theta = \dot{\theta}_d - \dot{\theta} \in \mathbb{R}$, $\ddot{e}_\theta = \ddot{\theta}_d - \ddot{\theta} \in \mathbb{R}$, and $k_{p,\theta} > 0$ and $k_{d,\theta} > 0$ two gains.

Let $\mu = \begin{bmatrix}\mu_x & \mu_y & \mu_z\end{bmatrix}^T \in \mathbb{R}^3$ be the acceleration of the birotor, expressed in Σ_i, where the magnitude is the desired thrust u_d produced by the remaining propellers, while the orientation $R_d \in SO(3)$ is given by the desired pitch and the current measured roll and yaw as follows:

$$R_d = \begin{bmatrix} c_{\theta_d} c_\psi & s_\phi s_{\theta_d} c_\psi - c_\phi s_\psi & c_\phi s_{\theta_d} c_\psi + s_\phi s_\psi \\ c_{\theta_d} s_\psi & s_p his_{\theta_d} s_\psi + c_\phi c_\psi & c_\phi s_{\theta_d} s_\psi - s_\phi c_\psi \\ -s_{\theta_d} & s_\phi c_{\theta_d} & c_\phi c_{\theta_d} \end{bmatrix}.$$

The definition of such an acceleration is thus given by

$$\mu = \frac{u_d}{m} R_d e_3 + g e_3. \tag{2.9}$$

Substituting $\theta = \theta_d - e_\theta$ into (2.2a) and taking into account the definition in (2.9) yield

$$\ddot{p}_b = \mu + \frac{u}{m}\delta, \tag{2.10}$$

where $\delta = \left[\delta_x \ \delta_y \ \delta_z\right]^T \in \mathbb{R}^3$ is the following interconnection vector

$$\delta_x = 2\cos(\phi)\cos(\psi)\sin(e_\theta/2)\cos(\theta_d - e_\theta/2), \tag{2.11a}$$

$$\delta_y = 2\cos(\phi)\sin(\psi)\sin(e_\theta/2)\cos(\theta_d - e_\theta/2), \tag{2.11b}$$

$$\delta_z = -2\cos(\phi)\sin(e_\theta/2)\sin(\theta_d - e_\theta/2). \tag{2.11c}$$

The acceleration μ can be designed as follows:

$$\mu = \ddot{p}_d + K_p e_p + K_d \dot{e}_p, \tag{2.12}$$

where $K_p, K_d \in \mathbb{R}^{3\times3}$ are positive definite gain matrices, $p_d, \dot{p}_d, \ddot{p}_d \in \mathbb{R}^3$ represent the desired position trajectory for the birotor, and $e_p = p_d - p_b$, $\dot{e}_p = \dot{p}_d - \dot{p}_b$, $\ddot{e}_p = \ddot{p}_d - \ddot{p}_b \in \mathbb{R}^3$ are the related tracking errors. Substituting (2.12) into (2.10) and (2.8) into (2.7) yields the following closed-loop equations:

$$\ddot{e}_p + K_d \dot{e}_p + K_p e_p = -\frac{u}{m}\delta, \tag{2.13a}$$

$$\ddot{e}_\theta + k_{d,\theta}\dot{e}_\theta + k_{p,\theta} e_\theta = 0. \tag{2.13b}$$

It remains to derive how to retrieve the desired total thrust u_d of the birotor and the desired pitch angle θ_d. Once the errors e_p and $\dot{e}_p$ are computed, knowing the planned acceleration $\ddot{p}_d$, the virtual control input μ can be computed from (2.12). The desired total thrust and the desired pitch angle can be computed by inverting (2.9) as follows:

$$u_d = m\sqrt{\mu_x^2 + \mu_y^2 + (\mu_z - g)^2}, \tag{2.14a}$$

$$\theta_d = \tan^{-1}\left(\frac{\mu_x c_\psi + \mu_y s_\psi}{\mu_z - g}\right), \tag{2.14b}$$

where ψ is measured from the inertial measurement unit (IMU). A second-order low-pass digital filter can be employed to reduce noise and compute both first and second derivatives of θ_d, to compute the pitch tracking errors e_θ and $\dot{e}_\theta$ in turn. The control input τ_θ is then computed from (2.6), with $\bar{\tau}_\theta$ obtained as from (2.8). Finally,

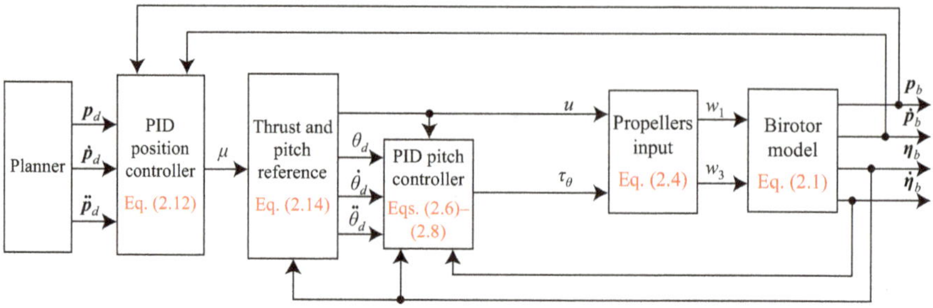

Figure 2.2 Block scheme of the PID-based control architecture. In red, the corresponding equations in the paper related to each block

the propeller inputs of the birotor can be computed inverting (2.4a) and (2.4c). The proposed controller is drawn in Figure 2.2.

Notice that two PD controllers are employed in (2.8) and (2.12). As noticed in [28], an integral action can be added without destroying the stability properties. The PD gains play like programmable stiffness and damping parameters, giving a physical interpretation to their choice.

Similar considerations can be made in the case of damage to motor 1 and/or motor 3. In such a case, the desired angle becomes the roll that can be computed by inverting $\mu = -(u/m)R_{d_r}e_3 + ge_3$, with $R_c \in SO(3)$ the equivalent of R_d by substituting θ_d with the measured θ and ϕ with the desired $\phi_d = \sin^{-1}\left((m/u_d)(\mu_y c_\psi - \mu_x s_\psi)\right)$.

2.4.2 Backstepping control scheme

The derivation of the proposed backstepping control scheme starts from the definition of the virtual acceleration μ as in (2.9). The desired total thrust and the desired pitch can be then retrieved from (2.14a) and (2.14b). Again, the goal is to design the vector μ suitably. The approach is different from the previous control scheme. First, an altitude control is derived. Afterward, the planar position control is designed. Finally, the control to achieve the desired pitch is addressed.

The following PD controller can be employed to control the altitude of the UAV

$$\mu_z = \ddot{z}_d + k_{d,z}\dot{e}_z + k_{p,z}e_z, \tag{2.15}$$

with $z_d \in \mathbb{R}$ the desired altitude in Σ_i, $e_z = z_d - z$, $\dot{e}_z = \dot{z}_d - \dot{z}$, $k_{p,z}, k_{d,z} > 0$ some gains.

For the planar control, it is worth noticing that the considered birotor, since it is an underactuated system continuously spinning around the vertical axis of Σ_b, can only rotate around the y_b-axis of Σ_b. Therefore, the projection of the birotor vertical axis z_b into the x_iy_i-plane of Σ_i is a rotating vector with rate $\dot{\psi}$. Figure 2.3 sketches the above concept. The following kinematic constraint can be thus introduced

$$\left[\dot{x}\ \dot{y}\right]^T = \beta\upsilon, \tag{2.16}$$

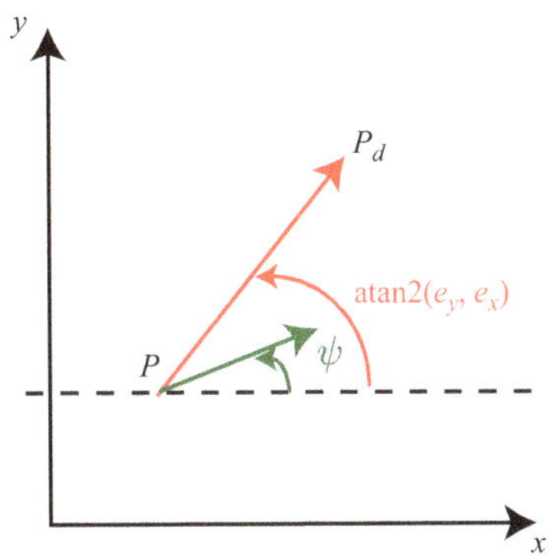

Figure 2.3 *The x_iy_i-plane of Σ_i is here represented. The point P represents the current position p_b of the birotor projected in such a plane. The point P_d represents the desired position p_d of the birotor on the same plane. The green vector is the current heading vector of the birotor in the x_iy_i-plane of Σ_i, which continuously rotates as the yaw angle. The red vector is the planar error creating an angle of atan2(e_y,e_x) with respect to the x_i-axis*

with $\beta = \left[\cos\psi \ \sin\psi\right]^T$, and $\upsilon \in \mathbb{R}$ the magnitude of the projection vector. The term υ is employed as a virtual input to design the desired planar velocities in Σ_i. The desired planar accelerations of the vector μ are obtained by differentiating (2.16) with respect to time as follows:

$$\left[\mu_x \ \mu_y\right]^T = \dot{\beta}\upsilon + \beta\alpha, \tag{2.17}$$

with $\alpha = \dot{\upsilon} \in \mathbb{R}$. The virtual inputs υ and α must be now designed to zero the error in the horizontal plane of Σ_i. Consider a regulation problem with $\dot{x}_d = \dot{y}_d = 0$. The planar errors are defined as $e_x = x_d - x$ and $e_y = y_d - y$. By deriving these two quantities and taking into account (2.16) yield

$$\left[\dot{e}_x \ \dot{e}_y\right]^T = -\beta\upsilon, \tag{2.18a}$$

$$\dot{\upsilon} = \alpha. \tag{2.18b}$$

Through a backstepping approach [8,29], the following virtual input can be designed to provide at least marginal stability in zeroing the planar errors:

$$\upsilon = k_\upsilon\sqrt{e_x^2 + e_y^2}\cos\left(\text{atan2}(e_y, e_x) - \psi\right), \tag{2.19}$$

with $k_\upsilon > 0$, and

$$\begin{aligned}\alpha = {} & (\dot{e}_x(k_\upsilon + k_\alpha) + e_x)c_\psi + (\dot{e}_y(k_\upsilon + k_\alpha) + e_y)s_\psi \\ & + k_\upsilon k_\alpha\sqrt{e_x^2 + e_y^2}\cos\left(\text{atan2}(e_y, e_x) - \psi(t)\right),\end{aligned} \tag{2.20}$$

with $k_\alpha > 0$.

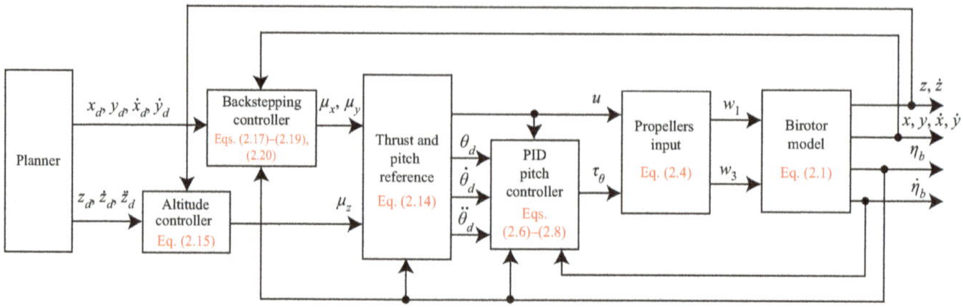

Figure 2.4 Block scheme of the backstepping-based control architecture. In red, the corresponding equations in the paper related to each block

Finally, the low-level attitude control for the birotor is designed as for the PID-based controller. Then, control law (2.6) is first introduced to get (2.7). The virtual control input $\bar{\tau}_\theta$ can be designed as in (2.8) to entirely control the pitch angle.

The derived backstepping controller is represented in Figure 2.4. To recap, as first, the position errors components e_x, e_y, e_z are computed, as well as the related time derivatives $\dot{e}_x, \dot{e}_y, \dot{e}_z$. Knowing the feedforward acceleration $\ddot{z}_d$, it is possible to compute the control input μ_z from (2.15). Taking into account both (2.19) and (2.20), the other two components of the virtual control input μ are retrieved from (2.17). The desired total thrust u_d and the pitch angle θ_d are then computed as in (2.14a) and (2.14b). A second-order low-pass digital filter is employed to reduce noise and compute both first and second derivatives of θ_d. Afterward, the pitch tracking errors e_θ and $\dot{e}_\theta$ are computed. The control input τ_θ is then computed from (2.6), with $\bar{\tau}_\theta$ obtained from (2.8). Finally, the propellers' speeds for the birotor are given by (2.4a) and (2.4c). An integral action might be added in (2.15) and (2.8) to increase tracking accuracy without destroying stability properties.

2.5 Numerical simulations

In this section, the two above-described controllers are compared in a case study simulating an emergency landing. As stated in Section 2.2, the employed controllers do not consider the switch from a nominal operating condition, in which all the motors are active, to the emergency landing procedure. Only the control of the birotor is evaluated through numerical simulations. The comparison between the results obtained with the two described control methodologies is thus analyzed. The next subsection introduces the parameters employed in the two controllers and the choices made for the simulation as well as the software environment. Afterward, the case study is presented and analyzed.

2.5.1 Description

The proposed control methodologies are tested in a MATLAB®/Simulink® environment running on a standard PC (2.6 GHz Intel Core i5, 8 GB 1,600 MHz DDR3). The two schemes depicted in Figures 2.2 and 2.4 have been implemented.

To simulate the dynamic of the birotor, the more accurate dynamic model (2.1a)–(2.1c), including air drag, is implemented. The dynamic parameters of the employed UAV are taken from a real AscTec Pelican quadrotor [30]. In detail, the following parameters are considered: $m = 1.25$ kg, $I_b = \text{diag}\{3.4, 3.4, 4.7\}$ kg^2, $l = 0.21$ m, $\rho_u = 1.8 \cdot 10^{-5}$ N s^2/rad^2, $\rho_c = 8 \cdot 10^{-7}$ N ms^2/rad^2. Besides, to consider saturations of the actuators, a maximum speed of about $\omega_i = 940$ rad/s (about 6,000 rpm) is considered. It was verified in practice that the birotor at steady state has a constant rotation of about 7 rad/s around its z_b-axis. Therefore, the friction coefficients in (2.1a)–(2.1c) are set to $F_p = \text{diag}\{1.25, 1.25, 5\}$ and $F_o = \text{diag}\{1, 1, 2\}$. Finally, the controllers are discretized at 100 Hz, to further test the robustness of the proposed controllers.

The gains for the PID-based controller are experimentally tuned as $K_p = \text{diag}\{6.25, 16, 25\}$, $K_d = \text{diag}\{6, 8, 10\}$, $k_{p,\theta} = 900$, and $k_{d,\theta} = 60$.

The gains for the backstepping controller are experimentally tuned as $k_\upsilon = 2$, $k_\alpha = 0.1$, $k_{p,\theta} = 144$, $k_{d,\theta} = 26.4$, $k_{p,z} = 16$, and $k_{d,z} = 8$.

In the following, a comparative case study is addressed. Given the desired trajectory, the two controllers are compared through a numerical simulation starting from the same initial conditions.

2.5.2 Case study

The considered case study is the tracking of a possible landing trajectory. The birotor starts from the position $p_b = \begin{bmatrix}1 & 1.5 & 4\end{bmatrix}^T$ m, which is the same desired initial position for the landing trajectory at $t = 0$ s, with $t > 0$ the time variable, and an initial yaw velocity of 3 rad/s. The desired goal position for the birotor is $\begin{bmatrix}0 & 0 & 0.5\end{bmatrix}^T$ m with respect to Σ_i at $t = 20$ s. The initial and final linear velocities and accelerations are zero. A seventh-order polynomial is employed to generate the desired trajectory and guarantee the conditions defined above. The birotor desired trajectory stays for other 10 s in a steady-state condition equal to the final one.

The time histories of the position error are depicted in Figure 2.5. The pictures show the position error for each of the Σ_i axes for the two implemented controllers. Both approaches successfully achieve the sought task. However, the backstepping approach needs more time to converge to the desired trajectory. The backstepping approach has a maximum error of about 0.15 m along the x-axis, 0.25 m along the y-axis, and 0.09 m along the vertical axis of Σ_i. The tridimensional Cartesian path followed by the birotor in both cases is depicted in Figure 2.6. The pitch error of the inner loop of both controllers is depicted in Figure 2.5(d). Also, in this case, the backstepping controller shows a more significant error of around 1 degree, which worsens the overall performance of the task.

The time histories of the commanded velocities for the birotor active propellers are illustrated in Figure 2.7. The propeller speeds do not saturate in both cases. Nevertheless, the behavior related to the backstepping controller appears overwrought.

It is worth recalling that the results show in any case robustness of both the control designs. The friction is indeed not included in the mathematical model employed for the control designs, even though it brings benefits to the stability analysis of

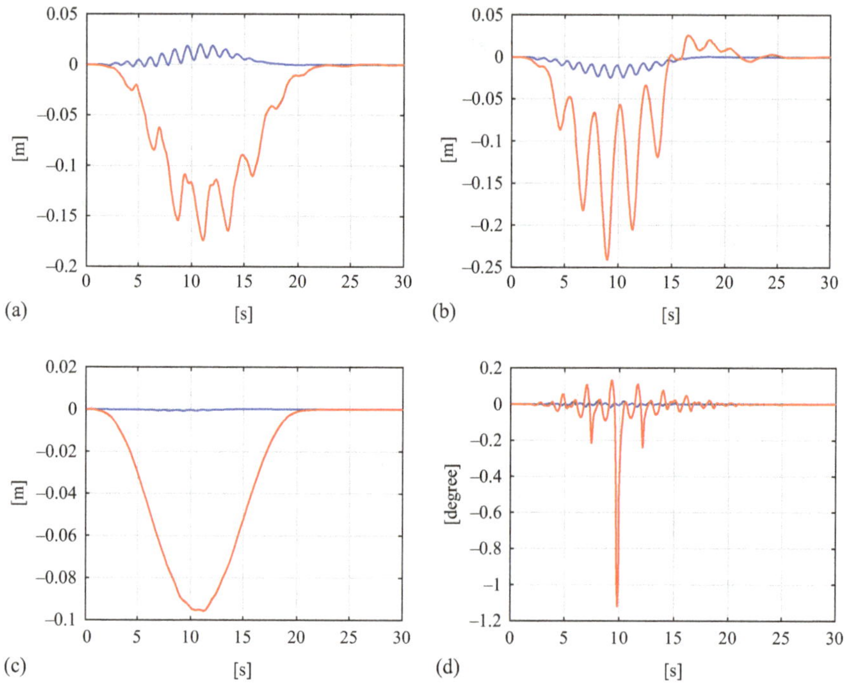

Figure 2.5 Position and pitch error for the considered case study. In blue, the results obtained with the PID-based controller. In red, the results obtained with the backstepping controller: (a) x-axis position error, (b) y-axis position error, (c) z-axis position error, and (d) pitch error

the closed-loop system. Besides, the controllers are discretized at 100 Hz as in the common practice. The final idea is that both the controllers fulfill the sought task. However, at least for the analyzed case study, the PID-based control seems to be more robust and employing less power for the motors.

2.6 Conclusion

This chapter provided a tutorial solution, based on [7,8], to solve the tracking control problem for a birotor UAV. Such a birotor configuration could be employed in those cases where a quadrotor UAV has one damaged propeller. Two control schemes have been proposed to deal with the contemplated problem: namely, a PID-based control scheme and a backstepping approach. It was shown that, with both approaches, the birotor could reach any position in the Cartesian space, and thus it could follow a safe emergency trajectory.

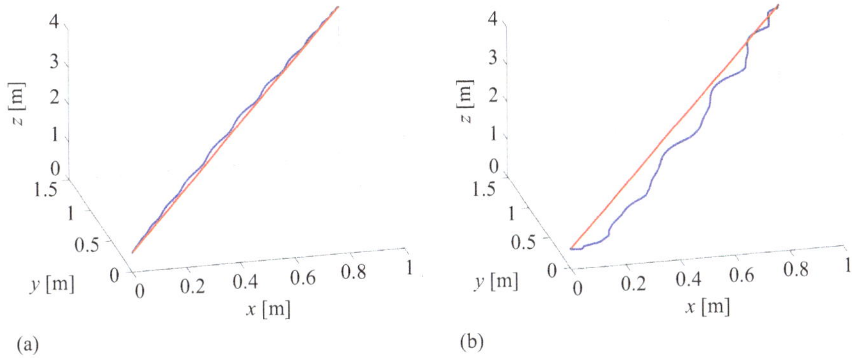

Figure 2.6 Tridimensional Cartesian path followed by the birotor with the two implemented controllers: (a) the path followed by employing the PID-based controller and (b) the path followed by employing the backstepping controller. In red, the desired path. In blue, the path followed by the birotor for the given controller

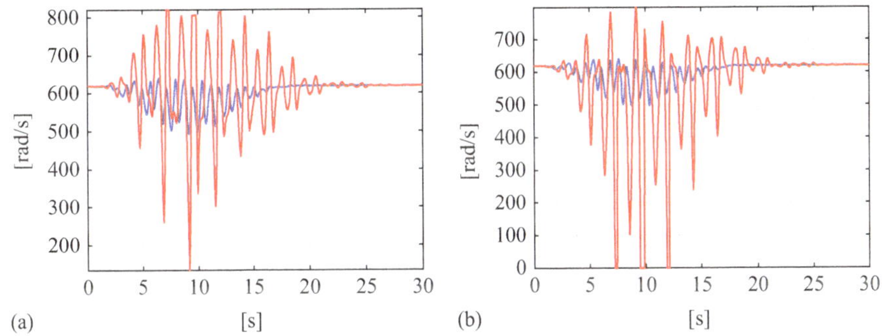

Figure 2.7 Commanded velocities of the active birotor propellers. In blue, the results obtained with the PID-based controller. In red, the results obtained with the backstepping controller: (a) motor 1 and (b) motor 3

Acknowledgments

The research leading to these results has been partially supported by the HYFLIERS project (Horizon 2020 Grant Agreement No. 779411) and the AERIAL-CORE project (Horizon 2020 Grant Agreement No. 871479). The authors are the sole responsible of its contents.

References

[1] Amazon Prime Air. http://www.amazon.com/b?node=8037720011.

[2] The White House. https://obamawhitehouse.archives.gov/the-press-office/2016/08/02/fact-sheet-new-commitments-accelerate-safe-integration-unmanned-aircraft.
[3] IEEE Spectrum. http://spectrum.ieee.org/aerospace/aviation/flying-selfie-bots-tagalong-video-drones-are-here/.
[4] Ruggiero F, Lippiello V, Ollero A. Aerial manipulation: A literature review. IEEE Robotics and Automation Letters. 2018;3(3):1957–1964.
[5] Blanke M, Staroswiecki M, Wu NE. Concepts and methods in fault-tolerant control. In: 2001 American Control Conference. vol. 4. Arlington, VA; 2001. p. 2606–2620.
[6] Zhang Y, Jiang J. Bibliographical review on reconfigurable fault-tolerant control systems. Annual Reviews in Control. 2008;32(2):229–252.
[7] Lippiello V, Ruggiero F, Serra D. Emergency landing for a quadrotor in case of a propeller failure: A PID approach. In: 2014 IEEE RSJ International Conference on Robotics and Automation. Chicago, IL, USA; 2014. p. 4782–4788.
[8] Lippiello V, Ruggiero F, Serra D. Emergency landing for a quadrotor in case of a propeller failure: A backstepping approach. In: 12th IEEE International Symposium on Safety, Security and Rescue Robots. Toyako-Cho, Hokkaido, Japan; 2014.
[9] Freddi A, Longhi S, Monteriu A. Actuator fault detection system for a mini-quadrotor. In: 2010 IEEE International Symposium on Industrial Electronics. Bari, Italy; 2010. p. 2055–2060.
[10] Sharifi F, Mirzaei M, Gordon BW, *et al.* Fault tolerant control of a quadrotor UAV using sliding mode control. In: 2010 Conference on Control and Fault-Tolerant Systems. Nice, France; 2010. p. 239–244.
[11] Sadeghzadeh I, Mehta A, Chamseddine A, *et al.* Active fault tolerant control of a quadrotor UAV based on gain scheduled PID control. In: 25th IEEE Canadian Conference on Electrical and Computer Engineering. Montreal, QC; 2012. p. 1–4.
[12] Saied M, Lussier B, Fantoni I, *et al.* Passive fault-tolerant control of an Octorotor using super-twisting algorithm: Theory and experiments. In: 2016 3rd Conference on Control and Fault-Tolerant Systems. Barcelona, Spain; 2016. p. 361–366.
[13] Ranjbaran M, Khorasani K. Fault recovery of an under-actuated quadrotor aerial vehicle. In: 49th IEEE Conference on Decision and Control. Atlanta, GA, USA; 2010. p. 4385–4392.
[14] Khebbache H, Sait B, Yacef F, *et al.* Robust stabilization of a quadrotor aerial vehicle in presence of actuator faults. International Journal of Information Technology, Control and Automation. 2012;2(2):1–13.
[15] Zhang Y, Chamseddine A. Fault tolerant flight control techniques with application to a quadrotor UAV testbed. In: Lombaerts T, editor. Automatic Flight Control Systems – Latest Developments. London: InTech; 2012. p. 119–150.
[16] Freddi A, Lanzon A, Longhi S. A feedback linearization approach to fault tolerance in quadrotor vehicles. In: 18th IFAC World Congress. Milan, Italy; 2011. p. 5413–5418.

[17] Kataoka Y, Sekiguchi K, Sampei M. Nonlinear control and model analysis of trirotor UAV model. In: 18th IFAC World Congress. Milan, Italy; 2011. p. 10391–10396.

[18] Lanzon A, Freddi A, Longhi S. Flight control of a quadrotor vehicle subsequent to a rotor failure. Journal of Guidance, Control, and Dynamics. 2014;37(2):580–591.

[19] Mueller MW, D'Andrea R. Stability and control of a quadrocopter despite the complete loss of one, two, or three propellers. In: 2014 IEEE International Conference on Robotics and Automation. Hong Kong, China; 2014. p. 45–52.

[20] Vey D, Lunze J. Structural reconfigurability analysis of multirotor UAVs after actuator failures. In: 2015 IEEE 54th Annual Conference on Decision and Control. Osaka, Japan; 2015. p. 5097–5104.

[21] Sun S, Sijbers L, Wang X, *et al.* High-speed flight of quadrotor despite loss of single rotor. IEEE Robotics and Automation Letters. 2018;4(4):3201–3207.

[22] Avant T, Lee U, Katona B, *et al.* Dynamics, hover configurations, and rotor failure restabilization of a morphing quadrotor. In: 2018 Annual American Control Conference. Wisconsin Center, Milwaukee, WI, USA; 2018. p. 4855–4862.

[23] Morozov YV. Emergency control of a quadrocopter in case of failure of two symmetric propellers. Automation and Remote Control. 2018;79(3):92–110.

[24] Merheb AR, Noura H, Bateman F. Emergency control of AR drone quadrotor UAV suffering a total loss of one rotor. IEEE/ASME Transactions on Mechatronics. 2017;22(2):961–971.

[25] Barghandan S, Badamchizadeh MA, Jahed-Motlagh MR. Improved adaptive fuzzy sliding mode controller for robust fault tolerant of a quadrotor. International Journal of Control, Automation and Systems. 2017;15(1):427–441.

[26] Saied M, Lussier B, Fantoni I, *et al.* Fault diagnosis and fault-tolerant control of an octorotor UAV using motors speeds measurements. IFAC-PapersOnLine. 2017;50(1):5263–5268.

[27] Lu P, van Kampen EJ. Active fault-tolerant control for quadrotors subjected to a complete rotor failure. In: 2015 IEEE/RSJ International Conference on Intelligent Robots and Systems. Hamburg, Germany; 2015. p. 4698–4703.

[28] Nonami K, Kendoul F, Suzuki S, *et al.* Autonomous flying robots. Unmanned Aerial Vehicles and Micro Aerial Vehicles. Berlin, Heidelberg, Germany: Springer-Verlag; 2010.

[29] Khalil HK. Nonlinear systems. Upper Saddle River, NJ, USA: Prentice Hall; 2002.

[30] Ruggiero F, Cacace J, Sadeghian H, *et al.* Impedance control of VToL UAVs with a momentum-based external generalized forces estimator. In: 2014 IEEE International Conference on Robotics and Automation. Hong Kong, China; 2014. p. 2093–2099.

Chapter 3

Observer-based LPV control design of quad-TRUAV under rotor-tilt axle stuck fault

Zhong Liu[1,2,3], Didier Theilliol[4], Liying Yang[1,2], Yuqing He[1,2] and Jianda Han[1,2,5,6]

Tilt rotor unmanned aerial vehicle (TRUAV) is one of hybrid and transformable aircraft with abilities of hovering and high-speed cruise and attracts much more attention in recent years. Considering the variable and nonlinear dynamics, linear parameter varying (LPV) methods provide convenient tools for model representation and controller synthesis of TRUAV, but its control in the transition procedure from hovering to high-speed cruise is still a challenge. Moreover, there are few references focusing on its fault-tolerant control (FTC) strategies, especially under actuator faults because of complex control coupling from rotors and aerodynamics. Aiming at these open control problems of TRUAV, the transition control of a quad-TRUAV is presented in this chapter with LPV method, and the FTC strategy is considered further for rotor-tilt axle stuck fault.

In the proposed control method, some virtual control inputs are introduced into the quad-TRUAV nonlinear model, and an equivalent LPV model is obtained for controller design. To estimate matched perturbations in this LPV-controlled plant, an observer-based LPV controller with a proportional integral observer (PIO) is synthesized. An inverse procedure is considered further for applied control inputs based on above virtual control inputs and state estimations. Under rotor-tilt axle stuck fault, a degraded model is taken into account for attitude stability mainly. The inverse procedure is reconstructed with the degraded dynamics, and the cruise ability of TRUAV is still kept by the redesigned reference value of angle of attack due to the similar

[1]State Key Laboratory of Robotics, Shenyang Institute of Automation, Chinese Academy of Sciences, Shenyang, P.R. China
[2]Institutes for Robotics and Intelligent Manufacturing, Chinese Academy of Sciences, Shenyang, P.R. China
[3]University of Chinese Academy of Sciences, Beijing, P.R. China
[4]Faculte des Sciences et Techniques, CRAN UMR 7039, CNRS, University of Lorraine, Vandœuvre-lès-Nancy, France
[5]Institute of Robotics and Automatic Information System, College of Artificial Intelligence, Nankai University, Tianjin, P.R. China
[6]Tianjin Key Laboratory of Intelligent Robotics, Nankai University, Tianjin, P.R. China

control effect from rotor-tilt angle and angle of attack. In the end, the effectiveness of the proposed control method and FTC is illustrated by numerical results.

3.1 Introduction

With the development of autonomous control and electronic technologies, unmanned aerial vehicles (UAVs) have been researched deeply and applied widely in the fields of agriculture and military [1]. As the typical representations of UAVs with conventional structures, rotorcraft UAVs (RUAVs) and fixed-wing UAVs (FWUAVs) also play irreplaceable roles for a long time. These conventional structures are beneficial to establish classical and common control methods for RUAVs and FWUAVs but induce some limitations for wider applications. RUAVs own ability of hovering, so they are suitable for operating tasks [2]. However, their endurance is worse than FWUAVs and could not cruise with a very high speed. On the contrary, FWUAVs are with long endurance and high cruise speed but could not hover at a fixed position. Consequently, the flexibility of FWUAVs for environments is worse than RUAVs [3]. To break these limitations due to essentially conventional structures, lots of hybrid UAVs with novel structures are designed [4] such as the convertiplane UAV, tail-sitter UAV and TRUAV displayed in Figure 3.1. These hybrid UAVs are all with abilities of hovering and high-speed cruise and also accompanied with different control problems [5–8], where the quad-TRUAV shown in Figure 3.1(d) is focused on mainly in this chapter.

Compared to convertiplane UAVs and tail-sitter UAVs, TRUAVs are with variable structures and dynamics. They are equipped with several rotors, which could

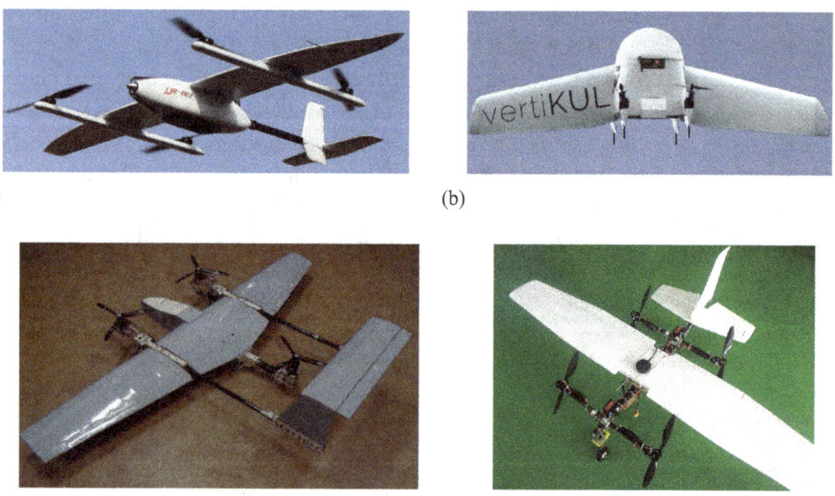

(a) (b) (c) (d)

Figure 3.1 Some hybrid UAV platforms: (a) convertiplane UAV [5], (b) tail-sitter UAV [6], (c) tri-TRUAV [7], (d) quad-TRUAV tilt from vertical position to horizontal position regarded as helicopter mode and airplane mode, respectively. Due to the dynamics similar with RUAV and FWUAV, the control of TRUAV in helicopter mode and airplane mode is mature based on existing knowledge, even though there are sort of structural differences [8]. Especially for more and more symmetric structures of some TRUAVs with more than three rotors, such as quad-TRUAVs in Figure 3.1(d) and presented in [9], the control difficulty in helicopter mode is relaxed further compared with dual-TRUAV in [10]. However, due to integrated abilities of hovering and high-speed cruise, the transition procedure between helicopter mode and airplane mode is always accompanied with varying and nonlinear dynamics. Consequently, pure linear control methods for fixed rotor-tilt angles [11] could not keep ideal control performance in whole transition procedure. In addition, typical nonlinear control methods are also hard to be applied directly due to the complex control coupling from rotors and aerodynamics. The simplified TRUAV model is considered sometimes for special nonlinear controller design [12].

Focusing on the nonlinear dynamics of TRUAVs from helicopter mode to airplane mode or inversely, two classical gain scheduling (GS) control structures are usually considered especially for applications [7,13–16]. In [13,14], two set of linear controllers are designed for helicopter mode and airplane mode, and a smooth weight with respect to rotor-tilt angle is calculated further to interpolate these two controllers for transition procedure. Moreover, divide and conquer GS method that is applied in [15,16], in which linear controllers for different rotor-tilt angles are switched directly according to flight velocity. This classical idea is also applied in the Pixhawk flight controller, and the control effectiveness has been validated [7]. Considering linear controllers that only ensure local stability facing nonlinear-controlled plants, a special flight envelop called tilt corridor is necessary in GS methods for the stability of whole transition procedure [8], which limits the maximum and minimum flight velocities with respect to a fixed rotor-tilt angle. Actually, tilt corridor indicates the potential relationship between flight velocity and TRUAV structure. However, the abovementioned GS methods mainly aim at applications, and heuristic knowledge is applied for desired control performance. Few references propose stability conditions of GS methods for TRUAV transition control. For the moment, stability analysis for the transition procedure is still an open challenge.

As the main flag value of TRUAV structure, rotor-tilt angle is crucial for acceleration and deceleration in transition procedure. In real TRUAV platforms, rotor-tilt axle is usually driven by servo through link mechanisms or chains. If it is stuck at one fixed angle, the transition procedure would be broken off, which might be fatal for safe flight. To ensure the stability or acceptable control performance under different faults, FTC methods for aerial vehicles and other systems have been researched for a long time, and many active and passive FTC strategies have been presented, see [17,18]. However, little attention has been paid for FTC problems of TRUAV, and rotor-tilt axle stuck fault is also neglected as a special actuator fault only for TRUAVs.

Based on the previous work [19], this chapter aims at the transition control and FTC strategy under rotor-tilt axle stuck fault of the quad-TRUAV shown in Figure 3.1(d). Several virtual control inputs represented as functions of applied control inputs and states are introduced into the quad-TRUAV nonlinear model to formulate an

LPV-controlled plant, which is a special nonlinear system with parameter-dependent state-space matrices [20]. An LPV controller is proposed to be synthesized with virtual control method, and an inverse procedure is designed further for applied control inputs. In the proposed control method, rotor-tilt angle is regarded as a control input with similar control effect as angle of attack, and asymptotic stability of the quad-TRUAV in transition procedure could be ensured in theory. Under rotor-tilt axle stuck fault, the quad-TRUV loses one control input, and a degraded dynamic model is considered for attitude stability mainly. The inverse procedure is also reconstructed for cruise ability with the redesigned reference value of angle of attack.

Note that virtual control inputs are functions of applied control inputs and states. To obtain applied control inputs, state values are necessary for designed inverse procedure. As only state estimations are available in application, some matched perturbations are induced into the LPV-controlled plant. Consequently, PIO [21] is applied to estimate these perturbations and states simultaneously. Compared with [19], the state estimator is designed further in this chapter, and an observer-based LPV controller [22] is synthesized finally, which owns potential FTC ability for bias actuator faults of virtual control inputs.

This chapter is organized as follows: Section 3.2 introduces the quad-TRUAV structure and establishes its longitudinal nonlinear model; Section 3.3 reformulates this nonlinear model into equivalent LPV representation and considers the closed-loop system with observer-based LPV controller; then, a main theorem with linear matrix inequality (LMI) conditions is presented in Section 3.4 to synthesize this LPV controller for virtual control inputs, and the inverse procedure for applied control inputs is designed based on virtual control inputs and state estimations; Section 3.5 introduces typical actuator stuck fault and further proposes the FTC strategy for rotor-tilt axle stuck fault; some numerical results are displayed in Section 3.6; Section 3.7 concludes the whole chapter.

3.2 Quad-TRUAV and nonlinear modeling

As the controlled plant of this chapter, the quad-TRUAV platform shown in Figure 3.1(d) is equipped with four tiltable rotors that could tilt together. Ailerons on wings, an elevator on horizontal tail and a rudder on vertical tail provide extra aerodynamic forces and moments for flight control. In helicopter mode with rotor-tilt angle as $\pi/2$ rad, the quad-TRUAV could hover at one fixed position or cruise with a low speed; in airplane mode with rotor-tilt angle as 0, the quad-TRUAV could cruise with a high speed; in mode transition procedure, the quad-TRUAV would accelerate with decreased rotor-tilt angle to ensure that wings generate enough lift or decelerate with increased rotor-tilt angle for helicopter mode.

To formulate the nonlinear model of this quad-TRUAV, some typical coordinate systems are defined in Figure 3.2, such as north-east-down coordinate system ($O_ex_ey_ez_e$), body-axis coordinate system ($O_bx_by_bz_b$) and wind-axis coordinate system ($O_wx_wy_wz_w$). Then, some states in these coordinate systems could also be defined, including flight velocity V, angle of attack α, pitch angle θ and pitch rate q. Similar

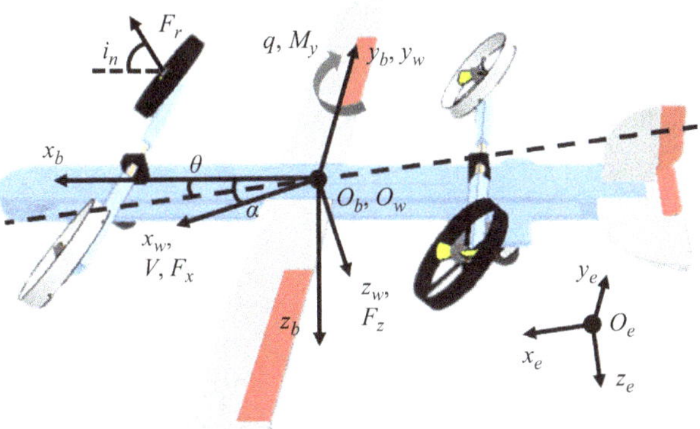

Figure 3.2 Definitions of coordinate systems and states

with the nonlinear modeling of fixed-wing aerial vehicle [23], following dynamic and kinematic equations are formulated:

$$\begin{aligned}
\dot{V} &= \frac{F_x}{m}, \\
\dot{h} &= V\sin(\theta - \alpha), \\
\dot{\alpha} &= \frac{F_z}{mV} + q, \\
\dot{q} &= \frac{M_y}{I_y}, \\
\dot{\theta} &= q,
\end{aligned} \tag{3.1}$$

where h is fight height, F_x and F_z are the resultant forces in the x- and z-directions of $O_w x_w y_w z_w$, M_y is the relevant moment in the y-direction of $O_b x_b y_b z_b$, m is the mass, and I_y is the moment of inertia in the y-direction.

Forces and moment in (3.1) could be represented as follows:

$$F_x = F_{rx} + F_{ax} + F_{gx}, \quad F_z = F_{rz} + F_{az} + F_{gz}, \quad M_y = M_{ry} + M_{ay}, \tag{3.2}$$

where rotors generate F_{rx}, F_{rz} and M_{ry}, aerodynamics induces F_{ax}, F_{az} and M_{ay}, and gravity only gives rise to component forces F_{gx} and F_{gz}. As for rotors driven by brushless direct current motors, generated thrusts are proportional to squares of rotor speeds [24]. By transforming thrusts into $O_w x_w y_w z_w$, the following component forces and moment are generated:

$$\begin{bmatrix} F_{rx} \\ F_{rz} \end{bmatrix} = \begin{bmatrix} \cos\alpha & \sin\alpha \\ -\sin\alpha & \cos\alpha \end{bmatrix} \begin{bmatrix} \cos i_n \\ -\sin i_n \end{bmatrix} 2C_t\rho AR^2(\Omega_F^2 + \Omega_B^2),$$

$$M_{ry} = 2C_t\rho AR^2 x_r \sin i_n(\Omega_F^2 - \Omega_B^2).$$

where C_t is the thrust coefficient, ρ is the air density, A is the rotor area, R is the rotor radius, i_n is the rotor-tilt angle, Ω_F and Ω_B are the forward and backward rotor speeds, and x_r is the longitudinal distance from the rotor to center of gravity. Moreover, aerodynamics induces drag, lift and pitch moment with the following forms [16,23]:

$$\begin{aligned}
F_{ax} &= -\frac{1}{2}\rho V^2 s \cdot C_D = -\frac{1}{2}\rho V^2 s \cdot (C_{D0} + C_{D1}\alpha + C_{D2}\alpha^2 + C_{D\delta}\delta_e),\\
F_{az} &= -\frac{1}{2}\rho V^2 s \cdot C_L = -\frac{1}{2}\rho V^2 s \cdot \left(C_{L0} + C_{L1}\alpha + \frac{\bar{c}}{2V}C_{Lq}q + C_{L\delta}\delta_e\right),\\
M_{ay} &= \frac{1}{2}\rho V^2 s\bar{c} \cdot C_m = \frac{1}{2}\rho V^2 s\bar{c} \cdot \left(C_{m0} + C_{m1}\alpha + \frac{\bar{c}}{2V}C_{mq}q + C_{m\delta}\delta_e\right),
\end{aligned}$$

where s is the wing area, $\bar{c}$ is the average chord length of wing, C_D, C_L and C_m are the aerodynamic coefficients of drag, lift and pitch moment, and δ_e is the elevator deflection. As for gravity, there are the following component forces:

$$\begin{bmatrix} F_{gx} \\ F_{gz} \end{bmatrix} = \begin{bmatrix} \cos\alpha & \sin\alpha \\ -\sin\alpha & \cos\alpha \end{bmatrix} \begin{bmatrix} -mg\sin\theta \\ mg\cos\theta \end{bmatrix} = \begin{bmatrix} -mg\sin(\theta - \alpha) \\ mg\cos(\theta - \alpha) \end{bmatrix},$$

where g is the acceleration of gravity.

Introducing (3.2) into (3.1), the nonlinear model of the quad-TRUAV could be formulated, where applied control input vector $u = \begin{bmatrix} i_n & \Omega_F & \Omega_B & \delta_e \end{bmatrix}^T$. For the following LPV control analysis, ignore the kinematic equation about h, and introduce a new state $\gamma = \theta - \alpha$. The state vector of the quad-TRUAV nonlinear model is defined as $x = \begin{bmatrix} V & \gamma & \alpha & q \end{bmatrix}^T$, and γ could be regarded as the virtual control input of h [19]. Measurement output vector is defined as $y = x$ with noise-free representation.

3.3 LPV control analysis

With the nonlinear model established previously, this section would reformulate (3.1) into polytopic LPV form with some virtual control inputs. Then, closed-loop system with an observer-based LPV controller would be established further to analyze the control structure for the quad-TRUAV.

3.3.1 Polytopic LPV representation

According to the specific form of (3.2), the nonlinear model of the quad-TRUAV owns complex control coupling from rotors and aerodynamics. To avoid considering the control coupling directly, the following virtual control inputs are considered:

$$v_V = F_V - g\sin\gamma - \frac{\rho s}{2m}C_{D1} \cdot V^2\alpha - \frac{\rho s}{2m}C_{D2} \cdot V^2\alpha^2, \tag{3.3}$$

$$v_\alpha = F_\alpha + g\frac{\cos\gamma}{V} - \frac{\rho s}{2m}C_{L0}V, \tag{3.4}$$

$$v_q = M_q + \frac{\rho s\bar{c}}{2I_y}C_{m0}V^2 + \frac{\rho s\bar{c}}{2I_y}C_{m1} \cdot V^2\alpha, \tag{3.5}$$

where

$$F_V = \frac{2\rho AR^2}{m} C_t \cdot \cos(i_n + \alpha)(\Omega_F^2 + \Omega_B^2) - \frac{\rho s}{2m} C_{D\delta} \cdot V^2 \delta_e, \tag{3.6}$$

$$F_\alpha = -\frac{2\rho AR^2}{mV} C_t \cdot \sin(i_n + \alpha)(\Omega_F^2 + \Omega_B^2) - \frac{\rho s}{2m} C_{L\delta} \cdot V \delta_e, \tag{3.7}$$

$$M_q = \frac{2\rho AR^2 x_r}{I_y} C_t \cdot \sin i_n (\Omega_F^2 - \Omega_B^2) + \frac{\rho s \bar{c}}{2I_y} C_{m\delta} \cdot V^2 \delta_e \tag{3.8}$$

include all control effects from different parts, and rotor-tilt angle i_n and angle of attack α take similar effect for the control of V and α. With above virtual control inputs, two simplified models with linear forms are established as follows:

$$\dot{V} = \left(0 - V \frac{\rho s}{2m} C_{D0}\right) V + v_V, \tag{3.9}$$

$$\begin{bmatrix} \dot{\gamma} \\ \dot{\alpha} \\ \dot{q} \end{bmatrix} = \left(\begin{bmatrix} 0 & 0 & a_{13} \\ 0 & 0 & 1 - a_{13} \\ 0 & 0 & 0 \end{bmatrix} + V \begin{bmatrix} 0 & a_{12} & 0 \\ 0 & -a_{12} & 0 \\ 0 & 0 & \frac{\rho s \bar{c}^2}{4I_y} C_{mq} \end{bmatrix} \right) \begin{bmatrix} \gamma \\ \alpha \\ q \end{bmatrix} + \begin{bmatrix} -1 & 0 \\ 1 & 0 \\ 0 & 1 \end{bmatrix} \begin{bmatrix} v_\alpha \\ v_q \end{bmatrix}, \tag{3.10}$$

where $a_{13} = (\rho s \bar{c}/4m) C_{Lq}$ and $a_{12} = (\rho s/2m) C_{L1}$. Note that flight velocities are separated from other states, which aims at FTC strategy under rotor-tilt axle stuck fault.

Obviously, system matrices of (3.9) and (3.10) are affinely parameter-dependent on V, and input matrices are all constant. This special form makes it possible to reformulate (3.9) and (3.10) into the following typical polytopic LPV form by sector nonlinearity method [25]:

$$\begin{aligned} \dot{x}(t) &= A(p)x(t) + Bv(t) + B_w w(t), \\ y(t) &= Cx(t) + D_w w(t), \end{aligned} \tag{3.11}$$

where $x(t) \in \mathrm{R}^{n_x}$, virtual control input vector $v(t) \in \mathrm{R}^{n_v}$, disturbance input vector $w(t) \in \mathrm{R}^{n_w}$, $A(p) = \sum_{i=1}^{n} p_1 A_i$ with varying parameters $0 \le p_i \le 1$ and $\sum_{i=1}^{n} p_i = 1$, p_i is the simplified representation of $p_i(V)$, and C is the identity matrix. Actually, separated subsystems (3.9) and (3.10) should be transformed into two polytopic LPV systems for two LPV controllers. To simplify representation in the following closed-loop analysis, integral polytopic LPV system (3.11) is considered for convenience, whose $n_x = 4$, $n_w = 4$, $n = 2$, $v = \left[\, v_V \,\vdots\, v_\alpha \;\; v_q \,\right]^T$, and state-space matrices are all with block diagonal structures.

Further consider virtual control inputs $v(t)$. According to (3.3)–(3.8), virtual control inputs could be represented as the functions of states $x(t)$ and applied control inputs $u(t)$ as the following algebraic relation:

$$v(t) = R_{vu}(x(t), u(t)). \tag{3.12}$$

With measurement noises, $x(t)$ is not available. Applying state estimations $\hat{x}(t)$,

$$\tilde{u}(t) = R_{vu}^{-1}(\hat{x}(t), v(t))$$

is available in application, where $R_{vu}^{-1}(\cdot,\cdot)$ is the inverse function of $R_{vu}(\cdot,\cdot)$. Take $\tilde{u}(t)$ into (3.11) in fault-free case without actuator fault:

$$\begin{aligned}\dot{x}(t) &= A(p)x(t) + BR_{vu}(x(t), \tilde{u}(t)) + B_w w(t)\\ &= A(p)x(t) + B\tilde{v}(t) + B_w w(t)\\ &= A(p)x(t) + Bv(t) + B\delta(t) + B_w w(t),\end{aligned}$$

where $\tilde{v}(t) \triangleq R_{vu}(x(t), \tilde{u}(t))$, and $\delta(t) \triangleq \tilde{v}(t) - v(t) = \begin{bmatrix}\delta_V & \delta_\alpha & \delta_q\end{bmatrix}^T$ is matched perturbation vector that could be compensated. In some other references, $\delta(t)$ is also regarded as bias actuator faults [21] or model errors from model simplification [26]. Consequently, the LPV-controlled plant of the quad-TRUAV is as follows:

$$\begin{aligned}\dot{x}(t) &= A(p)x(t) + Bv(t) + B\delta(t) + B_w w(t),\\ y(t) &= Cx(t) + D_w w(t).\end{aligned} \tag{3.13}$$

To ensure that varying parameters $p_i (i = 1, \ldots, n)$ are measurable, without considering the measurement noise about V, the first line of D_w needs to be zero row vector.

3.3.2 Closed-loop analysis with observer-based LPV control

With LPV-controlled plant (3.13), a state feedback LPV controller is proposed to be designed for closed-loop analysis. For this purpose, LPV PIO is necessary for estimations $\hat{x}(t)$ and $\hat{\delta}(t)$. Based on these estimated values, the observer-based LPV controller is considered in this subsection with $\hat{\delta}(t)$ as the compensation for matched perturbations.

For LPV controller design, to ensure partial measurement outputs $C_r y(t)$ track reference signal $r(t) \in \mathrm{R}^{n_r}$, define $e(t) = \int [r(t) - C_r y(t)]\, \mathrm{d}t$ as tracking error integration [27], and

$$\dot{e}(t) = -C_r C x(t) + r(t) - C_r D_w w(t). \tag{3.14}$$

With LPV controller gains $K_1(p) \in \mathrm{R}^{n_u \times n_r}$ and $K_2(p) \in \mathrm{R}^{n_u \times n_x}$, introduce the following control input vector into (3.13):

$$v(t) = \begin{bmatrix} K_1(p) & K_2(p) \end{bmatrix} \begin{bmatrix} e(t) \\ \hat{x}(t) \end{bmatrix} - \hat{\delta}(t). \tag{3.15}$$

Together with (3.14), there is the following closed-loop system:

$$\begin{aligned}\begin{bmatrix}\dot{e}(t)\\ \dot{x}(t)\end{bmatrix} &= \begin{bmatrix} 0 & -C_r C \\ BK_1(p) & A(p) + BK_2(p)\end{bmatrix}\begin{bmatrix} e(t) \\ x(t)\end{bmatrix} + \begin{bmatrix} 0 \\ BK_2(p)e_x(t) - Be_\delta(t)\end{bmatrix}\\ &\quad + \begin{bmatrix} r(t) - C_r D_w w(t) \\ B_w w(t)\end{bmatrix},\end{aligned} \tag{3.16}$$

where $e_x(t) = \hat{x}(t) - x(t)$ and $e_\delta(t) = \hat{\delta}(t) - \delta(t)$.

To estimate $x(t)$ and $\delta(t)$ simultaneously, the following PIO [21] could be considered with observer gains $L_1(p) \in \mathrm{R}^{n_x \times n_y}$ and $L_2(p) \in \mathrm{R}^{n_u \times n_y}$:

$$\begin{aligned} \dot{\hat{x}}(t) &= A(p)\hat{x}(t) + Bv(t) + B\hat{\delta}(t) + L_1(p)\left[C\hat{x}(t) - y(t)\right], \\ \dot{\hat{\delta}}(t) &= L_2(p)\left[C\hat{x}(t) - y(t)\right]. \end{aligned} \tag{3.17}$$

Together with original-controlled plant (3.13), there is the following closed-loop system:

$$\begin{bmatrix} \dot{e}_x(t) \\ \dot{e}_\delta(t) \end{bmatrix} = \begin{bmatrix} A(p) + L_1(p)C & B \\ L_2(p)C & 0 \end{bmatrix} \begin{bmatrix} e_x(t) \\ e_\delta(t) \end{bmatrix} + \begin{bmatrix} -B_w w(t) - L_1(p)D_w w(t) \\ -L_2(p)D_w w(t) - \dot{\delta}(t) \end{bmatrix}. \tag{3.18}$$

With the above analysis, assuming that $\dot{\delta}(t)$ is bounded and $\dot{\delta}(+\infty) = 0$ for limited energy, closed-loop systems (3.16) and (3.18) could be formulated as follows:

$$\begin{aligned} \begin{bmatrix} \dot{\tilde{x}}(t) \\ \dot{\overline{e}}(t) \end{bmatrix} &= \begin{bmatrix} \tilde{A}(p) + \tilde{B}K(p) & \tilde{B}K(p)\tilde{E} + \tilde{D} \\ 0 & \overline{A}(p) + L(p)\overline{C} \end{bmatrix} \begin{bmatrix} \tilde{x}(t) \\ \overline{e}(t) \end{bmatrix} + \begin{bmatrix} \tilde{B}_w \\ \overline{B}_w + L(p)\overline{D}_w \end{bmatrix} \omega(t), \\ z(t) &= \begin{bmatrix} 0 & C_e \end{bmatrix} \begin{bmatrix} \tilde{x}(r) \\ \overline{e}(t) \end{bmatrix} + F_w \omega(t), \end{aligned} \tag{3.19}$$

where

$$\tilde{x}(t) = \begin{bmatrix} e(t) \\ x(t) \end{bmatrix}, \ \overline{e}(t) = \begin{bmatrix} e_x(t) \\ e_\delta(t) \end{bmatrix}, \ \tilde{A}(p) = \begin{bmatrix} 0 & -C_r C \\ 0 & A(p) \end{bmatrix}, \ \tilde{B} = \begin{bmatrix} 0 \\ B \end{bmatrix},$$

$$K(p) = \left[\, K_1(p) \ \ K_2(p) \,\right], \ \tilde{E} = \begin{bmatrix} 0 & 0 \\ I & 0 \end{bmatrix}, \ \tilde{D} = \begin{bmatrix} 0 & 0 \\ 0 & -B \end{bmatrix}, \ \tilde{B}_w = \begin{bmatrix} I & -C_r D_w & 0 \\ 0 & B_w & 0 \end{bmatrix},$$

$$\overline{A}(p) = \begin{bmatrix} A(p) & B \\ 0 & 0 \end{bmatrix}, \ L(p) = \begin{bmatrix} L_1(p) \\ L_2(p) \end{bmatrix}, \ \overline{C} = \begin{bmatrix} C & 0 \end{bmatrix}, \ \overline{B}_w = \begin{bmatrix} 0 & -B_w & 0 \\ 0 & 0 & -I \end{bmatrix},$$

$$\overline{D}_w = \begin{bmatrix} 0 & -D_w & 0 \end{bmatrix}, \omega(t) = \begin{bmatrix} r(t) \\ w(t) \\ \dot{\delta}(t) \end{bmatrix}$$

and $z(t) \in \mathrm{R}^{n_x + n_u}$ is the performance output vector. Matrices C_e and F_w could be set for special performance outputs.

With the above closed-loop analysis based on virtual control inputs and observer-based LPV controller, the control structure for the quad-TRUAV in fault-free case is constructed in Figure 3.3. In this control structure, observer-based LPV control is designed for asymptotic stability of closed-loop system (3.19) with virtual control inputs $v(t)$, in which the velocity controller and the attitude controller for γ, α and q are synthesized severally based on (3.9) and (3.10). The inverse procedure needs to be considered further for applied control inputs. As only $\hat{x}(t)$ is available in application, it is $\tilde{u}(t)$ rather than $u(t)$ that is obtained from inverse procedure. Corresponding contents will be introduced in the next section at length. In the faulty case under rotor-tilt axle stuck fault, the inverse procedure would be reconstructed with fault

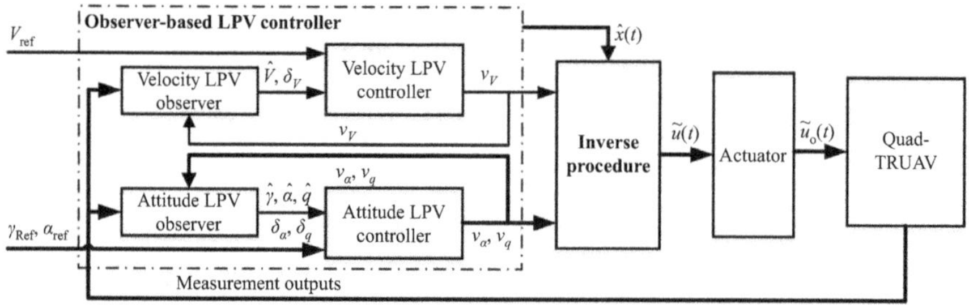

Figure 3.3 The control structure for the quad-TRUAV in fault-free case

information. This FTC strategy requires velocity controller and attitude controller are separated, which will be introduced deeply in Section 3.5.

3.4 Observer-based LPV control for the quad-TRUAV

With the control structure shown in Figure 3.3, this section would mainly focus on the fault-free case. The observer-based LPV controller with PIO would be synthesized first to ensure closed-loop stability of (3.19) with virtual control inputs. Then, the inverse procedure based on (3.3)–(3.8) would be designed further for applied control inputs of the quad-TRUAV.

3.4.1 Observer-based LPV controller synthesis

To design LPV controller (3.15) and PIO (3.17) for controlled plant (3.13), the following theorem is proposed for controller and observer gains.

Theorem 3.1. The closed-loop system (3.19) is asymptotically stable, and the estimation error of the LPV observer is with H_∞ disturbance attenuation τ, if there exist positive scalars ϕ_1, ϕ_2 and ϕ_x, invertible matrices W_1 and W_2, and matrices X, Y, $M(p) = \sum_{i=1}^{n} p_i M_i$ and $N(p) = \sum_{i=1}^{n} p_i N_i$ that satisfy the following conditions:

$$X > 0,\ Y > 0, \tag{3.20}$$

$$\mathcal{G}_i < 0,\ 0 \leq i \leq n, \tag{3.21}$$

where

$$\mathcal{G}_i = \begin{bmatrix} -\phi_1 \mathrm{He}(W_1) & 0 & \mathcal{G}_{(1,3)} & 0 & \phi_1 \tilde{E} & 0 & 0 \\ * & -\phi_1 \mathrm{He}(W_2) & 0 & 0 & \mathcal{G}_{(2,5)} & N_i \overline{D}_w & 0 \\ * & * & \mathcal{G}_{(3,3)} & \phi_x X & \tilde{D} & \tilde{B}_w & 0 \\ * & * & * & -2\phi_x I & 0 & 0 & 0 \\ * & * & * & * & \mathcal{G}_{(5,5)} & \mathcal{G}_{(5,6)} & C_e^T \\ * & * & * & * & * & -\tau I & F_w^T \\ * & * & * & * & * & * & -\tau I \end{bmatrix}$$

with $\mathrm{He}[\cdot] = [\cdot] + [\cdot]^T$, $\mathcal{G}_{(1,3)} = M_i^T \tilde{B}^T - \phi_1 W_1 + \phi_1 X$, $\mathcal{G}_{(2,5)} = N_i \overline{C} - \phi_2 W_2 + \phi_2 Y$, $\mathcal{G}_{(3,3)} = \mathrm{He}(\tilde{A}_i X + \tilde{B} M_i)$, $\mathcal{G}_{(5,5)} = \mathrm{He}(Y\overline{A}_i + N_i \overline{C})$ and $\mathcal{G}_{(5,6)} = Y\overline{B}_w + N_i \overline{D}_w$. Calculated LPV controller and observer gains are $K(p) = M(p)W_1^{-1}$ and $L(p) = W_2^{-T} N(p)$.

Proof. With Lyapunov function $\mathcal{V}(\tilde{x}(t), \overline{e}(t)) = \tilde{x}^T(t)X^{-1}\tilde{x}(t) + \overline{e}^T(t)Y\overline{e}(t)$, consider new vectors $\xi(t) = X^{-1}\tilde{x}(t)$, $\zeta_1(t) = -\xi(t) + W_1^{-1}[X\xi(t) + \tilde{E}\overline{e}(t)]$ [22] and $\zeta_2(t) = (W_2^{-1}Y - I)\overline{e}(t)$ to avoid bilinear matrix inequality problem of observer-based control design. Obviously,

$$\Psi_1 = (W_1 - X)\xi(t) + W_1\zeta_1(t) - \tilde{E}\overline{e}(t) = 0, \tag{3.22}$$

$$\Psi_2 = W_2\zeta_2(t) + (W_2 - Y)\overline{e}(t) = 0, \tag{3.23}$$

$$\Psi_x = \tilde{x}(t) - X\xi(t) = 0. \tag{3.24}$$

Let $K(p) = M(p)W_1^{-1}$ and $L(p) = W_2^{-T}N(p)$, and closed-loop system (3.19) becomes

$$\dot{\tilde{x}}(t) = [\tilde{A}(p)X + \tilde{B}M(p)]\xi(t) + \tilde{B}M(p)\zeta_1(t) + \tilde{D}\overline{e}(t) + \tilde{B}_w\omega(t),$$

$$\dot{\overline{e}}(t) = [\overline{A}(p) + W_2^{-T}N(p)\overline{C}]\overline{e}(t) + [\overline{B}_w + W_2^{-T}N(p)\overline{D}_w]\omega(t).$$

To ensure closed-loop stability and H_∞ disturbance attenuation, the following condition can be considered with relationships (3.22)–(3.24):

$$\begin{aligned} \frac{\mathrm{d}\mathcal{V}(\tilde{x}(t), \overline{e}(t))}{\mathrm{d}t} + \frac{z^T(t)z(t)}{\tau} &- \tau\omega^T(t)\omega(t) - 2\phi_1\zeta_1^T(t)\Psi_1(t) \\ &- 2\phi_2\zeta_2^T(t)\Psi_2(t) - 2\phi_x\tilde{x}^T(t)\Psi_x(t) < 0. \end{aligned} \tag{3.25}$$

Further expanding the left-hand side (LHS) of (3.25),

$$\begin{aligned} \mathrm{LHS}_{(3.25)} = \mathrm{He}\Big[&-\zeta_1^T(t)\phi_1\Psi_1(t) - \zeta_2^T(t)\phi_2\Psi_2(t) + \xi^T(t)\dot{\tilde{x}}(t) - \tilde{x}^T(t)\phi_x\Psi_x(t) \\ &+ \overline{e}^T(t)Y\dot{\overline{e}}(t) - \omega^T(t)\frac{\tau}{2}\omega(t) + \frac{z^T(t)z(t)}{2\tau}\Big]. \end{aligned}$$

With the above representation, the necessary and sufficient condition for (3.25) is as follows:

$$
\begin{aligned}
\mathrm{He}\ \Big\{ & -\zeta_1^T(t)\phi_1 W_1\zeta_1(t) - \zeta_1^T(t)\phi_1(W_1 - X)\xi(t) + \zeta_1^T(t)\phi_1\tilde{E}\bar{e}(t) \\
& - \zeta_2^T(t)\phi_2 W_2\zeta_2(t) - \zeta_2^T(t)\phi_2(W_2 - Y)\bar{e}(t) + \zeta_2^T(t)N(p)\overline{C}\bar{e}(t) + \zeta_2^T N(p)\overline{D}_w\omega(t) \\
& + \xi^T(t)\tilde{B}M(p)\zeta_1(t) + \xi^T(t)[\tilde{A}(p)X + \tilde{B}M(p)]\xi(t) + \xi^T(t)\tilde{D}\bar{e}(t) + \xi^T(t)\tilde{B}_w\omega(t) \\
& + \tilde{x}^T(t)\phi_x X\xi(t) - \tilde{x}^T(t)\phi_x\tilde{x}(t) \\
& + \bar{e}^T(t)[Y\overline{A}(p) + N(p)\overline{C}]\bar{e}(t) + \bar{e}^T(t)[Y\overline{B}_w + N(p)\overline{D}_w]\omega(t) \\
& + \omega^T(t)\left(-\frac{\tau}{2}\right)\omega(t) \\
& + \tau^{-1}z^T(t)C_e\bar{e}(t) + \tau^{-1}z^T(t)F_w\omega(t) + \tau^{-1}z^T(t)\left(-\frac{\tau}{2}I\right)\tau^{-1}z(t)\Big\} < 0.
\end{aligned}
$$

To represent into LMI form, define

$$\kappa = \begin{bmatrix} \zeta_1^T(t) & \zeta_2^T(t) & \xi^T(t) & \tilde{x}^T(t) & \bar{e}^T(t) & \omega^T(t) & \tau^{-1}z^T(t) \end{bmatrix}^T,$$

and (3.25) is equivalent to $\kappa^T(t)G(p)\kappa(t) < 0$, where

$$
G(p) = \begin{bmatrix}
-\phi_1\mathrm{He}(W_1) & 0 & G_{(1,3)} & 0 & \phi_1\tilde{E} & 0 & 0 \\
* & -\phi_1\mathrm{He}(W_2) & 0 & 0 & G_{(2,5)} & N(p)\overline{D}_w & 0 \\
* & * & G_{(3,3)} & \phi_x X & \tilde{D} & \tilde{B}_w & 0 \\
* & * & * & -2\phi_x I & 0 & 0 & 0 \\
* & * & * & * & G_{(5,5)} & G_{(5,6)} & C_e^T \\
* & * & * & * & * & -\tau I & F_w^T \\
* & * & * & * & * & * & -\tau I
\end{bmatrix}
$$

with $G_{(1,3)} = M^T(p)\tilde{B}^T - \phi_1 W_1 + \phi_1 X$, $G_{(2,5)} = N(p)\overline{C} - \phi_2 W_2 + \phi_2 Y$, $G_{(3,3)} = \mathrm{He}[\tilde{A}(p)X + \tilde{B}M(p)]$, $G_{(5,5)} = \mathrm{He}[Y\overline{A}(p) + N(p)\overline{C}]$ and $G_{(5,6)} = Y\overline{B}_w + N(p)\overline{D}_w$.

To ensure (3.25), there should be the following sufficient condition:

$$G(p) = \sum_{i=1}^{n} p_i\mathcal{G}_i < 0.$$

Consequently, condition (3.21) could be established due to $0 \le p_i \le 1$. And condition (3.20) is required primarily by the Lyapunov function. □

With above theorem, the following LMI minimization problem could be applied for minimum H_∞ disturbance attenuation:

$$
\begin{aligned}
& \min \tau \\
\text{s.t.} & \qquad\qquad\qquad\qquad\qquad\qquad (3.26) \\
& (3.20) \text{ and } (3.21).
\end{aligned}
$$

In addition, scalars ϕ_1, ϕ_2 and ϕ_x in Theorem 3.1 are applied to take relationships (3.22)–(3.24) into condition (3.25), which is beneficial to reduce conservatism from new vectors $\xi(t)$, $\zeta_1(t)$ and $\zeta_2(t)$. To determine suitable values of these scalars, line search [22] is a classical and direct method.

3.4.2 Inverse procedure design

With the abovementioned observer-based LPV controller, virtual control inputs $v(t)$ could be obtained with (3.15) and (3.17). For applied control inputs, inverse procedure design focuses on the algebraic relation between $v(t)$ and $u(t)$ to solve the equation $R_{vu}(x(t), u(t)) = v(t)$ with respect to $u(t)$. As pure mathematical analysis, $x(t)$ is assumed to be available in this subsection. For application with the control structure shown in Figure 3.3, $x(t)$ should be replaced by $\hat{x}(t)$.

According to (3.3)–(3.5), F_V, F_α and M_q could be represented as the following forms based on v_V, v_α and v_q:

$$\begin{aligned} F_V &= v_V + g\sin\gamma + \frac{\rho s}{2m}C_{D1}\cdot V^2\alpha + \frac{\rho s}{2m}C_{D2}\cdot V^2\alpha^2, \\ F_\alpha &= v_\alpha - g\frac{\cos\gamma}{V} + \frac{\rho s}{2m}C_{L0}V, \\ M_q &= v_q - \frac{\rho s\bar{c}}{2I_y}C_{m0}V^2 - \frac{\rho s\bar{c}}{2I_y}C_{m1}\cdot V^2\alpha. \end{aligned} \tag{3.27}$$

The above variables include all control effects from applied control inputs, and their algebraic forms (3.6)–(3.8) would be analyzed mainly.

According to the form of M_q in (3.8), rotor speed difference $\Omega_F^2 - \Omega_B^2$ and deflection of elevator δ_e take control effect concurrently in transition procedure. To deal with this control coupling, introduce rotor control weight η. With the form of M_q and $\eta \in [0, 1]$,

$$\Omega_F^2 - \Omega_B^2 = \frac{\eta M_q}{(2\rho AR^2 x_r/I_y)C_t\cdot\sin i_n}, \tag{3.28}$$

$$\delta_e = \frac{(1-\eta)M_q}{(\rho s\bar{c}/2I_y)C_{m\delta}\cdot V^2}, \tag{3.29}$$

where i_n in (3.28) is still unknown, and (3.29) assigns the value of elevator deflection based on η. The value of η should represent the weight of control effect from rotors according to TRUAV structure or flight velocity [19]. A similar weight value is also applied in GS method, but there is not a theoretical form of this value. For inverse procedure design here,

$$\eta = 1 - \frac{V^2}{V_c}, \tag{3.30}$$

where V_c is the regular cruise speed in airplane mode.

According to the forms of F_V and F_α in (3.6) and (3.7), the following equations could be formulated with the value of δ_e from (3.29):

$$
\begin{aligned}
\cos(i_n+\alpha)\left(\Omega_F^2+\Omega_B^2\right) &= \frac{F_V+(\rho s/2m)C_{D\delta}\cdot V^2\delta_e}{(2\rho AR^2/m)C_t} \triangleq \overline{F}_V, \\
\sin(i_n+\alpha)\left(\Omega_F^2+\Omega_B^2\right) &= \frac{F_\alpha+(\rho s/2m)C_{L\delta}\cdot V\delta_e}{-(2\rho AR^2/mV)C_t} \triangleq \overline{F}_\alpha.
\end{aligned}
\tag{3.31}
$$

It is obvious that $i_n+\alpha=\arctan\left(\overline{F}_\alpha/\overline{F}_V\right)$. In transition procedure with a variable rotor-tilt angle, i_n could compensate the control effect of α as follows [19]:

$$
i_n=\arctan\left(\frac{\overline{F}_\alpha}{\overline{F}_V}\right)-\alpha. \tag{3.32}
$$

To ensure that the quad-TRUAV reaches airplane mode ($i_n=0$) with velocity reference V_{ref}, the angle of attack α should also be with a special reference value α_{ref}. For this reference value, the equilibrium point of nonlinear model (3.1) with $V=V_{\text{ref}}$ should be calculated. Consequently, consider the stable case with $i_n=0$ and $V=V_{\text{ref}}$, equilibrium points of γ and q must be equal to 0 to ensure stable values of flight height and pitch angle. Based on the nonlinear model of the quad-TRUAV, equilibrium points of angle of attack (α_0) and other control inputs (Ω_{F0}, Ω_{B0} and δ_{e0}) should satisfy the following equations:

$$
\begin{aligned}
&\frac{\rho s}{2m}C_{D0}V_{\text{ref}}^2+\frac{\rho s}{2m}C_{D1}V_{\text{ref}}^2\alpha_0+\frac{\rho s}{2m}C_{D2}V_{\text{ref}}^2\alpha_0^2-\frac{2\rho AR^2}{m}C_t\cos\alpha_0\left(\Omega_{F0}^2+\right.\\
&\left.\Omega_{B0}^2\right)+\frac{\rho s}{2m}C_{D\delta}V_{\text{ref}}^2\delta_{e0}=0,\\
&\frac{\rho s}{2m}C_{L0}V_{\text{ref}}+\frac{\rho s}{2m}C_{L1}V_{\text{ref}}\alpha_0-\frac{g}{V_{\text{ref}}}+\frac{2\rho AR^2}{mV_{\text{ref}}}C_t\sin\alpha_0\left(\Omega_{F0}^2+\Omega_{B0}^2\right)\\
&+\frac{\rho s}{2m}C_{L\delta}V_{\text{ref}}\delta_{e0}=0,\\
&\frac{\rho s\overline{c}}{2I_y}C_{m0}V_{\text{ref}}^2+\frac{\rho s\overline{c}}{2I_y}C_{m1}V_{\text{ref}}^2\alpha_0+\frac{\rho s\overline{c}}{2I_y}C_{m\delta}V_{\text{ref}}^2\delta_{e0}=0,
\end{aligned}
\tag{3.33}
$$

which could be solved for α_0 as the desired reference value α_{ref}.

With the value of i_n from (3.32), $\Omega_F^2-\Omega_B^2$ could be calculated by (3.28). Together with the following equation based on (3.31),

$$
\Omega_F^2+\Omega_B^2=\sqrt{\overline{F}_V^2+\overline{F}_\alpha^2}, \tag{3.34}
$$

the values of forward and backward rotor speeds could also be assigned.

In general, (3.28)–(3.32) and (3.34) together with reference value α_{ref} from (3.33) constitute the inverse procedure for applied control inputs. For application with the control structure shown in Figure 3.3, V, γ and α in these equations should be replaced by their estimations $\hat{V}$, $\hat{\gamma}$ and $\hat{\alpha}$. The stability of the quad-TRUAV in the transition procedure could be ensured asymptotically by these applied control inputs.

3.5 Fault-tolerant design

3.5.1 *Actuator stuck fault*

According to [28], actuator stuck fault is defined as follows:

$$\tilde{u}_o(t) = \Phi(t)\tilde{u}(t) + [I - \Phi(t)]u_F.$$

This equation corresponds to "Actuator" block presented in Figure 3.3, in which u_F is a constant vector that represents the magnitude of stuck failure. In addition, diagonal matrix $\Phi(t) = \text{diag}(\phi_1(t), \ldots, \phi_j(t), \ldots)$, $\phi_j(t) = 1$ means that the jth actuator is fault-free, whereas $\phi_j(t) = 0$ means that the jth actuator is faulty and lock-in-place.

Under rotor-tilt axles stuck, the control structure designed in Figure 3.3 is inapplicable anymore. For the following FTC strategy, assume that fault information could be estimated or measured accurately and timely by fault detection and diagnosis (FDD) module. According to the preliminary results in [19], the proposed FTC strategy requires real-time performance of FDD methods.

3.5.2 *Degraded model method for FTC*

After rotor-tilt axle stuck fault, the algebraic relation (3.12) becomes

$$v(t) = R_{vu}\big(x(t), u(t)|_{i_n = i_{nF}}\big),$$

where i_{nF} is the fixed and faulty rotor-tilt angle. With the abovementioned equation, one control input of the quad-TRUAV is lost. In this case, a degraded model could be applied to ensure partial control performance [29].

To determine the degraded model, the dynamics of the controlled plant should be analyzed. Similar with conventional UAVs, inner-loop control of the quad-TRUAV about attitudes is fundamental for the stability, and the UAV would not be with diverging states with stable flight height and angle of attack. With this common sense about the TRUAV dynamics, only subsystem (3.10) about attitudes needs to be kept as the degraded model, and the subsystem about flight velocity (3.9) could be in an open loop directly [19].

Based on above analysis, flight velocity control is ignored directly after rotor-tilt axle stuck fault. Consequently, virtual control input v_V is unnecessary, and this is the reason why the velocity controller and the attitude controller are separated. With control inputs v_α and v_q from the attitude LPV controller, the inverse procedure should be reconstructed with measured or estimated fault magnitude i_{nF}. Corresponding FTC structure under rotor-tilt axles stuck fault is presented in Figure 3.4.

To reconstruct the inverse procedure, consider (3.4), (3.5), (3.7) and (3.8) once more. F_α and M_q could still be calculated by (3.27) according to the forms of (3.4) and (3.5). With δ_e and η from (3.29) and (3.30), there are the following rotor speeds for FTC with α_F as an adjustable reference value under stuck fault:

$$\Omega_F^2 + \Omega_B^2 = \frac{F_\alpha + (\rho s/2m)C_{L\delta} \cdot V\delta_e}{-(2\rho AR^2/mV)C_t \cdot \sin(i_{nF} + \alpha)}, \tag{3.35}$$

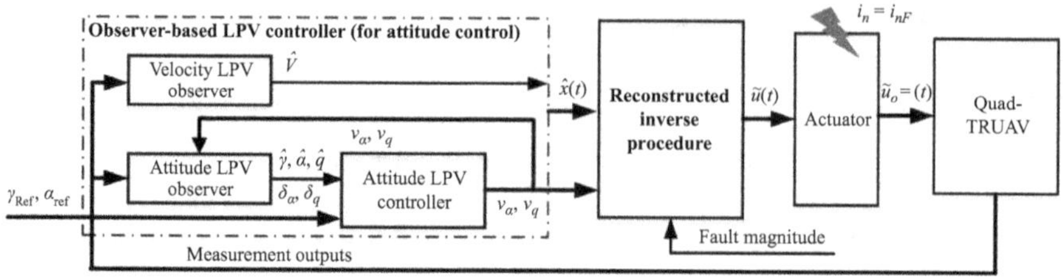

Figure 3.4 The control structure for the quad-TRUAV in faulty case

$$\Omega_F^2 - \Omega_B^2 = \frac{\eta M_q}{(2\rho AR^2 x_r/I_y)C_t \cdot \sin i_{nF}}, \tag{3.36}$$

$$\alpha_{\text{ref}} = \alpha_F. \tag{3.37}$$

Note that, without flight velocity control, V_{ref} could not be reached directly with above FTC strategy. However, because i_n and α own the same control effect for velocity control, reference value α_F provides potential redundancy for velocity tracking. To determine this reference value, the equilibrium point of nonlinear model (3.1) with $V = V_{\text{ref}}$ and $i_n = i_{nF}$ should be calculated. By considering the stable case similar with (3.33), the following equations are formulated:

$$\begin{aligned}
&\frac{\rho s}{2m}C_{D0}V_{\text{ref}}^2 + \frac{\rho s}{2m}C_{D1}V_{\text{ref}}^2\alpha_0 + \frac{\rho s}{2m}C_{D2}V_{\text{ref}}^2\alpha_0^2 - \frac{2\rho AR^2}{m}C_t \cos\left(i_{nF}+\right.\\
&\left.\alpha_0\right)\left(\Omega_{F0}^2 + \Omega_{B0}^2\right) + \frac{\rho s}{2m}C_{D\delta}V_{\text{ref}}^2\delta_{e0} = 0,\\
&\frac{\rho s}{2m}C_{L0}V_{\text{ref}} + \frac{\rho s}{2m}C_{L1}V_{\text{ref}}\alpha_0 - \frac{g}{V_{\text{ref}}} + \frac{2\rho AR^2}{mV_{\text{ref}}}C_t \sin\left(i_{nF} + \alpha_0\right)\left(\Omega_{F0}^2+\right.\\
&\left.\Omega_{B0}^2\right) + \frac{\rho s}{2m}C_{L\delta}V_{\text{ref}}\delta_{e0} = 0,\\
&\frac{\rho s\bar{c}}{2I_y}C_{m0}V_{\text{ref}}^2 + \frac{\rho s\bar{c}}{2I_y}C_{m1}V_{\text{ref}}^2\alpha_0 + \frac{2\rho AR^2 x_r}{I_y}C_t \cdot \sin i_{nF}\left(\Omega_{F0}^2 - \Omega_{B0}^2\right)+\\
&\frac{\rho s\bar{c}}{2I_y}C_{m\delta}V_{\text{ref}}^2\delta_{e0} = 0,
\end{aligned} \tag{3.38}$$

which could be solved for α_0 as the desired reference value α_F. However, the permissible range of angle of attack should be taken into consideration.

In general, (3.29), (3.30) and (3.35)–(3.37) together with α_F from (3.38) reconstruct the inverse procedure under rotor-tilt axle stuck fault. Moreover, V, γ and α in these equations should be replaced by $\hat{V}$, $\hat{\gamma}$ and $\hat{\alpha}$ for application.

3.6 Numerical results

For the following numerical results, parameters of the quad-TRUAV nonlinear model are listed as follows with International System of Units:

$$\begin{gathered}m = 2.71, I_y = 0.0816, \bar{c} = 0.2966, s = 0.452, R = 0.12, A = R^2\pi, x_r = 0.1847,\\ C_t = 0.0041, g = 9.8, \rho = 1.225, C_{D0} = 0.0111, C_{D1} = 2.6278 \times 10^{-4},\\ C_{D2} = 0.0035, C_{D\delta} = 0.0061, C_{L0} = 0.1982, C_{L1} = 0.1823, C_{Lq} = 1.5177,\\ C_{L\delta} = 0.0102, C_{m0} = 0.0189, C_{m1} = 4.333 \times 10^{-6}, C_{mq} = -1.0365,\\ C_{m\delta} = -0.1186.\end{gathered}$$

The polytopic LPV system (3.13) is with the following state-space matrices:

$$A_1 = \left[\begin{array}{c:ccc} 0 & 0 & 0 & 0 \\ \hdashline 0 & 0 & 0 & 0.0230 \\ 0 & 0 & 0 & 0.9770 \\ 0 & 0 & 0 & 0 \end{array}\right], A_2 = \left[\begin{array}{c:ccc} -0.0340 & 0 & 0 & 0 \\ \hdashline 0 & 0 & 0.5588 & 0.0230 \\ 0 & 0 & -0.5588 & 0.9770 \\ 0 & 0 & 0 & -4.6403 \end{array}\right],$$

$$B = \left[\begin{array}{c:cc} 1 & 0 & 0 \\ \hdashline 0 & -1 & 0 \\ 0 & 1 & 0 \\ 0 & 0 & 1 \end{array}\right], B_w = I, C = I, D_w = \mathrm{diag}(0, 0.1, 0.1, 0.1), C_e = I, F_w = 0,$$

$$C_r = \left[\begin{array}{c:ccc} 1 & 0 & 0 & 0 \\ \hdashline 0 & 1 & 0 & 0 \\ 0 & 0 & 1 & 0 \end{array}\right] \text{ and } r = \left[\begin{array}{c} V_{\mathrm{ref}} \\ \hdashline \gamma_{\mathrm{ref}} \\ \alpha_{\mathrm{ref}} \end{array}\right].$$

And varying parameters $p_1 = (\overline{V} - V/\overline{V} - \underline{V}) = (30 - V/30 - 0)$ and $p_2 = 1 - p_1$, where 0 and 30 m/s are minimum and maximum flight velocities. For the above LPV system, the flight velocity controller and the attitude controller could be designed severally by LMI minimization problem (3.26) and MATLAB® LMI toolbox [30] with parameters $\phi_1 = 70$, $\phi_2 = 110$ and $\phi_x = 0.05$. For inverse procedure design, set regular cruise speed $V_c = 21.5$.

3.6.1 Fault-free results

Numerical results without rotor-tilt axle stuck fault are displayed in this subsection with initial conditions $x(0) = [2.1 \quad 0 \quad 0 \quad 0]^T$, $\hat{x}(0) = [2.2 \quad 0 \quad 0 \quad 0]^T$ and $\hat{\delta}(0) = [0 \quad 0 \quad 0]^T$.

Set $w(t) = \left[0.1 \sin 2\pi t \quad 0.001 \sin 2\pi t \quad 0.001 \sin 2\pi t \quad 0.01 \sin 2\pi t\right]^T$ as disturbance input vector. Figure 3.5(a) shows the curves of reference values, measurement outputs, and their estimations based on estimated states. Figure 3.5(b) further displays corresponding tracking errors and estimation errors. Along with V_{ref} varies from 2 to 22 m/s with fixed slope 2 m/s^2, α_{ref} varies from 0 to -0.008106 rad with a first-order inertia link $1/(s+1)$, in which the final value of α_{ref} is calculated from (3.33). The reference value of γ ensures tracking control of flight height as shown in

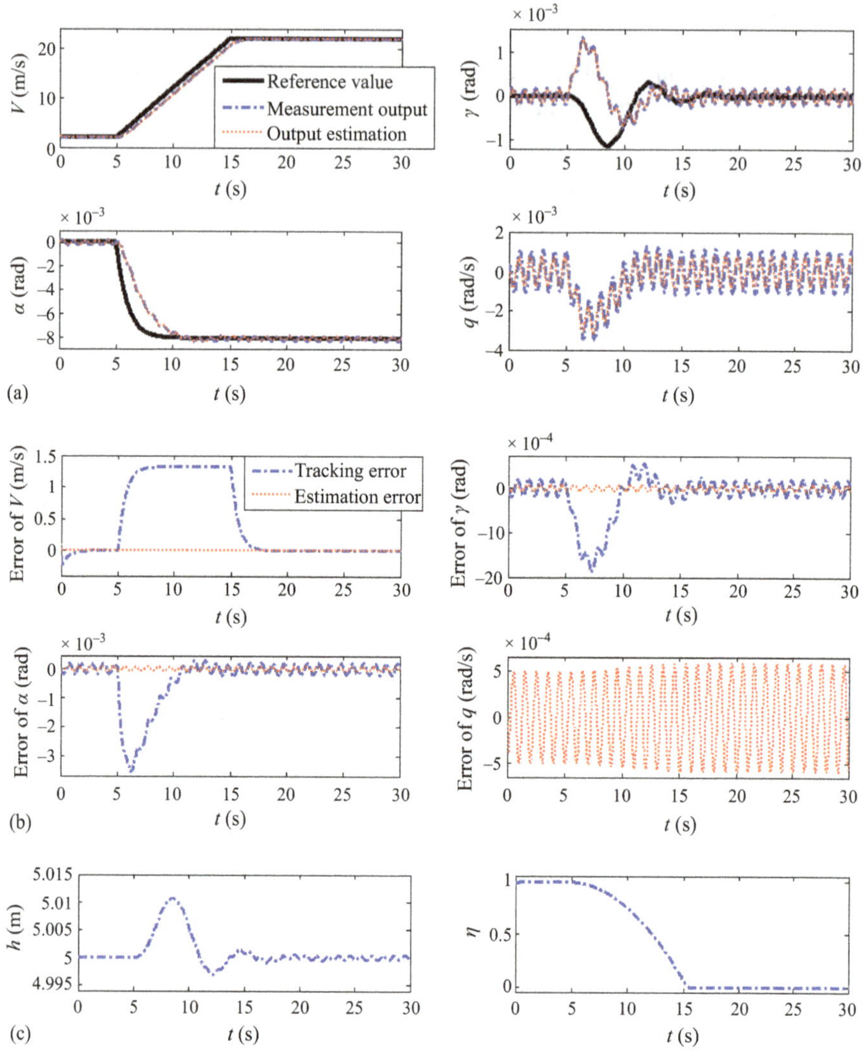

Figure 3.5 Measurement outputs and control weight: (a) measurement outputs and estimations, (b) tracking errors and estimation errors, (c) flight height and control weight

the first curve of Figure 3.5(c) based on the flight height controller designed in [19]. With the above curves, the tracking control of the LPV controller is displayed, and states are also estimated with the LPV observer. Figure 3.6 shows curves of matched perturbations and their estimations from designed PIO. These estimated values are applied for compensation with (3.15).

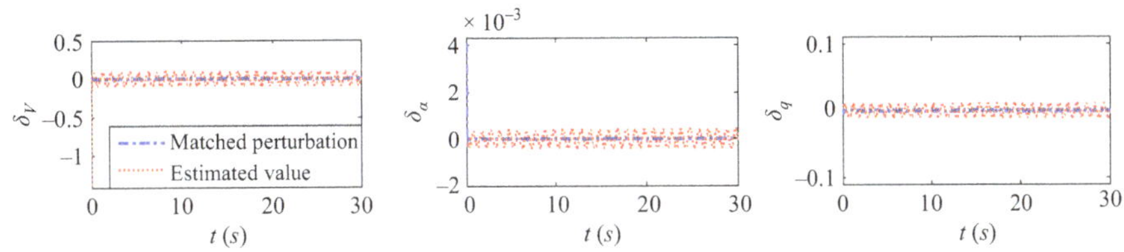

Figure 3.6 Matched perturbations and estimations

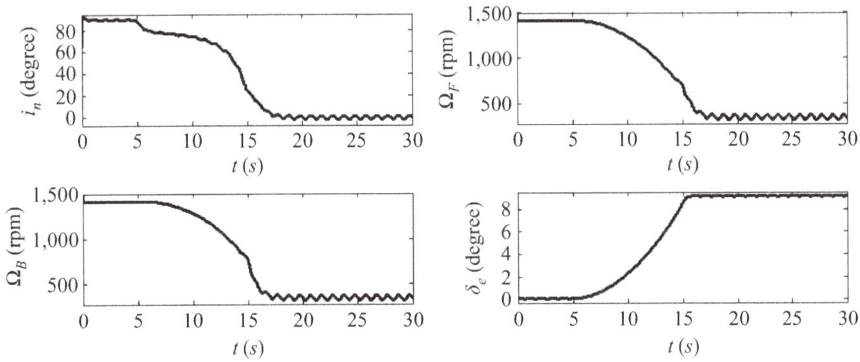

Figure 3.7 Applied control inputs

For inverse procedure, the second curve of Figure 3.5(c) displays the value of control weight calculated from (3.30), and Figure 3.7 shows curves of the applied control inputs. To accelerate in transition procedure, the rotor-tilt angle i_n has to be decreased for extra longitudinal force. Consequently, the variation of i_n from $\pi/2$ rad to 0 is achieved. With these applied control inputs, the stability of the quad-TRUAV in transition procedure is ensured asymptotically in theory.

3.6.2 FTC results under fault

Numerical results with rotor-tilt axle stuck fault are displayed in this subsection with the same initial conditions and disturbance input vector as fault-free results. Consider faulty rotor-tilt angle $i_{nF} = \pi/4$, and FTC strategy could obtain this fault information after 0.3 s time delay.

Under rotor-tilt axle stuck fault, if there is no FTC, the controlled plant would be companied with diverging states as shown in Figure 3.8. Applying proposed FTC strategy, Figure 3.9(a) displays stable measurement outputs and their estimations, Figure 3.9(b) presents tracking errors and estimation errors, and Figure 3.10 shows corresponding applied control inputs. Obviously, before the appearance of stuck fault, the applied control inputs are the same with the values in fault-free case. Then, the rotor-tilt angle reaches $\pi/4$ rad and is stuck at this position. As velocity control is

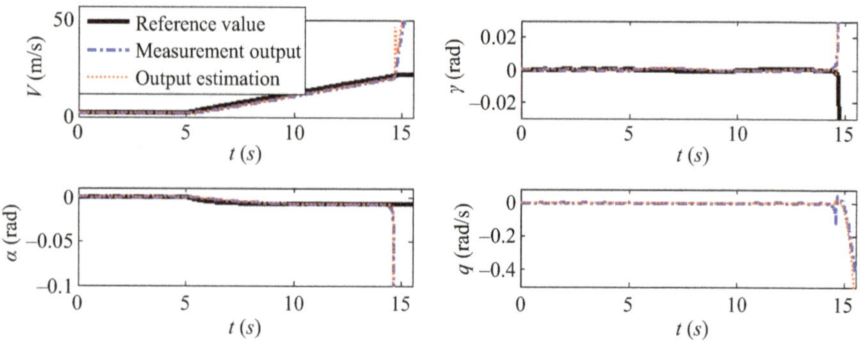

Figure 3.8 Measurement outputs without FTC under fault

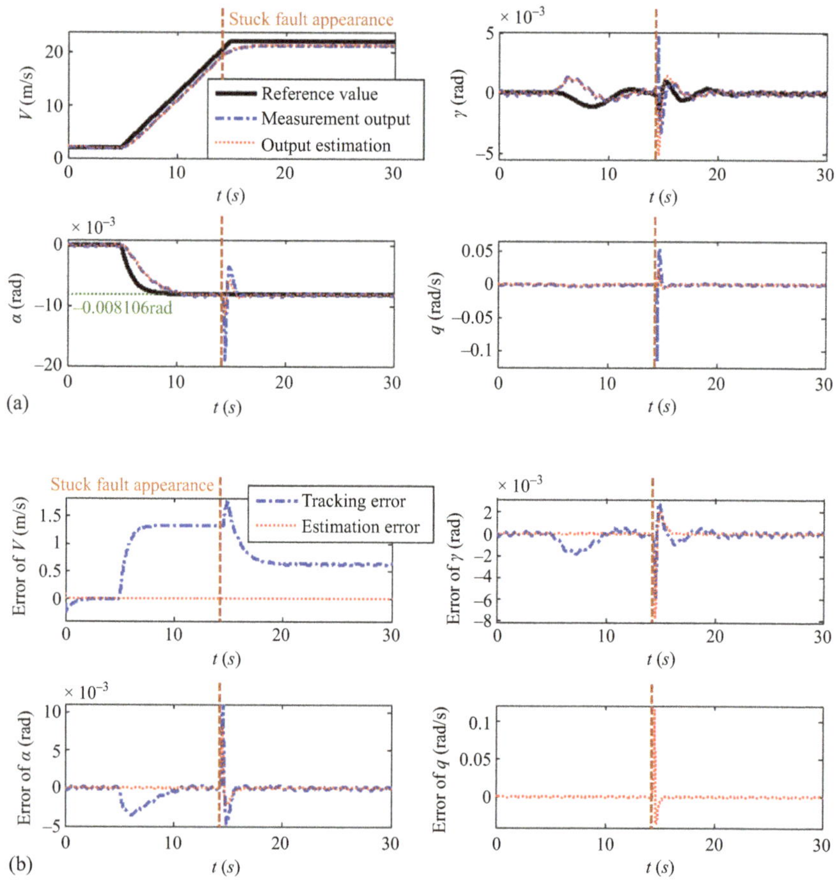

Figure 3.9 Measurement outputs and errors with FTC under fault: (a) measurement outputs and estimations and (b) tracking errors and estimation errors

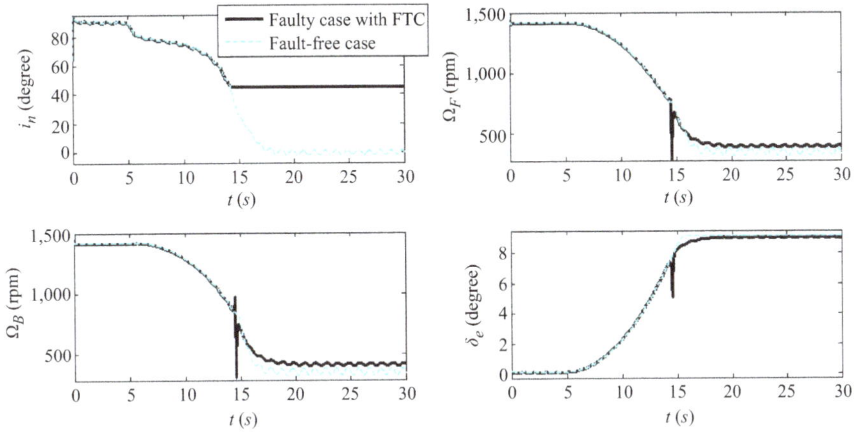

Figure 3.10 Applied control inputs with FTC under fault (compared with the values in fault-free case)

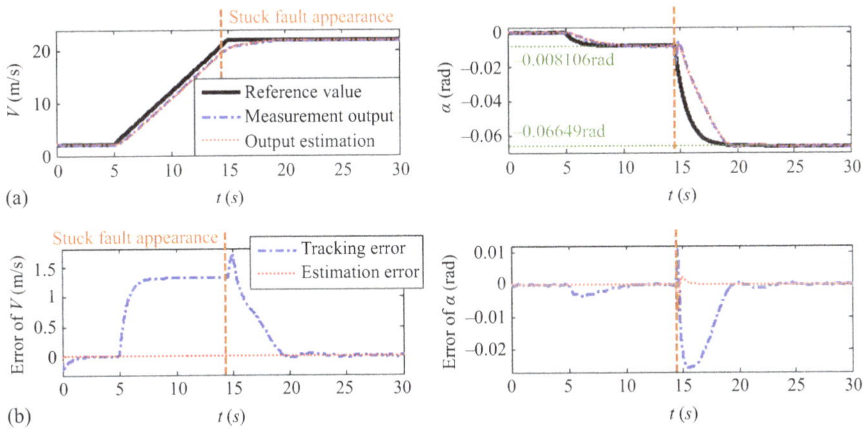

Figure 3.11 Velocity tracking control under fault: (a) velocity tracking control based on angle of attack and (b) tracking errors and estimation errors

ignored in FTC strategy, and the final value of α_{ref} is still -0.008106 rad as fault-free case, flight velocity could not reach the desired reference value. Other measurement outputs and some applied control inputs are with transient fluctuations after rotor-tilt axle stuck fault, which are caused by the delay time of fault information.

To ensure tracking control of flight velocity, (3.38) is considered for the new reference value of angle of attack. After rotor-tilt axle stuck fault, the redesigned final value of α_{ref} is equal to -0.06649 rad according to (3.38). In this way, flight velocity could reach reference value 22 m/s again as shown in Figure 3.11, and the quad-TRUAV still owns the cruise ability based on the reference value α_F. However, the permissible range of angle of attack should be taken into account.

3.7 Conclusions

This chapter introduces the control method of the quad-TRUAV in the transition procedure under rotor-tilt axle stuck fault. By introducing virtual control inputs, the nonlinear model of the quad-TRUAV is transformed into the LPV form, and observer-based LPV control is proposed to ensure the closed-loop stability of controlled plant. For applied control inputs, the inverse procedure is designed further, and the stability of transition procedure is ensured asymptotically in theory. After rotor-tilt axle stuck fault, the inverse procedure is reconstructed with degraded dynamics by ignoring velocity control. With the redesigned reference value of angle of attack, the cruise ability of TRUAV under stuck fault is still kept.

Acknowledgments

This work was supported by the National Key R&D Program of China (2018YFB1307500), National Natural Science Foundation of China (U1608253), and National Natural Science Foundation of Guangdong Province (2017B010116002).

References

[1] Wikimedia Foundation, Inc. List of Unmanned Aerial Vehicle Applications [Internet]. America: Wikimedia Foundation, Inc.; [cited 2018 Sep 14]. Available from: https://en.wikipedia. org/wiki/List_of_unmanned_aerial_vehicle_applications.

[2] Song D, Meng X, Qi J, Han J. Strategy of dynamic modeling and predictive control on 3-DoF rotorcraft aerial manipulator system. Robot. 2015;37(2):152–60.

[3] Kim HJ, Kim M, Lim H, *et al.* Fully autonomous vision-based net-recovery landing system for a fixed-wing UAV. IEEE–ASME Trans Mechatron. 2013;18(4):1320–33.

[4] Saeed AS, Younes AB, Islam S, Dias J, Seneviratne L, Cai G. A review on the platform design, dynamic modeling and control of hybrid UAVs. Proceedings of International Conference on Unmanned Aircraft Systems; 2015 Jun 9–12; Denver, CO, USA. 2015. p. 806–15.

[5] Chengdu JOUAV Automation Tech Co., Ltd. CW-007 [Internet]. Chengdu: Chengdu JOUAV Automation Tech Co., Ltd; c2015- [cited 2018 Sep 14]. Available from: http://www.jouav.com/index.php/Jouav/index/CW_007.htmll =en-us.

[6] Hochstenbach M, Notteboom C, Theys B. Design and control of an unmanned aerial vehicle for autonomous parcel delivery with transition from vertical

take-off to forward Flight – VertiKUL, a quadcopter tailsitter. Int J Micro Air Veh. 2015;7(4):395–406.
[7] Chen C, Zhng J, Zhang D, Shen L. Control and flight test of a tilt-rotor unmanned aerial vehicle. Int J Adv Rob Syst. 2017;1:1–12.
[8] Liu Z, He Y, Yang L, Han J. Control techniques of tilt rotor unmanned aerial vehicle systems: a review. Chin J Aeronaut. 2017;30(1):135–48.
[9] Flores G, Lugo I, Lozano R. 6-DOF hovering controller design of the quad tiltrotor aircraft: simulations and experiments. Proceedings of 53rd IEEE Conference on Decision and Control; 2014 Dec 15–17; Los Angeles, CA, USA. 2014. p. 6123–8.
[10] Choi SW, Kang Y, Chang S, Koo S, Kim JM. Development and conversion flight test of a small tiltrotor unmanned aerial vehicle. J Aircr. 2010;47(2):730–2.
[11] Oner KT, Cetinsoy E, Sirimoglu E, *et al.* Mathematical modeling and vertical flight control of a tilt-wing UAV. Turk J Electr Eng Comput Sci. 2012;20(1):149–57.
[12] Flores-Colunga GR, Lozano-Leal R. A nonlinear control law for hover to level flight for the quad tilt-rotor UAV. Proceedings of 19th World Congress The International Federation of Automatic Control; 2014 Aug 24–29; Cape Town, South Africa. 2014. p. 11055–9.
[13] Muraoka K, Okada N, Kubo D, Sato M. Transition flight of quad tilt wing VTOL UAV. Proceedings of 28th International Congress of the Aeronautical Sciences; 2012.
[14] Verling S, Zilly J. Modeling and control of a VTOL glider [dissertation]. Swiss Federal Institute of Technology Zurich. Zurich, Switzerland; 2002.
[15] Dickeson JJ, Miles D, Cifdaloz O, Wells VL, Rodriguez AA. Robust LPV H_∞ gain-scheduled hover-to-cruise conversion for a tilt-wing rotorcraft in the presence of CG variations. Proceedings of 46th IEEE Conference on Decision and Control; 2007 Dec 12–14; New Orleans, LA, USA. 2007. p. 2773–8.
[16] Zhao W, Underwood C. Robust transition control of a Martian coaxial tiltrotor aerobot. Acta Astronaut. 2014;99:111–29.
[17] Qi X, Qi J, Didier T, *et al.* A review on fault diagnosis and fault tolerant control methods for single-rotor aerial vehicles. J Intell Rob Syst. 2014;73:535–55.
[18] Zhang Y, Jiang J. Bibliographical review on reconfigurable fault-tolerant control systems. Annu Rev Control. 2008;32:229–52.
[19] Liu Z, Didier T, Yang L, He Y, Han J. Mode transition and fault tolerant control under rotor-tilt axle stuck fault of quad-TRUAV. Proceedings of 10th Fault Detection, Supervision and Safety for Technical Process; 2018 Aug 29–31; Warsaw, Poland. 2018. p. 991–7.
[20] Rugh WJ, Shamma JS. Research on gain scheduling. Automatica. 2000; 36:1401–25.
[21] Koenig D, Mammar S. Design of proportional-integral observer for unknown input descriptor systems. IEEE Trans Autom Control. 2003;47(12):2057–62.
[22] Koroglu H. Improved conditions for observer-based LPV control with guaranteed L2-gain performance. 2014 American Control Conference; 2014 Jun 4–6; Portland, OR, USA. 2014. p. 3760–5.

[23] Marcos A, Balas GJ. Development of linear-parameter-varying models for aircraft. J Guid Control Dyn. 2004;27(2):218–28.

[24] Pounds P, Mahony R, Corke P. Modelling and control of a large quadrotor robot. Control Eng Pract. 2010;18:691–9.

[25] Tanaka K, Wang H. Fuzzy Control Systems Design and Analysis: A Linear Matrix Inequality Approach. New York, NY: John Wiley & Sons, Inc.; 2001.

[26] Song Q, Han J. UKF-based active modeling and model-reference adaptive control for mobile robots. Robot. 2005;27(3):226–30.

[27] Lopez-Estrada FR, Ponsart JC, Didier T, Zhang Y, Astorga-Zaragoza CM. LPV model-based tracking control and robust sensor fault diagnosis for a quadrotor UAV. J Intell Rob Syst. 2016;84(1–4):163–77.

[28] Qi X, Didier T, Qi J, Zhang Y, Han J. Self-healing control against actuator stuck failures under constraints: application to unmanned helicopters. Adv Intell Comput Diagn Control. 2016;386:193–207.

[29] Souanef T, Fichter W. Fault tolerant L1 adaptive control based on degraded models. Advances in Aerospace Guidance, Navigation and Control. 2015.

[30] Gahinet P, Nemirovski A, Laub AJ, Chilali M. LMI Control Toolbox for Use With MATLAB. Natick, MA, USA: The MathWorks, Inc.; 1994.

Chapter 4

An unknown input observer-based framework for fault and icing detection and accommodation in overactuated unmanned aerial vehicles

Andrea Cristofaro[1], Damiano Rotondo[2], and Tor Arne Johansen[3]

The use of unmanned aerial vehicles (UAVs) as support to operations in remote areas and harsh environments, such as marine operations in the Arctic, is becoming more and more crucial. These systems are expected to face very critical weather conditions, and for this reason they are naturally prone to the occurrence of icing. The formation of ice layers on airfoils decreases the lift and, simultaneously, increases the drag and the mass of the vehicle, thus requiring additional engine power and implying a premature stall angle. By adopting some tools from the control allocation framework, this chapter aims at presenting an unknown input observer (UIO) approach for fault and icing diagnosis and accommodation in UAVs equipped with a redundant suite of effectors and actuators. The chapter is structured as follows. At first, the UAV model and the basic setups are given, and then the fault and icing effects on the UAV dynamics are discussed. A short overview on the design of UIOs is proposed, and a main section is dedicated to present the icing diagnosis and accommodation tasks. Furthermore, by means of appropriate scheduling parameters, the UIO-based diagnosis scheme can be extended to the nonlinear aircraft dynamics by considering the framework of linear parameter varying (LPV) systems. A collection of simulation examples concludes the chapter and illustrates the practical application of the proposed architecture.

4.1 Introduction

Detection and accommodation of ice adhesion on wings, control surfaces and sensors is a challenging and primary issue for UAVs, since ice accretion modifies the shape of the aircraft and alters the pressure measurements, this causing adverse changes on

[1]Department of Computer, Control and Management Engineering (DIAG), Sapienza University of Rome, Rome, Italy
[2]Department of Electrical and Computer Engineering (IDE), University of Stavanger (UiS), Stavanger, Norway
[3]Department of Engineering Cybernetics, Center for Autonomous Marine Operations and Systems (AMOS), NTNU, Trondheim, Norway

aerodynamic forces and reducing the maneuvering capabilities. The phenomenon of icing, which can be regarded as a structural fault, is a well-recognized problem in aviation research since the early 1900s [1]. Inflight icing is typically caused by the impact of supercooled water droplets (SWDs). Under certain atmospheric conditions, water droplets remain cooled and do not freeze until they reach very low temperatures; however, if a droplet impacts on the aircraft surface, it freezes immediately and accretes ice [2]. The rate and the severity of icing are determined by several factors, such as shape and roughness of the impacting surface, vehicle speed, air temperature and relative humidity. The consequences of icing are even more severe for small unmanned aircrafts due to their simple architecture and limited payload, this making them mostly unsuitable for the typical anti-icing and deicing devices that are used in large airplanes. Small UAVs are also more prone to icing than most other aircrafts since they often operate at low altitude where high humidity and SWDs are encountered more frequently. Larger aircrafts tend to operate at high altitude (except for takeoff and landing) where there are fewer risks of icing. Some advanced deicing systems for UAVs have been proposed recently based on layers of coating material made of carbon nanotubes [3,4]. However, since these are very power consuming, in order to guarantee the efficiency of the system it is very important to rely on fault/icing detection schemes with fast and accurate responses. On the other hand, the availability of redundant control surfaces is a key advantage toward safe aircraft maneuvering and stability in spite of icing.

UIOs [5] are a versatile and useful tool for generating robust detection filters, as they can be made insensitive to certain input space directions by assigning specific directions to the residuals as long as some structural algebraic conditions on the system are fulfilled. This is particularly interesting in connection with the control allocation framework [6,7]. The work on UIO-based icing diagnosis has been initiated in [8,9], where the problem of icing detection has been addressed considering, respectively, the lateral and the longitudinal model of the vehicle. Further improvements have been obtained using multiple-models [10] and LPV methods [11,12].

4.2 Vehicle model

The UAV nonlinear model consists of three equations for the airspeed components $(\tilde{u}, \tilde{v}, \tilde{w})$, which represent the velocity relative to the wind, three equations for the Euler angles $(\tilde{\phi}, \tilde{\theta}, \tilde{\psi})$, which define the attitude of the vehicle, and three equations for the angular rates $(\tilde{p}, \tilde{q}, \tilde{r})$ [13]:

$$
\begin{aligned}
m\dot{\tilde{u}} &= m(\tilde{r}\tilde{v} - \tilde{q}\tilde{w} - g\sin\tilde{\theta}) + \mathscr{A}_x + \mathscr{T} \\
m\dot{\tilde{v}} &= m(\tilde{p}\tilde{w} - \tilde{r}\tilde{u} + g\cos\tilde{\theta}\sin\tilde{\phi}) + \mathscr{A}_y \\
m\dot{\tilde{w}} &= m(\tilde{q}\tilde{u} - \tilde{p}\tilde{v} + g\cos\tilde{\theta}\cos\tilde{\phi}) + \mathscr{A}_z
\end{aligned}
$$

$$
\begin{aligned}
\dot{\tilde{\phi}} &= \tilde{p} + \tilde{q}\sin\tilde{\phi}\tan\tilde{\theta} + \tilde{r}\cos\tilde{\phi}\tan\tilde{\theta} \\
\dot{\tilde{\theta}} &= \tilde{q}\cos\tilde{\phi} - \tilde{r}\sin\tilde{\phi} \\
\dot{\tilde{\psi}} &= \tilde{q}\sin\tilde{\phi}\sec\tilde{\theta} + \tilde{r}\cos\tilde{\phi}\sec\tilde{\theta}
\end{aligned}
$$

$$\begin{aligned}\dot{\tilde{p}} &= \Gamma_1 \tilde{p}\tilde{q} - \Gamma_2 \tilde{q}\tilde{r} + \mathscr{M}_p \\ \dot{\tilde{q}} &= \Gamma_5 \tilde{p}\tilde{r} - \Gamma_6(\tilde{p}^2 - \tilde{r}^2) + \mathscr{M}_q \\ \dot{\tilde{r}} &= \Gamma_7 \tilde{p}\tilde{q} - \Gamma_1 \tilde{q}\tilde{r} + \mathscr{M}_r\end{aligned}$$

where m is the vehicle mass, g is the gravitational acceleration, $\mathscr{A}_i$ are the aerodynamical forces (lift and drag), $\mathscr{T}$ is the thrust force, $\mathscr{M}_i$ are the aerodynamical torques and Γ_i are the coefficients obtained as combinations of the main inertia coefficients I_{xx}, I_{yy}, I_{zz} and I_{xz}. The velocities $\tilde{u}, \tilde{v}, \tilde{w}$ are expressed in the body frame, i.e., along longitudinal, lateral and vertical body direction, respectively, and they represent the aircraft speed relative to the wind.

The aerodynamical forces and torques are expressed by the quasilinear relationships

$$\begin{aligned}\mathscr{T} &= \tfrac{\rho S_{prop} C_{prop}}{2m}\left(k_m^2 \tilde{\tau}_t - V_a^2\right) \\ \mathscr{A}_x &= \tfrac{\rho \tilde{V}_a^2 S}{2m}\left(C_X(\tilde{\alpha}) + C_{X_q}(\tilde{\alpha})\tfrac{c\tilde{q}}{2\tilde{V}_a} + C_{X_{\delta_e}}(\tilde{\alpha})\tilde{\tau}_e\right) \\ \mathscr{A}_y &= \tfrac{\rho \tilde{V}_a^2 S}{2m}\left(C_{Y_0} + C_{Y_\beta}\beta + C_{Y_p}\tfrac{b\tilde{p}}{2\tilde{V}_a} + C_{Y_r}\tfrac{b\tilde{r}}{2\tilde{V}_a} + C_{Y_{\delta_a}}\tilde{\tau}_a + C_{Y_{\delta_r}}\tilde{\tau}_r\right) \\ \mathscr{A}_z &= \tfrac{\rho \tilde{V}_a^2 S}{2m}\left(C_Z(\tilde{\alpha}) + C_{Z_q}(\tilde{\alpha})\tfrac{c\tilde{q}}{2\tilde{V}_a} + C_{Z_{\delta_e}}(\tilde{\alpha})\tilde{\tau}_e\right) \\ \mathscr{M}_p &= \tfrac{1}{2}\rho \tilde{V}_a^2 Sb\left(C_{p_0} + C_{p_\beta}\tilde{\beta} + C_{p_p}\tfrac{b\tilde{p}}{2\tilde{V}_a} + C_{p_r}\tfrac{b\tilde{r}}{2\tilde{V}_a} + C_{p_{\delta_a}}\tilde{\tau}_a + C_{p_{\delta_r}}\tilde{\tau}_r\right) \\ \mathscr{M}_q &= \tfrac{\rho \tilde{V}_a^2 Sc}{2I_{yy}}\left(C_{m_0} + C_{m_\alpha}\tilde{\alpha} + C_{m_q}\tfrac{c\tilde{q}}{2\tilde{V}_a} + C_{m_{\delta_e}}\tilde{\tau}_e\right) \\ \mathscr{M}_r &= \tfrac{1}{2}\rho \tilde{V}_a^2 Sb\left(C_{r_0} + C_{r_\beta}\tilde{\beta} + C_{r_p}\tfrac{b\tilde{p}}{2\tilde{V}_a} + C_{r_r}\tfrac{b\tilde{r}}{2\tilde{V}_a} + C_{r_{\delta_a}}\tilde{\tau}_a + C_{r_{\delta_r}}\tilde{\tau}_r\right)\end{aligned}$$

where ρ is the air density, S_{prop} is the area of the propeller and C_{prop} is its aerodynamical coefficient, k_m is the constant that specifies the efficiency of the motor, S is the wing surface area, m is the vehicle mass, c is the mean aerodynamic chord and b is the wingspan of the UAV.

The overall inputs entering the system are the propeller angular speed $\tilde{\tau}_t$ (assumed to be positive) and the surface deflections $\tilde{\tau}_e$, $\tilde{\tau}_a$ and $\tilde{\tau}_r$ producing torques. The nondimensional coefficients C_i are usually referred to as *stability* and *control derivatives*. Some of them are nonlinear functions of the angle-of-attack $\tilde{\alpha}$, defined as

$$\tilde{\alpha} = \arctan\left(\frac{\tilde{u}}{\tilde{w}}\right)$$

according to

$$\begin{aligned}C_X(\tilde{\alpha}) &= \left(C_{L_0} + C_{L_\alpha}\tilde{\alpha}\right)\sin\tilde{\alpha} - \left(C_{D_0} + C_{D_\alpha}\tilde{\alpha}\right)\cos\tilde{\alpha} \\ C_{X_q}(\tilde{\alpha}) &= C_{L_q}\sin\tilde{\alpha} - C_{D_q}\cos\tilde{\alpha} \\ C_{X_{\delta_e}}(\tilde{\alpha}) &= C_{L_{\delta_e}}\sin\tilde{\alpha} - C_{D_{\delta_e}}\cos\tilde{\alpha} \\ C_Z(\tilde{\alpha}) &= -\left[\left(C_{D_0} + C_{D_\alpha}\tilde{\alpha}\right)\sin\tilde{\alpha} + \left(C_{L_0} + C_{L_\alpha}\tilde{\alpha}\right)\cos\tilde{\alpha}\right] \\ C_{Z_q}(\tilde{\alpha}) &= -\left(C_{D_q}\sin\tilde{\alpha} + C_{L_q}\cos\tilde{\alpha}\right) \\ C_{Z_{\delta_e}}(\tilde{\alpha}) &= -\left(C_{D_{\delta_e}}\sin\tilde{\alpha} + C_{L_{\delta_e}}\cos\tilde{\alpha}\right)\end{aligned}$$

Finally, $\tilde{V}_a$ and $\tilde{\beta}$ are the total airspeed and the sideslip angle, respectively, defined as

$$\tilde{V}_a = \sqrt{\tilde{u}^2 + \tilde{v}^2 + \tilde{w}^2}, \qquad \tilde{\beta} = \arcsin\left(\frac{\tilde{u}}{\tilde{V}_a}\right)$$

4.2.1 Linearization

The design of a complete nonlinear controller for the UAV systems is a very ambitious and challenging objective, since there are several requirements to be met and a number of different configurations and scenarios to be faced with specifically tuned performances. However, it is a common procedure to select a finite number of operating conditions, and to develop linearized control schemes for each of them: as long as the vehicle configuration lies in the hull of a given operating point, the corresponding linear controller is applied. In this regard, let us consider a suitable trim condition $x^* := (u^*, v^*, w^*, \phi^*, \theta^*, \psi^*, p^*, q^*, r^*, \tau_t^*, \tau_a^*, \tau_e^*, \tau_r^*)$ and linearize the system about such operating point. Introducing the incremental variables

$$\begin{aligned} &u := \tilde{u} - u^*, \ v := \tilde{v} - v^*, \ w := \tilde{w} - w^* \\ &\phi := \tilde{\phi} - \phi^*, \ \theta := \tilde{\theta} - \theta^*, \ \psi := \tilde{\psi} - \psi^* \\ &p := \tilde{p} - p^*, \ q := \tilde{q} - q^*, \ r := \tilde{r} - r^* \\ &\tau_t := \tilde{\tau}_t - \tau_t^*, \ \tau_a := \tilde{\tau}_a - \tau_a^*, \ \tau_e := \tilde{\tau}_e - \tau_e^*, \ \tau_r := \tilde{\tau}_r - \tau_r^*. \end{aligned}$$

one obtains a 6-DOF linear system describing the linearized coupled longitudinal/lateral dynamics of the aircraft:

$$\dot{\mathbf{x}} = \mathbf{A}\mathbf{x} + \mathbf{B}\tau \tag{4.1}$$

with $\mathbf{x} := [u, v, w, \phi, \theta, \psi, p, q, r]^T$, $\tau = [\tau_t, \tau_e, \tau_a, \tau_r]^T$ and plant matrices

$$\mathbf{A} = \begin{bmatrix} X_u & X_v & X_w & 0 & X_\theta & 0 & 0 & X_q & X_r \\ Y_u & Y_v & Y_w & Y_\phi & Y_\theta & 0 & Y_p & 0 & Y_r \\ Z_u & Z_v & Z_w & Z_\phi & Z_\theta & 0 & Z_p & Z_q & 0 \\ 0 & 0 & 0 & \Phi_\phi & \Phi_\theta & 0 & \Phi_p & \Phi_q & \Phi_r \\ 0 & 0 & 0 & \Theta_\phi & 0 & 0 & 0 & \Theta_q & \Theta_r \\ 0 & 0 & 0 & \Psi_\phi & \Psi_\theta & 0 & 0 & \Psi_q & \Psi_r \\ L_u & L_v & L_w & 0 & 0 & 0 & L_p & L_q & L_r \\ M_u & M_v & M_w & 0 & 0 & 0 & M_p & M_q & M_r \\ N_u & N_v & N_w & 0 & 0 & 0 & N_p & N_q & N_r \end{bmatrix}$$

$$\mathbf{B} = \begin{bmatrix} X_{\tau_t} & X_{\tau_e} & 0 & 0 \\ 0 & 0 & Y_{\tau_a} & Y_{\tau_r} \\ 0 & Z_{\tau_e} & 0 & 0 \\ 0 & 0 & 0 & 0 \\ 0 & 0 & 0 & 0 \\ 0 & 0 & 0 & 0 \\ 0 & 0 & L_{\tau_a} & L_{\tau_r} \\ 0 & M_{\tau_e} & 0 & 0 \\ 0 & 0 & N_{\tau_a} & N_{\tau_r} \end{bmatrix}$$

The structure of such matrices is clearly provided by the linearization of the kinematic relationships, and the coefficients are essentially determined by stability and control derivatives of the vehicle.

4.2.2 Measured outputs

The navigation system is supposed to be equipped with a sensor suite, including a pitot tube aligned with the longitudinal body axis, a GPS, an altimeter, gyroscopes and accelerometers. The following main outputs $\mathbf{y} \in \mathbb{R}^7$ are therefore considered: horizontal airspeed $y_1 = \tilde{u}$, attitude angles $(y_2, y_3, y_4) = (\tilde{\phi}, \tilde{\theta}, \tilde{\psi})$ and angular velocities $(y_5, y_6, y_7) = (\tilde{p}, \tilde{q}, \tilde{r})$; according to the linearization and the operating conditions, the output matrix $\mathring{\mathbf{C}} \in \mathbb{R}^{7\times 9}$ associated with (4.1) turns out to be

$$\mathring{\mathbf{C}} := \begin{bmatrix} 1 & 0_{1\times 2} & 0_{1\times 6} \\ 0_{6\times 1} & 0_{6\times 2} & I_{6\times 6} \end{bmatrix} \tag{4.2}$$

In addition, the GPS and the altimeter provide position measurements $(\tilde{x}_N, \tilde{x}_E, \tilde{x}_D)$ expressed in the inertial frame; the position coordinates are assigned by the kinematic equations:

$$\begin{bmatrix} \dot{\tilde{x}}_N \\ \dot{\tilde{x}}_E \\ \dot{\tilde{x}}_D \end{bmatrix} = \mathbf{R}\left(\tilde{\phi}, \tilde{\theta}, \tilde{\psi}\right) \begin{bmatrix} \tilde{u} \\ \tilde{v} \\ \tilde{w} \end{bmatrix} + \nu,$$

where the matrix $\mathbf{R}(\cdot,\cdot,\cdot)$ represents the rotation from body to inertial frame and $\nu = [\nu_N \ \nu_E \ \nu_D]^T$ is the wind speed (expressed in the inertial frame). We notice that, whenever an accurate wind speed estimator is available [14–16], the interpolation of the estimated wind speed with the average aircraft speed that can be computed through the GPS data provides also a measurement of the relative velocities v, w and hence in this case one can rely on an output matrix $\bar{\mathbf{C}} \in \mathbb{R}^{9\times 9}$.

We will refer to $\mathbf{C} = \bar{\mathbf{C}}$ as to *full information case* and to $\mathbf{C} = \mathring{\mathbf{C}}$ as to *partial information case*.

4.2.3 Control allocation setup

We put our focus on overactuated vehicles, this being a key feature while performing tasks such as fault accommodation and control reconfiguration. The redundancy of control surfaces can be expressed by a simple, linear effector model:

$$\tau = \mathbf{G}\delta, \quad \delta := \begin{bmatrix} \tau_t \\ \delta_1 \end{bmatrix} \tag{4.3}$$

where $\delta_1 \in \mathbb{R}^4$ is a vector incorporating left and right aileron deflections, and left and right rudder (or elevon) deflections:

$$\delta_1 = \begin{bmatrix} \delta_{al} \\ \delta_{ar} \\ \delta_{rl} \\ \delta_{rr} \end{bmatrix}$$

A typical class of vehicles equipped with a surface architecture of this kind is offered, for example, by the so-called V-tail aircrafts [13]. In particular, a pitching moment is induced when moving the ailerons jointly, while moving them alternatively produces a rolling moment. Similarly, yawing moment is induced by an alternative movement of elevons, while a joint movement produces a pitching moment. The matrix $\mathbf{G} \in \mathbb{R}^{4\times 5}$ is assigned by

$$\mathbf{G} = \begin{bmatrix} 1 & 0 & 0 & 0 & 0 \\ 0 & \varepsilon & \varepsilon & 1 & 1 \\ 0 & \frac{1}{2} & -\frac{1}{2} & 0 & 0 \\ 0 & 0 & 0 & 1 & -1 \end{bmatrix},$$

where the parameter $\varepsilon > 0$ is inverse proportional to the distance between the aircraft center of gravity (usually assumed aligned with wings) and the tail.
Concerning the low-level control loop, the system is supposed to be controlled by an autopilot responsible to produce the desired control effect τ_c that is typically provided by a suitable controller. The generation of the desired control effect is distributed over the redundant effectors according to (4.3); in particular the control allocation module is in charge to determine δ_c such that

$$\tau_c = \mathbf{G}\delta_c.$$

4.2.4 Wind disturbance

The airspeed dynamics is affected by the wind effect that can be expressed by the additional input:

$$-\mathbf{R}\left(\tilde{\phi}, \tilde{\theta}, \tilde{\psi}\right)\dot{\nu}$$

where $\dot{\nu} = [\dot{\nu}_N \ \dot{\nu}_E \ \dot{\nu}_D]^T$ is the wind acceleration expressed in the inertial frame. The wind ν is typically decomposed as the sum of a steady component (known or accurately estimated) ν^* with $\dot{\nu}^* = 0$ and a turbulence component ν': this leads to an input disturbance $\xi(t) \in \mathbb{R}^9$ given by

$$\xi(t) = \mathbf{N}(t) \begin{bmatrix} \dot{\nu}'_N \\ \dot{\nu}'_E \\ \dot{\nu}'_D \end{bmatrix}, \ \mathbf{N}(t) := \begin{bmatrix} -\mathbf{R}\left(\tilde{\phi}, \tilde{\theta}, \tilde{\psi}\right) \\ 0_{6\times 3} \end{bmatrix} \tag{4.4}$$

Summarizing, combining (4.1)–(4.4), we have to deal with the following uncertain linear plant:

$$\begin{aligned} \dot{\mathbf{x}} &= \mathbf{A}\mathbf{x} + \mathbf{B}\mathbf{G}\delta + \mathbf{N}\dot{\nu}' \\ \mathbf{y} &= \mathbf{C}\mathbf{x} \end{aligned} \tag{4.5}$$

where $\mathbf{N} = \mathbf{N}(t)$ is a time-varying input matrix. In order to model in a realistic way the wind gusts, the widely accepted Dryden wind turbulence model, also known as Dryden gusts, is used [17]. The Dryden model uses spatially varying stochastic processes to represent the components of the gusts, specifying their power spectral density.

4.3 Icing and fault model

The accretion of clear ice on the aircraft surfaces modifies stability and control derivatives according to the following linear model [18]:

$$C_{\sharp_i}^{ice} = (1 + \eta \mathscr{K}_{\sharp_i}) C_{\sharp_i}, \quad \begin{array}{l} \sharp = X, Y, Z, L, M, N \\ i = u, v, w, p, q, r, \tau_e, \tau_a, \tau_r \end{array} \tag{4.6}$$

where η is the icing severity factor and the coefficient $\mathscr{K}_{\sharp}$ depends on aircraft specifications [18]; the clean condition corresponds to $\eta = 0$, while the all iced condition occurs for $\eta = \eta_{\max}$ [19]. Such model has been developed on the basis of real data obtained from different icing encounters [18]. The overall effect of icing can be modeled as an additive disturbance term $\eta\omega$, where η is a scalar unknown quantity and the vector ω is assigned by

$$\omega = \mathbf{A}_{\mathscr{E}} \mathbf{x} + \mathbf{B}_{\mathscr{E}} \tau$$

with

$$\mathbf{A}_{\mathscr{E}} = \begin{bmatrix} \mathscr{E}_{X_u} & \mathscr{E}_{X_v} & \mathscr{E}_{X_w} & 0 & 0 & 0 & 0 & \mathscr{E}_{X_q} & 0 \\ \mathscr{E}_{Y_u} & \mathscr{E}_{Y_v} & \mathscr{E}_{Y_w} & 0 & 0 & 0 & \mathscr{E}_{Y_p} & 0 & \mathscr{E}_{Y_r} \\ \mathscr{E}_{Z_u} & \mathscr{E}_{Z_v} & \mathscr{E}_{Z_w} & 0 & 0 & 0 & 0 & \mathscr{E}_{Z_q} & 0 \\ 0 & 0 & 0 & 0 & 0 & 0 & 0 & 0 & 0 \\ 0 & 0 & 0 & 0 & 0 & 0 & 0 & 0 & 0 \\ 0 & 0 & 0 & 0 & 0 & 0 & 0 & 0 & 0 \\ \mathscr{E}_{L_u} & \mathscr{E}_{L_v} & \mathscr{E}_{L_w} & 0 & 0 & 0 & \mathscr{E}_{L_p} & 0 & \mathscr{E}_{L_r} \\ \mathscr{E}_{M_u} & \mathscr{E}_{M_v} & \mathscr{E}_{M_w} & 0 & 0 & 0 & 0 & \mathscr{E}_{M_q} & 0 \\ \mathscr{E}_{N_u} & \mathscr{E}_{N_v} & \mathscr{E}_{N_w} & 0 & 0 & 0 & \mathscr{E}_{N_p} & 0 & \mathscr{E}_{N_r} \end{bmatrix}$$

$$\mathbf{B}_{\mathscr{E}} = \begin{bmatrix} 0 & \mathscr{E}_{X_{\tau_e}} & 0 & 0 \\ 0 & 0 & \mathscr{E}_{Y_{\tau_a}} & \mathscr{E}_{Y_{\tau_r}} \\ 0 & \mathscr{E}_{Z_{\tau_e}} & 0 & 0 \\ 0 & 0 & 0 & 0 \\ 0 & 0 & 0 & 0 \\ 0 & 0 & 0 & 0 \\ 0 & 0 & \mathscr{E}_{L_{\tau_a}} & \mathscr{E}_{L_{\tau_r}} \\ 0 & \mathscr{E}_{M_{\tau_e}} & 0 & 0 \\ 0 & 0 & \mathscr{E}_{N_{\tau_a}} & \mathscr{E}_{N_{\tau_r}} \end{bmatrix}$$

where the coefficients $\mathscr{E}_{\sharp}$ are obtained from $\mathscr{K}_{\sharp}$ and $C_{\sharp}$ by performing linear combinations. The icing severity factor evolves according to the law,

$$\eta = f(\upsilon) \cdot \chi$$

where υ is the fraction of water freezing at a point on a surface to the water impinging on the surface,

$$\upsilon = \frac{\text{mass of water freezing}}{\text{mass of water impinging}},$$

and $\chi \geq 0$ is the accumulation parameter defined as the mass flux [20]

$$\dot{\chi} = \frac{e\lambda F_a}{\rho c}(1 - \iota_{airfoil}), \tag{4.7}$$

e being the collection efficiency, λ the liquid water content, F_a is the free stream velocity, ρ the ice density, c is the airfoil chord and $\iota_{airfoil} \in [0, 1]$ is the airfoil icing protection coefficient. Both the fraction υ and the ice density ρ depend on the air temperature and the relative humidity. In particular, when the temperature is below -10°C the factor υ satisfies $\upsilon \approx 1$, this corresponding to rime ice formation; on the other hand, if the temperature is between -10°C and 0°C, glaze ice typically appears with $\upsilon < 1$. On the other hand, due to aerodynamical cooling effects, icing can also occur when the outside air temperature is close to freezing point but still warmer than 0°. It has been observed experimentally [18] that the icing severity factor achieves its maximum $\eta_{\max}$ when the freezing fraction υ is close to the value $\upsilon_g = 0.2$, while it decreases to a steady value as υ approaches 1.
Icing can also impact on the aircraft maneuverability by decreasing the effectiveness of control surfaces. Assume that the effector position is driven by the dynamical relationships

$$\dot{\delta}_\flat = f_\flat(\delta_\flat, \gamma_\flat), \quad \flat = al, ar, rl, rr,$$

where $\gamma_\flat$ is the actuator input and the vector field $f(\cdot,\cdot)$ is supposed to be asymptotically stable for the free dynamics $\dot{\delta}_\flat = f_\flat(\delta_\flat, 0)$. The presence of ice may cause malfunctions of actuators and surface blockage; the effects can be generally modeled as the combination of loss of efficiency multiplicative factors $d_\flat(t)$ and additive factors $\varphi_\flat$:

$$\dot{\delta}_\flat = d_\flat f_\flat(\delta_\flat, \gamma_\flat) + \varphi_\flat, \quad \flat = al, ar, rl, rr \tag{4.8}$$

$$\begin{aligned} &d_\flat = 1 && t \leq t_0 \\ &d_\flat \in [0, 1) && t > t_0 \\ &\varphi_\flat \neq 0 && t \geq t_0' \end{aligned} \tag{4.9}$$

It must be noticed that the same model catches also the effects of electrical or mechanical faults that may occur in effectors and actuators. For this reason, throughout the chapter, model (4.8) and (4.9) will be referred to as a generic effector fault.

4.4 Unknown input observer framework

The approach presented in this survey is based on UIOs [5]; the main advantage of such observers is that, if some structural conditions are met, the parameters can be designed such that the resulting estimation error is independent of some inputs of the systems, even if these are not measured directly. In this regard, let us consider a general linear system of the form:

$$\begin{cases} \dot{\mathbf{x}} = \mathbf{A}\mathbf{x} + \mathbf{B}\mathbf{v} + \mathbf{X}\mathbf{v}_{\mathbf{un}} \\ \mathbf{y} = \mathbf{C}\mathbf{x} \end{cases}$$

where $\mathbf{v}$ is the nominal input, and $\mathbf{v_{un}}$ is an unknown additional input, e.g. a disturbance. The input matrix $\mathbf{X}$ is supposed to be known. The general structure of a UIO for such linear plant is the following:

$$\begin{cases} \dot{\mathbf{z}} = \mathbf{Fz} + \mathbf{SBv} + \mathbf{Ky} \\ \hat{\mathbf{x}} = \mathbf{z} + \mathbf{Hy} \end{cases}$$

where the matrices $\mathbf{F}, \mathbf{S}, \mathbf{K}$ and $\mathbf{H}$ are design parameters. It is worth to note that, in order to achieve a correct asymptotic state estimation, the matrix $\mathbf{F}$ has to be Hurwitz, i.e., all its poles must lie in the open left half-plane. The estimation error is defined as the difference between the true state $\mathbf{x}(t)$ and the estimated state $\hat{\mathbf{x}}(t)$:

$$\varepsilon(t) = \mathbf{x}(t) - \hat{\mathbf{x}}(t).$$

Exploiting the observer structure, the dynamics of the error is ruled by the following equation:

$$\begin{aligned} \dot{\varepsilon} &= [(\mathbf{I_{n\times n}} - \mathbf{HC})\mathbf{A} - \mathbf{KC} + \mathbf{FHC}]\mathbf{x} - \mathbf{F}\hat{\mathbf{x}} \\ &\quad + (\mathbf{I_{n\times n}} - \mathbf{HC})\mathbf{X}\mathbf{v_{un}} + (\mathbf{I} - \mathbf{HC} - \mathbf{S})\mathbf{Bv} \end{aligned}$$

Setting $\mathbf{K} = \mathbf{K_1} + \mathbf{K_2}$, if the following conditions are satisfied

$$\mathbf{S} = \mathbf{I_{n\times n}} - \mathbf{HC} \tag{4.10}$$

$$\mathbf{F} = \mathbf{SA} - \mathbf{K_1C}, \quad \sigma(\mathbf{F}) \in \mathbb{C}^- \tag{4.11}$$

$$\mathbf{K_2} = \mathbf{FH} \tag{4.12}$$

then the latter equation reduces to

$$\dot{\varepsilon} = \mathbf{F}\varepsilon + \mathbf{SX}\mathbf{v_{un}}$$

where $\sigma(\cdot)$ stands for the spectrum of a matrix and the set $\mathbb{C}^-$ in the left open complex half-plane. It is worth to note that a sufficient condition for $\mathbf{F}$ to be Hurwitz is the freedom to assign the eigenvalues of the matrix $\mathbf{SA} - \mathbf{K_1C}$ through the feedback gain $\mathbf{K_1}$; to this purpose we give the following statement.

Theorem 4.1. *([5]) Let $\mathbf{Q} \in \mathbb{R}^{n\times \ell}$ be a matrix satisfying the following conditions*

(C1) $rank(\mathbf{Q}) = rank(\mathbf{CQ})$;

(C2) *the pair* $(\mathbf{C}, \mathbf{A_Q})$ *is detectable, where*

$$\mathbf{A_Q} := \mathbf{A} - \mathbf{Q}((\mathbf{CQ})^{\mathrm{T}}\mathbf{CQ})^{-1}(\mathbf{CQ})^{\mathrm{T}}\mathbf{CA}.$$

Then the matrices $\mathbf{H}, \mathbf{S}, \mathbf{F}, \mathbf{K}$ *can be found such that equalities (4.10)–(4.12) hold true together with*

$$\mathbf{SQ} = 0. \tag{4.13}$$

A particular solution for $\mathbf{H}$ *is given by*

$$\mathbf{H_Q} = \mathbf{Q}((\mathbf{CQ})^{\mathrm{T}}\mathbf{CQ})^{-1}(\mathbf{CQ})^{\mathrm{T}}.$$

Conversely, if equalities (4.10)–(4.13) are satisfied, then $\mathbf{Q}$ *verifies* (C1) *and* (C2).

The previous theorem states necessary and sufficient conditions for the existence of an UIO such that (4.13) is satisfied. This latter condition is fundamental to make the estimation error insensitive to some of the effector/actuator faults or independent of additive uncertain inputs.

UIOs are well established and helpful tools for robust fault detection, and they have been studied already in the context of flight fault diagnosis [21,22]. An observer-satisfying property (4.13) is sometimes defined as a *zeroing fault input* UIO [7], and a bank of fault detection and isolation schemes can be easily designed thanks to such feature by selecting the **S** matrix in order to cancel out the whole input matrix **B** except, sequentially, its jth column as j varies in $\{1, 2, \dots, m\}$. By this procedure, the set of measurable signals $\{\rho^{(j)} = \mathbf{C}\varepsilon^{(j)}\}_{j=1}^{m}$ is provided, and the faults can be detected and isolated according to a simple logic,

$$\begin{cases} ||\rho^{(j)}(t)|| \leq \gamma^{(j)} \Leftarrow \text{ no faults} \\ ||\rho^{(j)}(t)|| > \gamma^{(j)} \Rightarrow \text{ presence of faults in the } j\text{th actuator} \end{cases}$$

where $\gamma^{(j)}$ are suitable thresholds depending on measurement noise levels and other bounded perturbations. An alternative design scheme is based on UIOs with *constrained output directions* [23,24]. Assume $m \leq p$ and consider the canonical basis of $\mathbb{R}^p$, namely $\mathbf{e}_1, \dots, \mathbf{e}_p$. Since by assumption the output matrix **C** is full-rank, there exists $\Omega \in \mathbb{R}^{n \times p}$ such that

$$\mathbf{C}\Omega = \mathbf{I}_{\mathbf{p}\times\mathbf{p}} = [\mathbf{e}_1 \ \cdots \ \mathbf{e}_p].$$

The general solution of such equation is given by

$$\Omega = \mathbf{C}^T(\mathbf{C}\mathbf{C}^T)^{-1} + [\mathbf{I}_{\mathbf{n}\times\mathbf{n}} - \mathbf{C}^T(\mathbf{C}\mathbf{C}^T)^{-1}\mathbf{C}]\Omega_*, \tag{4.14}$$

where $\Omega_* \in \mathbb{R}^{n \times p}$ is an arbitrary matrix. Denoting by $\Omega_1, \dots, \Omega_p$ the columns of the matrix Ω, the basic idea of the method is to design the observer parameters in order to guarantee that, if a fault occurs in the jth actuator, then the estimation error maintains the direction Ω_j during the system evolution, this corresponding to a fixed direction $\mathbf{e}_j$ for the residual. It is worth to note that a design constraint for the achievability of this condition is that directions $\Omega_1, \dots, \Omega_p$ need to correspond to eigenvectors of the observer matrix **F** [25].

4.5 Diagnosis and accommodation

4.5.1 Detection and isolation in UAVs using UIOs

This section is devoted to the design of a fault diagnosis scheme able to detect the icing and to identify whether the presence of ice on a particular control surface is causing a loss of effectiveness of the surface itself, or the ice accretion on the leading edge of the wings and tail is causing a change of the airfoils aerodynamical properties. The method is to design the FD scheme based on UIOs with constrained output directions [8,9,26]. Referring to the effector model (4.3), let us consider the decomposition

$$\mathbf{G} = [\mathbf{G}_1 \ \mathbf{G}_2 \ \mathbf{G}_3 \ \mathbf{G}_4 \ \mathbf{G}_5]$$

and set $\mathbf{W} = \mathbf{BG}$ with

$$\mathbf{W} = [\mathbf{W}_1\ \mathbf{W}_2\ \mathbf{W}_3\ \mathbf{W}_4\ \mathbf{W}_5].$$

Consider a generic linear UIO assigned by the equations

$$\begin{aligned}\dot{\mathbf{z}} &= \mathbf{Fz} + \mathbf{SB}\tau_c + \mathbf{Ky}\\ \hat{\mathbf{x}} &= \mathbf{z} + \mathbf{Hy}\end{aligned}$$

where τ_c is the nominal commanded input, and assume that the observer matrices are designed such that conditions (4.10) and (4.11) are met. The matrix $\mathbf{H} \in \mathbb{R}^{9\times p}$, with either $p = 9$ or $p = 7$, is a free design parameter that can be tuned in order to assign desired directions to residuals and to decouple them from unknown input disturbances. However, due to the rank deficiency of $\mathbf{W}$, it is not possible in these cases to address a decoupled distribution of the faults effects over a set of characteristic directions in the output space and it is then required to deal with linear combinations of such directions. The two cases of full and partial information have to be treated separately.

4.5.1.1 Full information case

Let us consider the case $\mathbf{C} = \bar{\mathbf{C}}$ first. Referring to the wind input matrix (4.4), let us select three linear independent and constant vectors $\mathbf{N}_1, \mathbf{N}_2, \mathbf{N}_3 \in \mathbb{R}^9$ with

$$span\{\mathbf{N}_1, \mathbf{N}_2, \mathbf{N}_3\} = span\mathbf{N}(t) \quad \forall t \geq 0.$$

The basic idea is to make the residuals independent on the three components of the wind force, and to assign a particular output directions to each of the four input vectors corresponding to control surfaces (not including the engine throttle input vector $\mathbf{W}_1$). This is expressed by the identities

$$\bar{\mathbf{C}}(\mathbf{I} - \mathbf{H}\bar{\mathbf{C}})\mathbf{N}_\mathrm{i} = 0, \quad i = 1, 2, 3 \tag{4.15}$$

$$\bar{\mathbf{C}}(\mathbf{I} - \mathbf{H}\bar{\mathbf{C}})\mathbf{W}_{\mathrm{i}+1} = \mathbf{e}_\mathrm{i},\ i = 1, 2, 3, 4, \tag{4.16}$$

where $\mathbf{e}_\mathrm{i}$ is the ith vector of the standard basis in the output space $\mathbb{R}^7$. On the other hand, the four conditions (4.16) cannot be imposed simultaneously due to the lack of rank of the matrix $\mathbf{G}$, and we have to limit to consider three of them. To design such matrix $\mathbf{H}$ one can proceed as follows. Select a set of three independent vectors in $\mathbb{R}^9$, namely $\{\mathbf{b}_1, \mathbf{b}_2, \mathbf{b}_3\}$, such that

$$\bar{\mathbf{C}}\mathbf{b}_\mathrm{i} = \mathbf{e}_\mathrm{i} \quad i = 1, 2, 3.$$

Setting $\Upsilon = [0\ 0\ 0\ \mathbf{b}_1\ \mathbf{b}_2\ \mathbf{b}_3]$ and $\Lambda_{234} = [\mathbf{N}_1\ \mathbf{N}_2\ \mathbf{N}_3\ \mathbf{W}_2\ \mathbf{W}_3\ \mathbf{W}_4]$ with $\Upsilon, \Lambda_{234} \in \mathbb{R}^{9\times 6}$ a simple solution is given by

$$\mathbf{H}_{234} = (\Lambda_{234} - \Upsilon)(\bar{\mathbf{C}}\Lambda_{234})^{-\mathbf{L}},$$

where $(\cdot)^{-L}$ stands for the left pseudoinverse of a matrix. Choosing a different combination, say $\Lambda_{345} = [\mathbf{N}_1\ \mathbf{N}_2\ \mathbf{N}_3\ \mathbf{W}_3\ \mathbf{W}_4\ \mathbf{W}_5]$, a second solution is found

$$\mathbf{H}_{345} = (\Lambda_{345} - \Upsilon)(\bar{\mathbf{C}}\Lambda_{345})^{-\mathbf{L}}.$$

Summarizing, we have two distinct UIOs, whose estimated states $\hat{\mathbf{x}}^{(1)}, \hat{\mathbf{x}}^{(2)}$ satisfy

$$\dot{\mathbf{x}} - \dot{\hat{\mathbf{x}}}^{(1)} = \mathbf{F}^{(1)}(\mathbf{x} - \hat{\mathbf{x}}^{(1)}) + \mathbf{S}^{(1)}(\mathbf{N}\dot{\nu} + \mathbf{BG}\tilde{\delta} + \eta\omega) \tag{4.17}$$

$$\dot{\mathbf{x}} - \dot{\hat{\mathbf{x}}}^{(2)} = \mathbf{F}^{(2)}(\mathbf{x} - \hat{\mathbf{x}}^{(2)}) + \mathbf{S}^{(2)}(\mathbf{N}\dot{\nu} + \mathbf{BG}\tilde{\delta} + \eta\omega) \tag{4.18}$$

where the matrices are defined as

$$\mathbf{F}^{(i)} = \mathbf{S}^{(i)}\mathbf{A} - \mathbf{K}_1^{(i)}\bar{\mathbf{C}}, \quad i = 1, 2$$
$$\mathbf{S}^{(1)} = \mathbf{I} - \mathbf{H}_{234}\bar{\mathbf{C}}, \quad \mathbf{S}^{(2)} = \mathbf{I} - \mathbf{H}_{345}\bar{\mathbf{C}}$$
$$\mathbf{S}^{(1)}[\mathbf{W}_2\ \mathbf{W}_3\ \mathbf{W}_4] = \mathbf{S}^{(2)}[\mathbf{W}_3\ \mathbf{W}_4\ \mathbf{W}_5] = [\mathbf{b}_1\ \mathbf{b}_2\ \mathbf{b}_3]$$

and $\tilde{\delta}$ is the deviation of the actual control inputs from the commanded inputs

$$\tilde{\delta} = \delta - \delta_c, \text{ with } \mathbf{G}\delta_{\mathbf{c}} = \tau_{\mathbf{c}}.$$

The key point of the construction is the selection of the matrix $\mathbf{K}_1^{(i)}$ $i = 1, 2$ in a way that $\mathbf{F}^{(i)}$ is Hurwitz with the triple $\{\mathbf{b}_1, \mathbf{b}_2, \mathbf{b}_3\}$ included in the set of its eigenvectors. This can easily be addressed by choosing an arbitrary Hurwitz matrix $\mathbf{M}^{(i)}$ having $\{\mathbf{b}_1, \mathbf{b}_2, \mathbf{b}_3\}$ as eigenvectors and setting

$$\mathbf{K}_1^{(i)} = (\mathbf{S}^{(i)}\mathbf{A} - \mathbf{M}^{(i)})\bar{\mathbf{C}}^{-1}. \tag{4.19}$$

The icing identification can be performed combining the two observers, i.e., defining a suitable logic to gather information from residual directions. Let us denote by $\bar{\Pi}_i$ the linear projection operator on the subspace $span\{\mathbf{e}_i\} \subset \mathbb{R}^9, i = 1, \ldots, 9$, e.g.,

$$\bar{\Pi}_1 = diag(1, 0, 0, 0, 0, 0, 0, 0, 0)$$
$$\bar{\Pi}_2 = diag(0, 1, 0, 0, 0, 0, 0, 0, 0)$$
$$\vdots$$
$$\bar{\Pi}_9 = diag(0, 0, 0, 0, 0, 0, 0, 0, 1).$$

Proposition 4.1. *Set* $\varepsilon^{(i)} = \bar{\mathbf{C}}(\mathbf{x} - \hat{\mathbf{x}}^{(i)})$, $i = 1, 2$. *Assume the estimator transients due to the initial conditions to be negligible, i.e.,* $\varepsilon^{(i)}(0) = 0$. *Then, we can state the following decision rule.*

- $\bar{\Pi}_1\varepsilon^{(1)} = \varepsilon^{(1)} \neq 0 \Rightarrow$ *faulty effector* δ_{al}
- $\left\{\begin{array}{l}\bar{\Pi}_2\varepsilon^{(1)} = \varepsilon^{(1)} \neq 0 \\ \bar{\Pi}_1\varepsilon^{(2)} = \varepsilon^{(2)} \neq 0\end{array}\right\} \Rightarrow$ *faulty effector* δ_{ar}
- $\left\{\begin{array}{l}\bar{\Pi}_3\varepsilon^{(1)} = \varepsilon^{(1)} \neq 0 \\ \bar{\Pi}_2\varepsilon^{(2)} = \varepsilon^{(2)} \neq 0\end{array}\right\} \Rightarrow$ *faulty effector* δ_{rl}
- $\bar{\Pi}_3\varepsilon^{(2)} = \varepsilon^{(2)} \neq 0 \Rightarrow$ *faulty effector* δ_{rr}
- $\bar{\Pi}_j\varepsilon^{(i)} \neq \varepsilon^{(i)}\ \forall i = 1, 2\ \forall j = 1, 2, 3 \Rightarrow$ *airfoil icing*

A formal proof of this result can be found in [27].

Remark 4.1. *The proposed decision rule can be extended to handle the case of multiple faults, i.e., the simultaneous failure of two effectors. The residuals turn out to be directed as combinations of two of the basis vectors, and hence the faulty devices can be still identified. However in this latter case, due rank deficiency, it is unlikely that exact control reconfiguration can be achieved by control allocation.*

Remark 4.2. *It is worth to note that only six of nine outputs are effectively used in the proposed scheme, i.e., 3 degrees of freedom for zeroing the wind effects, and 3 degrees of freedom for assigning output directions to the columns of the matrix* **W**. *Such dual redundancy may be used to enhance robustness of the algorithm in spite of sensor faults. As a matter of fact, icing may also occur on airspeed sensors, typically causing overestimation of velocity; for this reason, in the absence of probes equipped with heating devices, it may be safer to base the icing diagnosis method on a set of six outputs only, namely* $(\phi, \theta, \psi, p, q, r)$*, in order to exclude possible biased airspeed measurements.*

4.5.1.2 Partial information case

In the partial information case $\mathbf{C} = \mathring{\mathbf{C}}$, we have seven independent outputs, and hence there are seven directions that can be freely assigned. The construction is analogous to the full information case but, since $dim(span\{\mathring{\mathbf{C}}\mathbf{N}(\mathbf{t})\}) = 1$, there is no feasible way to design the observer being decoupled from wind accelerations in the lateral and vertical directions, namely $\dot{\nu}_y^{\vdash}, \dot{\nu}_z^{\vdash}$. Similarly to the previous design procedure, we set

$$\mathbf{H_{234}} = (\Lambda_{234} - \Upsilon)(\mathring{\mathbf{C}}\Lambda_{\mathbf{234}})^{-\mathbf{L}},$$
$$\mathbf{H_{345}} = (\Lambda_{345} - \Upsilon)(\mathring{\mathbf{C}}\Lambda_{\mathbf{345}})^{-\mathbf{L}}$$

where the original matrices Λ and Υ have been replaced by

$$\Lambda_{234} = [\mathbf{N_1}\ \mathbf{W_2}\ \mathbf{W_3}\ \mathbf{W_4}]$$
$$\Lambda_{345} = [\mathbf{N_1}\ \mathbf{W_3}\ \mathbf{W_4}\ \mathbf{W_5}]$$
$$\Upsilon = [\mathbf{0}\ \mathbf{b_1}\ \mathbf{b_2}\ \mathbf{b_3}],$$

with $\mathbf{N_1} \in span\{\mathring{\mathbf{C}}\mathbf{N}(\mathbf{t})\}$. The following assumption, which is an extension of (4.19), guarantees that the eigenstructures of the matrices $\mathbf{F}^{(i)}$ can be assigned properly [25].

Assumption 4.1. *A matrix* $\mathbf{K_1^{(i)}}, i = 1, 2,$ *can be designed such that* $\sigma(\mathbf{S^{(i)}A} - \mathbf{K_1^{(i)}}\mathring{\mathbf{C}}) \in \mathbb{C}^-$ *together with*

$$\mathbf{F}^{(i)}\mathbf{b_j} = (\mathbf{S^{(i)}A} - \mathbf{K_1^{(i)}}\mathring{\mathbf{C}})\mathbf{b_j} = \lambda_\mathbf{j}^{(i)}\mathbf{b_j} \tag{4.20}$$

for some $\lambda_j^{(i)} < 0$ *and* $j = 1, 2, 3$.

A sufficient condition for the existence of a gain matrix $\mathbf{K}_1^{(i)}$ with the desired properties can be found in [26].
Let us denote by $\mathring{\Pi}_i$ the linear projection operator on the subspace $span\{\mathbf{e}_i\} \subset \mathbb{R}^7, i = 1, \ldots, 7$, e.g.,

$$\begin{aligned}\mathring{\Pi}_1 &= diag(1,0,0,0,0,0,0)\\ \mathring{\Pi}_2 &= diag(0,1,0,0,0,0,0)\\ &\vdots\\ \mathring{\Pi}_7 &= diag(0,0,0,0,0,0,1).\end{aligned}$$

In addition, we denote by $\mathring{\Pi}_i^{\perp}$ the projection operator on the subspace orthogonal to $span\{\mathbf{e}_i\}$. Since in this case the wind disturbances are not completely decoupled, we need to introduce suitable thresholds in the partial information isolation scheme. Suppose that bounds on the wind accelerations are available, i.e., $|\dot{\nu}_y^{\vdash}| \leq \vartheta_y, \quad |\dot{\nu}_z^{\vdash}| \leq \vartheta_z$. Set $\vartheta = [0\ \vartheta_y\ \vartheta_z]^T$ and

$$\mu^{(i)}(t) = \int_0^t e^{\mathbf{F}^{(i)}(t-\varsigma)}\mathbf{S}^{(i)}\vartheta\, d\varsigma,$$

the latter being an upper bound for the forced response of residuals to wind disturbances.

The following result is a generalization of the rule stated in Proposition 4.1, where the identities have been replaced by inequalities in order to take into account model uncertainties and disturbances, these being no longer decoupled from residuals in the partial information case.

Proposition 4.2. *Set* $\varepsilon^{(i)} = \mathring{\mathbf{C}}(\mathbf{x} - \hat{\mathbf{x}}^{(i)}), i = 1, 2$. *Assume the estimator transients due to the initial conditions to be negligible, i.e.,* $\varepsilon^{(i)}(0) = 0$. *Then, we can state the following decision rule for* $\varepsilon^{(i)} \neq 0$:

- $\|\mathring{\Pi}_1\varepsilon^{(1)} - \varepsilon^{(1)}\| \leq \|\mathring{\Pi}_1^{\perp}\mu^{(1)}\| \Rightarrow$ *faulty effector* δ_{al}
- $\left.\begin{cases}\|\mathring{\Pi}_2\varepsilon^{(1)} - \varepsilon^{(1)}\| \leq \|\mathring{\Pi}_2^{\perp}\mu^{(1)}\| \\ \|\mathring{\Pi}_1\varepsilon^{(2)} - \varepsilon^{(2)}\| \leq \|\mathring{\Pi}_1^{\perp}\mu^{(2)}\|\end{cases}\right\} \Rightarrow$ *faulty effector* δ_{ar}
- $\left.\begin{cases}\|\mathring{\Pi}_3\varepsilon^{(1)} - \varepsilon^{(1)}\| \leq \|\mathring{\Pi}_3^{\perp}\mu^{(1)}\| \\ \|\mathring{\Pi}_2\varepsilon^{(2)} - \varepsilon^{(2)}\| \leq \|\mathring{\Pi}_2^{\perp}\mu^{(2)}\|\end{cases}\right\} \Rightarrow$ *faulty effector* δ_{rl}
- $\|\mathring{\Pi}_3\varepsilon^{(2)} - \varepsilon^{(2)}\| \leq \|\mathring{\Pi}_3^{\perp}\mu^{(2)}\| \Rightarrow$ *faulty effector* δ_{rr}
- $\|\mathring{\Pi}_j\varepsilon^{(i)} - \varepsilon^{(i)}\| > \|\mathring{\Pi}_j^{\perp}\mu^{(i)}\|\ \forall i = 1,2\ \forall j = 1,2,3 \Rightarrow$ *airfoil icing*

Remark 4.3. *In this case, the isolation of a fault is achieved when the difference between the residual $\varepsilon^{(i)}$ and its projection on the direction $\mathbf{e}_j$ remains bounded by the projection of the wind acceleration threshold $\mu^{(i)}$ on the orthogonal subspace $\{\mathbf{e}_j\}^{\perp}$. A similar logic can be adopted to handle other types of system perturbations, such as model uncertainties or measurement noise. Moreover, a frequency separation approach with the introduction of additional filters may be helpful, since wind acceleration and sensor noise are high-frequency disturbances while icing is characterized by a low frequency.*

4.5.2 Control allocation-based icing/fault accommodation

Once the icing has been detected, the control scheme is switched to some alarm mode. The alarm mode may be read as a reconfiguration in the case of effector icing [7,28], or as the activation of an automated deicing device in the case of accretion of ice on the leading edge of the wings [4]. Let us treat the two scenarios separately.

4.5.2.1 Effector icing: control reconfiguration

Suppose that one of the effectors has been identified as faulty or iced, say $\delta_{\flat^\star}$ with $\flat^\star \in \{al, ar, rl, rr\}$. Avoiding the use of the effector $\delta_{\flat^\star}$ is desirable in order to prevent loss of control, and hence the corresponding actuator input $v_{\flat^\star}$ is set to zero in (4.8). However, due to the presence of the factor $d_{\flat^\star}$, this does not ensure that the state of $\delta_{\flat^\star}$ converges to zero [29]; in particular, denoting by $\Psi_{\flat^\star}(t, t_d)$ the solution of the free evolution equation (4.8) for t larger than the fault detection time t_d, one has

$$\lim_{t\to+\infty} \Psi_{\flat^\star}(t, t_d) = \bar{\delta}_{\flat^\star} \in \mathbb{R}, \quad \delta^{\dagger}_{\flat^\star}(t) := \Psi_{\flat^\star}(t, t_d) - \bar{\delta}_{\flat^\star}.$$

With a slight abuse of notation, we indicate with $\mathbf{G}_{\flat^\star}$ the column of $\mathbf{G}$ corresponding to the faulty effector, and with $\breve{\delta} \in \mathbb{R}^4$, $\breve{\mathbf{G}}_{\flat^\star} \in \mathbb{R}^{4\times 4}$ the reduced-order control input and the matrix obtained from $\mathbf{G}$ by removing the column $\mathbf{G}_{\flat^\star}$, respectively.

The control allocation scheme is updated, and the new task is to generate a control action able to produce the desired virtual input using the safe effectors only:

$$\breve{\tau}_c = \breve{\mathbf{G}}_{\flat^\star}\breve{\delta},$$

where $\breve{\tau}_c$ has also been modified in order to compensate for the torque generated by the faulty device, i.e., $\breve{\tau}_c = \tau_c - \mathbf{G}_{\flat^\star}\Psi_{\flat^\star}(t, t_d)$. To take into account for possible physical or operational limitations $(\delta_{al}, \delta_{ar}, \delta_{rl}, \delta_{rr}) \in \mathscr{Y}$, a direct control allocation method is proposed to generate the updated virtual input. Denoting by $\tau(\alpha)$ the diagonal matrix $\tau(\alpha) = diag(1, \alpha, \alpha, \alpha)$ with $\alpha \in [0, 1]$, the allocation reduces to the optimization problem

$$\max_{\alpha\in[0,1]} : \exists\breve{\delta} \in \mathscr{Y}_{\flat^\star} \text{ with } \breve{\mathbf{G}}_{\flat^\star}\breve{\delta} = \tau(\alpha)\tau_c$$

or, equivalently $\max_{\alpha\in[0,1]} : \breve{\delta}_c = \breve{\mathbf{G}}_{\flat^\star}^{-1}\tau(\alpha)\tau_c \in \breve{\mathscr{Y}}_{\flat^\star}$, where $\breve{\mathscr{Y}}_{\flat^\star}$ is the reduced-order constraint set. The advantage of direct allocation is that, even though the inputs are saturated, their joint effect has exactly the same orientation as the desired virtual input with a possible downsized magnitude, thus reducing the risk of stall angles and other hazardous conditions for aircraft stability. On the other hand, in certain operational conditions, it could be preferable to give priority to generate a particulate torque rather than the others: in this case, the direct allocation can be modified choosing independently the rescaling factors, i.e., setting $\tau = (1, \alpha_1, \alpha_2, \alpha_3)$.

We notice that, if icing occurs on several effectors, the degrees of freedom in the control allocation scheme might be insufficient to guarantee a complete reconfiguration. However, performing an estimation of the loss of efficiency (see for instance [30]) allows one to compensate partially for the icing effect while keeping the faulty effectors in use.

4.5.2.2 Airfoil leading edge icing: automated deicing system

If ice accretion on airfoils is detected, the automated icing protection system must be turned on, this corresponding to $\iota_{iairfoil} > 0$ in (4.7). The icing protection system is mainly constituted of layers of coating material, a coating temperature sensor and a microcontroller, together with a thermocouple and an electric power source [4].

The coating efficiency is regulated by the microcontroller with a PID that uses the temperature as input. The ice layers detach when the coating temperature is above 0°: due to aerodynamic cooling effect, a safety margin is imposed to ensure complete ice melting, this corresponding to a positive reference temperature $T_\star$. In particular $\iota_{airfoil}$ is an increasing function of coating temperature T and ice thickness χ with $\iota_{airfoil}(T_\star, \chi) \geq 1$ for $\chi > 0$.

Remark 4.4. *The icing protection subsystem can also be used in anti-icing mode. Gathering atmospheric data such as relative humidity, the automated icing protection system can be turned on whenever the aircraft encounters potential icing conditions.*

4.6 Enhanced quasi-LPV framework

The UIO framework described in Section 4.4 and used for detection and isolation of faults and icing in Section 4.5 has the limitation of considering an LTI model of the UAV, which has been obtained by linearizing the vehicle nonlinear model as described in Section 4.2. Therefore, the developed approach is reliable only as long as the linearized model is consistent with the nonlinear one. A way to overcome this limitation, which allows maintaining the simplicity brought by a linear structure, is to use an LPV formulation to cope with the nonlinearities. Unlike linearization techniques, LPV methods do not involve any approximation, since they rely on an exact transformation of the original nonlinear system into a linear-like one, by incorporating all the original nonlinearities within some varying parameters that schedule the state space matrices [31]. The resulting model is referred to as *quasi-LPV*, due to the dependence of the varying parameters on endogenous signals.

4.6.1 Nonlinear embedding

More specifically, the UAV nonlinear model described in Section 4.2 can be brought to a quasi-LPV form using the nonlinear embedding in the parameters approach [32,33]:

$$\dot{\mathbf{x}} = \mathbf{A}(\mathbf{x})\mathbf{x} + \mathbf{B}(\mathbf{x})\tau + \mathbf{d}(\mathbf{x}) \tag{4.21}$$

with

$$\mathbf{A}(\mathbf{x}) = \begin{bmatrix} X_u(\mathbf{x}) & X_v(\mathbf{x}) & X_w(\mathbf{x}) & 0 & 0 & 0 & 0 & X_q(\mathbf{x}) & X_r(\mathbf{x}) \\ Y_u(\mathbf{x}) & Y_v(\mathbf{x}) & Y_w(\mathbf{x}) & 0 & 0 & 0 & Y_p(\mathbf{x}) & 0 & Y_r(\mathbf{x}) \\ Z_u(\mathbf{x}) & Z_v(\mathbf{x}) & Z_w(\mathbf{x}) & 0 & 0 & 0 & Z_p(\mathbf{x}) & Z_q(\mathbf{x}) & 0 \\ 0 & 0 & 0 & 0 & 0 & 0 & 1 & \Phi_q(\mathbf{x}) & \Phi_r(\mathbf{x}) \\ 0 & 0 & 0 & 0 & 0 & 0 & 0 & \Theta_q(\mathbf{x}) & \Theta_r(\mathbf{x}) \\ 0 & 0 & 0 & 0 & 0 & 0 & 0 & \Psi_q(\mathbf{x}) & \Psi_r(\mathbf{x}) \\ L_u(\mathbf{x}) & L_v(\mathbf{x}) & L_w(\mathbf{x}) & 0 & 0 & 0 & L_p(\mathbf{x}) & L_q(\mathbf{x}) & L_r(\mathbf{x}) \\ M_u(\mathbf{x}) & M_v(\mathbf{x}) & M_w(\mathbf{x}) & 0 & 0 & 0 & M_p(\mathbf{x}) & M_q(\mathbf{x}) & M_r(\mathbf{x}) \\ N_u(\mathbf{x}) & N_v(\mathbf{x}) & N_w(\mathbf{x}) & 0 & 0 & 0 & N_p(\mathbf{x}) & N_q(\mathbf{x}) & N_r(\mathbf{x}) \end{bmatrix}$$

$$\mathbf{B}(\mathbf{x}) = \begin{bmatrix} X_{\tau_t} & X_{\tau_e}(\mathbf{x}) & 0 & 0 \\ 0 & 0 & Y_{\tau_a}(\mathbf{x}) & Y_{\tau_r}(\mathbf{x}) \\ 0 & Z_{\tau_e}(\mathbf{x}) & 0 & 0 \\ 0 & 0 & 0 & 0 \\ 0 & 0 & 0 & 0 \\ 0 & 0 & 0 & 0 \\ 0 & 0 & L_{\tau_a}(\mathbf{x}) & L_{\tau_r}(\mathbf{x}) \\ 0 & M_{\tau_e}(\mathbf{x}) & 0 & 0 \\ 0 & 0 & N_{\tau_a}(\mathbf{x}) & N_{\tau_r}(\mathbf{x}) \end{bmatrix}$$

$$\mathbf{d}(\mathbf{x}) = \begin{bmatrix} -g\sin\tilde{\theta} & g\cos\tilde{\theta}\sin\tilde{\phi} & g\cos\tilde{\theta}\cos\tilde{\phi} & \mathbf{0}_{1\times 6} \end{bmatrix}^T$$

where the role of the coefficients is akin to the one in the linearized model (4.1), although their expressions differ.

4.6.2 LPV unknown input observer

Taking into account the wind, the actuator faults and the icing, the quasi-LPV UAV model can be brought to the general LPV form

$$\begin{cases} \dot{\mathbf{x}} = \mathbf{A}(\vartheta)\mathbf{x} + \mathbf{B}(\vartheta)\mathbf{v} + \mathbf{X}(\vartheta)\mathbf{v}_{\mathbf{un}} + \mathbf{d}(\vartheta) \\ \mathbf{y} = \mathbf{C}\mathbf{x} \end{cases}$$

with notation similar to the one used in Section 4.4, but with the relevant property that the matrices **A**, **B**, **X** depend on a varying parameter vector ϑ, which is measured or estimated. It is also assumed that its derivative $\dot{\vartheta}$ is measured or estimated.

The structure of a UIO for such LPV plant is the following:

$$\begin{cases} \dot{\mathbf{z}} = \mathbf{F}(\vartheta)\mathbf{z} + \mathbf{S}(\vartheta)\mathbf{B}(\vartheta)\mathbf{v} + \mathbf{K}(\vartheta)\mathbf{y} - \dot{\mathbf{H}}(\vartheta, \dot{\vartheta})\mathbf{y} + \mathbf{d}(\vartheta) - \mathbf{H}(\vartheta)\mathbf{C}\mathbf{d}(\vartheta) \\ \hat{\mathbf{x}} = \mathbf{z} + \mathbf{H}(\vartheta)\mathbf{y} \end{cases}$$

where $\dot{\mathbf{H}}(\vartheta, \dot{\vartheta})$ is the time derivative of $\mathbf{H}(\vartheta)$.

Then, the dynamics of the estimation error is described by

$$\begin{aligned} \dot{\varepsilon} = {} & [(\mathbf{I}_{\mathbf{n}\times\mathbf{n}} - \mathbf{H}(\vartheta)\mathbf{C})\mathbf{A}(\vartheta) - \mathbf{K}(\vartheta)\mathbf{C} + \mathbf{F}(\vartheta)\mathbf{H}(\vartheta)\mathbf{C}]\mathbf{x} - \mathbf{F}(\vartheta)\hat{\mathbf{x}} \\ & + (\mathbf{I}_{\mathbf{n}\times\mathbf{n}} - \mathbf{H}(\vartheta)\mathbf{C})\mathbf{X}(\vartheta)\mathbf{v}_{\mathbf{un}} + (\mathbf{I}_{\mathbf{n}\times\mathbf{n}} - \mathbf{H}(\vartheta)\mathbf{C} - \mathbf{S}(\vartheta))\mathbf{B}(\vartheta)\mathbf{v} \end{aligned}$$

which, through the choice

$$\mathbf{S}(\vartheta) = \mathbf{I}_{\mathbf{n}\times\mathbf{n}} - \mathbf{H}(\vartheta)\mathbf{C} \tag{4.22}$$

$$\mathbf{F}(\vartheta) = \mathbf{S}(\vartheta)\mathbf{A}(\vartheta) - \mathbf{K}_1(\vartheta)\mathbf{C} \tag{4.23}$$

$$\mathbf{K}_2(\vartheta) = \mathbf{F}(\vartheta)\mathbf{H}(\vartheta) \tag{4.24}$$

$$\mathbf{K}(\vartheta) = \mathbf{K}_1(\vartheta) + \mathbf{K}_2(\vartheta) \tag{4.25}$$

becomes

$$\dot{\varepsilon} = \mathbf{F}(\vartheta)\varepsilon + \mathbf{S}(\vartheta)\mathbf{X}(\vartheta)\mathbf{v}_{\mathbf{un}}$$

Notice that $\mathbf{F}(\vartheta)$ can be chosen as a constant matrix **F** through an appropriate choice of the matrix $\mathbf{K}_1(\vartheta)$, which allows assuring convergence to zero of the estimation error ε with $\mathbf{v}_{\mathbf{un}} = \mathbf{0}$ if $\sigma(\mathbf{F}) \in \mathbb{C}^-$. On the other hand, the matrix function $\mathbf{S}(\vartheta)$ can be chosen to constrain the range of $\mathbf{S}(\vartheta)\mathbf{X}(\vartheta)$, in such a way that different output directions of the residuals are assigned for the unknown inputs acting on the system, with the aim of identifying the cause for some detected system malfunctions.

4.6.3 Application to the UAV fault/icing diagnosis

Due to the superposition of effects and the lack of degrees of freedom in the UIO design, it is not possible to decouple completely the wind disturbance and icing effects from the actuator faults. However, it is still possible to design the UIO matrices such that a successful fault/icing diagnosis can be achieved. For the sake of simplicity, only the full information case will be detailed.

Let us notice that the following condition holds

$$\eta\omega \in span\{\mathbf{B}(\mathbf{x}), \mathbf{e}_2, \mathbf{e}_3\} \quad \forall t \geq 0.$$

which allows us defining our target as designing the UIO matrices with the following properties:

$$\mathbf{S}(\mathbf{x})[\mathbf{B}(\mathbf{x})\ \mathbf{e}_2\ \mathbf{e}_3\ \mathbf{e}_4\ \mathbf{e}_5\ \mathbf{e}_6] = \mathbf{I}_{9\times 9}$$

$$\mathbf{F}\mathbf{e}_i = \lambda_i^F \mathbf{e}_i \quad \forall i = 1, \ldots, 9$$

where $\lambda_i^F \in \mathbb{C}^-$, $i = 1, \ldots, 9$, are the desired eigenvalues of the matrix $\mathbf{F}$.

It is easy to check that the resulting matrix $\mathbf{S}(\mathbf{x})$ has the following structure:

$$\mathbf{S}(\mathbf{x}) = \begin{pmatrix} \frac{1}{X_{\tau_t}} & 0 & 0 & 0 & 0 & 0 & 0 & s_{18}(\mathbf{x}) & 0 \\ 0 & 0 & 0 & 0 & 0 & 0 & 0 & s_{28}(\mathbf{x}) & 0 \\ 0 & 0 & 0 & 0 & 0 & 0 & s_{37}(\mathbf{x}) & 0 & s_{39}(\mathbf{x}) \\ 0 & 0 & 0 & 0 & 0 & 0 & s_{47}(\mathbf{x}) & 0 & s_{49}(\mathbf{x}) \\ 0 & 1 & 0 & 0 & 0 & 0 & s_{57}(\mathbf{x}) & 0 & s_{59}(\mathbf{x}) \\ 0 & 0 & 1 & 0 & 0 & 0 & 0 & s_{68}(\mathbf{x}) & 0 \\ 0 & 0 & 0 & 1 & 0 & 0 & 0 & 0 & 0 \\ 0 & 0 & 0 & 0 & 1 & 0 & 0 & 0 & 0 \\ 0 & 0 & 0 & 0 & 0 & 1 & 0 & 0 & 0 \end{pmatrix}$$

where the elements $s_{ij}(\mathbf{x})$ are the functions of the elements appearing in $\mathbf{B}(\mathbf{x})$.

Under the assumption that at a given time either a single fault or icing could act on the system (no simultaneous multiple faults and icing), the following decision rule can be employed for faults/icing diagnosis.

Proposition 4.3. *Set* $\varepsilon = \mathbf{x} - \hat{\mathbf{x}}$ *and assume estimator transients due to initial conditions to be negligible, i.e.,* $\varepsilon(0) = 0$*, and the wind to be steady, i.e.,* $\nu = \nu^*$ *with* $\dot{\nu}^* = 0$*. Then we can state the following decision rule.*

- $\bar{\Pi}_i \varepsilon = 0 \quad \forall i = 1, \ldots, 9 \Rightarrow$ *no faults*
- $\left\{\begin{matrix} \bar{\Pi}_1 \varepsilon \neq 0 \\ \bar{\Pi}_i \varepsilon = 0 \quad \forall i = 2, \ldots, 9 \end{matrix}\right\} \Rightarrow$ *faulty thrust*
- $\left\{\begin{matrix} \bar{\Pi}_2 \varepsilon \neq 0 \\ \bar{\Pi}_i \varepsilon = 0 \quad \forall i = 1, 3, \ldots, 9 \end{matrix}\right\} \Rightarrow$ *faulty elevator*
- $\left\{\begin{matrix} \bar{\Pi}_3 \varepsilon \neq 0 \\ \bar{\Pi}_i \varepsilon = 0 \quad \forall i = 1, 2, 4, \ldots, 9 \end{matrix}\right\} \Rightarrow$ *faulty aileron*
- $\left\{\begin{matrix} \bar{\Pi}_4 \varepsilon \neq 0 \\ \bar{\Pi}_i \varepsilon = 0 \quad \forall i = 1, \ldots, 3, 5, \ldots, 9 \end{matrix}\right\} \Rightarrow$ *faulty rudder*
- *else* $\Rightarrow$ *airfoil icing*

Remark 4.5. *A threshold-based logic similar to the one described in Section 4.4 should be considered due to the presence of the wind turbulence input disturbance* $\xi(t)$ *described by* (4.4). *It is worth noticing that*

$$\mathbf{S}(\mathbf{x})\mathbf{N}(t) = [\times \ 0\,0\,0 \ \times \ \times \ 0\,0\,0]$$

with $\times$ *indicating nonzero elements, which means that (at least in theory) the wind turbulence should affect only* $\bar{\Pi}_i\varepsilon$, $i = 1, 4, 6$. *However, due to undesired effects such as the presence of sensor noise and parametric uncertainties, the wind turbulence would affect the other residuals as well, although at a much smaller extent.*

4.7 Illustrative example: the Aerosonde UAV

The model of a typical small unmanned aircraft, the Aerosonde UAV (AAI Corporation, Textron Inc.) has been used to illustrate the use of the UIO framework for icing and fault diagnosis. A linear system describing the vehicle dynamics about the trim condition

$$\begin{aligned} u^* &= 22.95\ \mathrm{m/s},\ v^* = 0.5\ \mathrm{m/s}\ w^* = 2.3\ \mathrm{m/s} \\ \phi^* &= 0\ \mathrm{rad},\ \theta^* = 0.2\ \mathrm{rad},\ \psi^* = 0\ \mathrm{rad} \\ p^* &= 0\ \mathrm{rad/s},\ q^* = 0\ \mathrm{rad/s},\ r^* = 0\ \mathrm{rad/s} \end{aligned}$$

can be easily obtained taking a first-order approximation. Assuming the air density $\rho = 1.2682\ \mathrm{kg/m^3}$, the system matrices **A**, **B** can be computed using the control and stability derivatives for the Aerosonde UAV that are reported in [13]. The icing impact coefficients $\mathscr{E}$ in the matrices $A_{\mathscr{E}}$ and $B_{\mathscr{E}}$ can be estimated noticing that, in total icing conditions, the change in lift and drag coefficients has been experimentally observed to obey the rule [18]:

- 10% reduction of coefficients C_{Z_α}, $C_{Z_{\delta_e}}$, C_{m_α}, $C_{m_{\delta_e}}$, C_{p_β}, C_{p_p}, $C_{p_{\delta_a}}$
- 8% reduction of coefficients $C_{Y_{\delta_r}}$, $C_{p_{\delta_r}}$, C_{r_r}, $C_{r_{\delta_r}}$
- 20% reduction of coefficients C_{Y_β}, C_{r_β}

The system is supposed to be controlled by an autopilot and, in the considered scenario, its objective is keeping the airspeed constant while slowly changing the pitch (a ramp has been considered as reference). Wind disturbances ν have been included in the simulations, with a maximum admissible acceleration $\|\dot{\nu}\| \le 8\ \mathrm{m/s^2}$.

For the sake of simplicity, only the results for the partial information case have been reported, this being the most practically significant scenario. The bank of UIOs has been designed, and to show the efficiency of the methods, sensor noise has been included in the simulation study. As a matter of fact, the gain matrices $\mathbf{K_1}^{(i)}$ can be chosen as to satisfy Assumption 4.1. In the first example, a fault has been supposed to affect the left rudder δ_{rl} for $t \ge 40$ s, this causing a progressive loss of efficiency of the device. Figures 4.1 and 4.2 show the behavior of residuals $\varepsilon^{(1)}$, $\varepsilon^{(2)}$. Despite the presence of noise and wind disturbances, in each residual a single component turns out to be distinctly affected by the control surfaces failure, i.e., direction $\mathbf{e}_3$ for residual $\varepsilon^{(1)}$ and direction $\mathbf{e}_2$ for residual $\varepsilon^{(2)}$: according to the decision rule, the fault can be therefore, correctly identified. Control reconfiguration is activated for $t \geq 60$ s and the nominal control action is recovered, as clearly illustrated in Figure 4.3, where the pitch behavior is depicted. For the sake of completeness, the resulting lateral airspeed v is also reported in Figure 4.4, although its dynamics is only marginally affected by the fault.

The second example corresponds to incremental icing with the severity factor η slowly varying from 0 to 0.2: Figures 4.5 and 4.6 illustrate the behavior of residuals

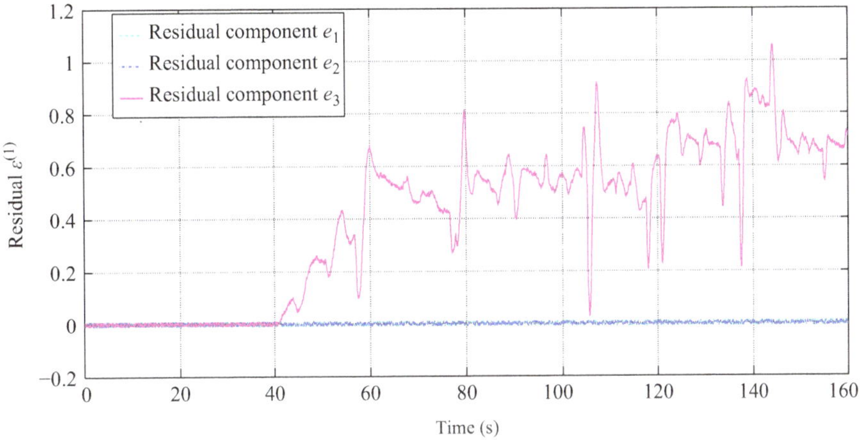

Figure 4.1 Faulty rudder δ_{rl}: components $\mathbf{e}_1$,$\mathbf{e}_2$ and $\mathbf{e}_3$ of residual $\varepsilon^{(1)}$ (partial information case)

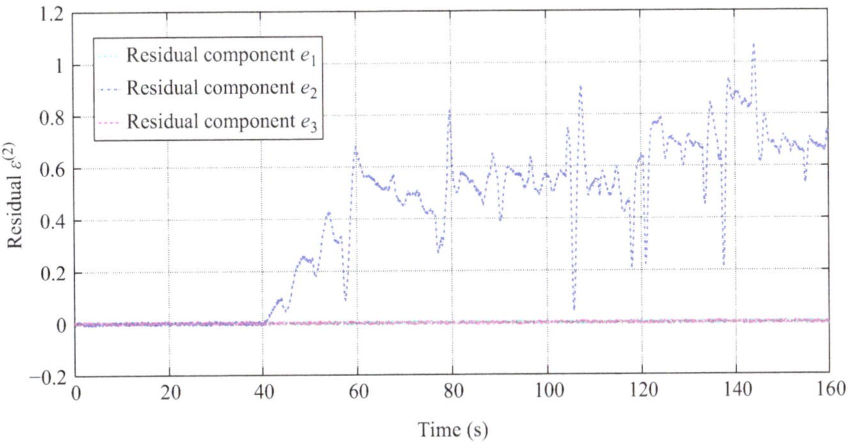

Figure 4.2 Faulty rudder δ_{rl}: components $\mathbf{e}_1$, $\mathbf{e}_2$ and $\mathbf{e}_3$ of residual $\varepsilon^{(2)}$ (partial information case)

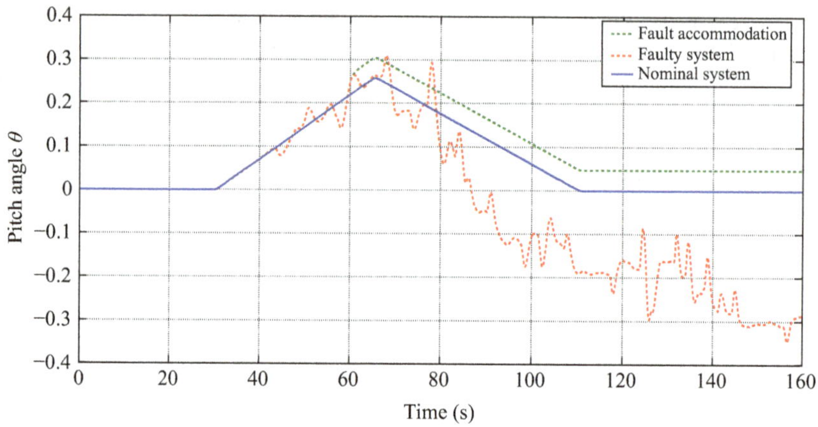

Figure 4.3 Pitch angle θ: reconfigured system, faulty system, nominal system

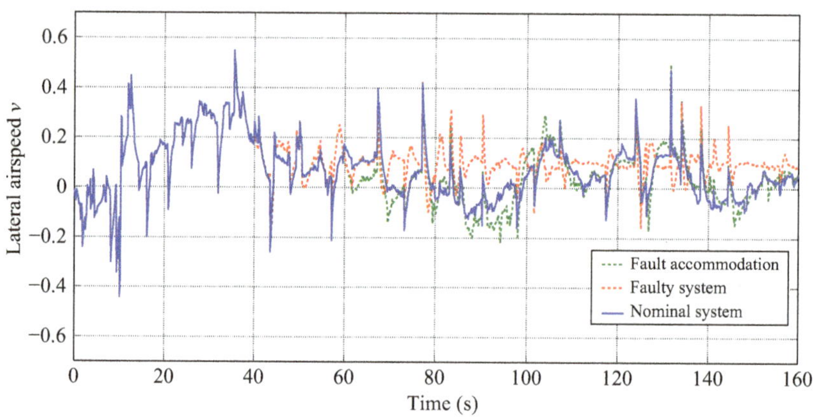

Figure 4.4 Lateral airspeed v: reconfigured system, faulty system, nominal system

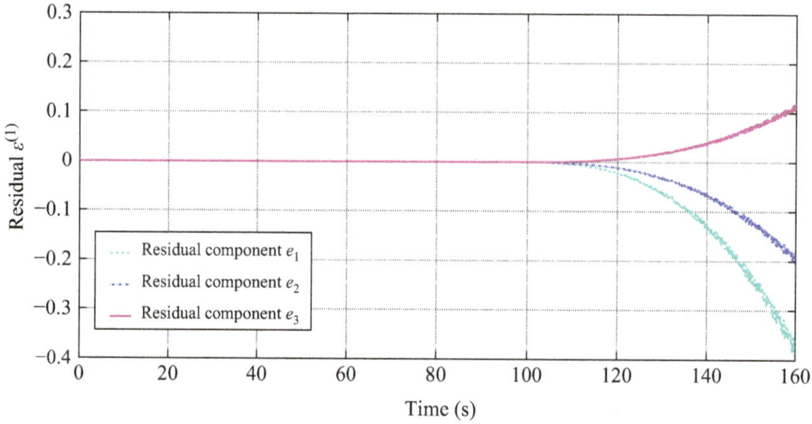

Figure 4.5 Incremental airfoil icing: components $\mathbf{e}_1$, $\mathbf{e}_2$ and $\mathbf{e}_3$ of residual $\varepsilon^{(1)}$

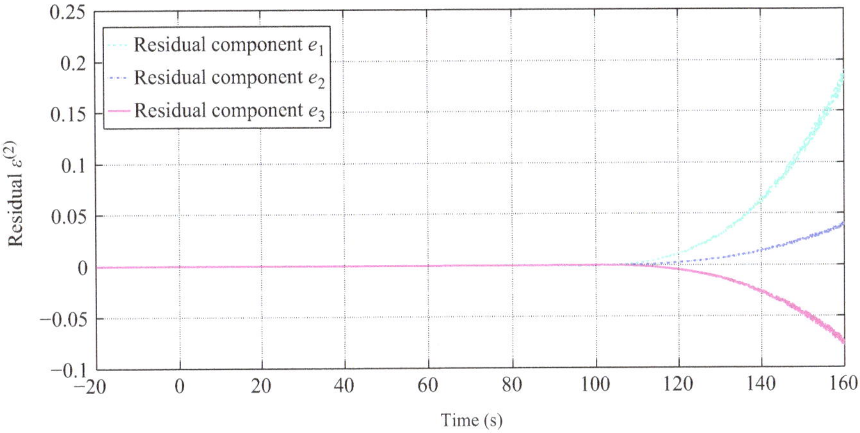

Figure 4.6 Incremental airfoil icing: components $\mathbf{e}_1$, $\mathbf{e}_2$ and $\mathbf{e}_3$ of residual $\varepsilon^{(2)}$

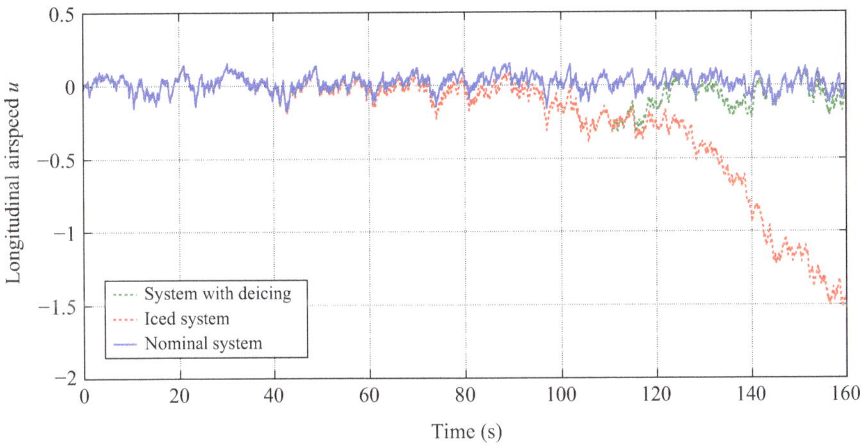

Figure 4.7 Horizontal airspeed u: system with ice accretion, nominal system (ice-free), activation of automated deicing system

$\varepsilon^{(1)}$ and $\varepsilon^{(2)}$ (with partial information): each of the three components $\mathbf{e}_1, \mathbf{e}_2$ and $\mathbf{e}_3$ is significantly affected by the system perturbation, this allowing to identify the anomalous effect caused by the ice accretion on airfoils. Finally the deicing routine has been activated, i.e., $\iota_{airfoil} > 0$ for $t \geq 120$ s, and the results of icing accommodation on the longitudinal airspeed and on the pitch are shown in Figures 4.7 and 4.8, respectively: the icing severity factor is reduced until a good performance of the system is recovered.

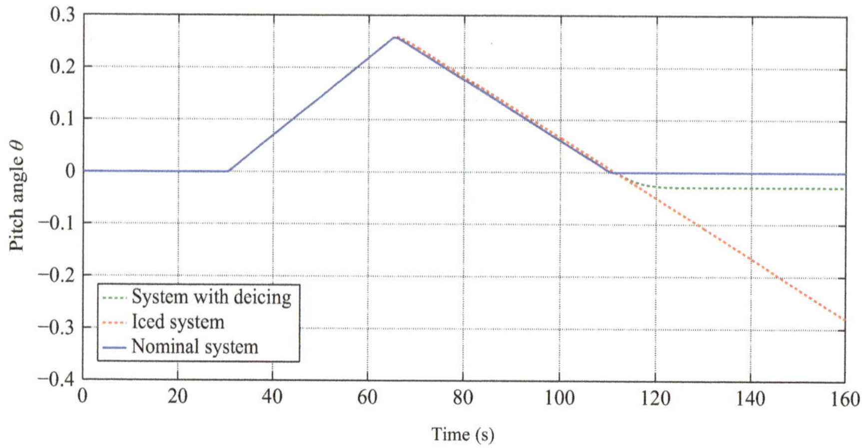

Figure 4.8 Pitch angle θ: system with ice accretion, nominal system (ice-free), activation of automated deicing system

References

[1] Caliskan F, Hajiyev C. A review of in-flight detection and identification of aircraft icing and reconfigurable control. Progress in Aerospace Sciences. 2013;60:12–34.

[2] Myers TG, Hammond DW. Ice and water film growth from incoming supercooled droplets. International Journal of Heat and Mass Transfer. 1999;42:2233–2242.

[3] Bone S, Duff M. Carbon nanotubes to de-ice UAVs. Technical Report. 2012. http://13614282187/eng12/Author/data/2122docx.

[4] Sørensen KL, Helland AS, Johansen TA. Carbon nanomaterial-based wing temperature control system for in-flight anti-icing and de-icing of unmanned aerial vehicles. In: IEEE Aerospace Conference. 2015.

[5] Chen J, Patton RJ, Zhang HY. Design of unknown input observers and robust detection filters. International Journal of Control. 1996;63:85–105.

[6] Johansen TA, Fossen TI. Control allocation: A survey. Automatica. 2013; 49:1087–1103.

[7] Cristofaro A, Johansen TA. Fault tolerant control allocation using unknown input observers. Automatica. 2014;50(7):1891–1897.

[8] Tousi M, Khorasani K. Robust observer-based fault diagnosis for an unmanned aerial vehicle. In: Systems Conference (SysCon), 2011 IEEE International; 2011. p. 428–434.

[9] Cristofaro A, Johansen TA. An unknown input observer approach to icing detection for unmanned aerial vehicles with linearized longitudinal motion. In: American Control Conference (ACC); 2015. p. 207–213.

[10] Cristofaro A, Johansen TA, Aguiar AP. Icing detection and identification for unmanned aerial vehicles: Multiple model adaptive estimation. In: 2015 European Control Conference (ECC); 2015. p. 1645–1650.
[11] Rotondo D, Cristofaro A, Johansen TA, *et al.* Icing detection in unmanned aerial vehicles with longitudinal motion. In: 2015 IEEE Conference on Control Applications (CCA) – Part of the 2015 IEEE Multi-Conference on Systems and Control (MSC); 2015. p. 984–989.
[12] Seron MM, Johansen TA, De Dona' JA, *et al.* Detection and estimation of icing in unmanned aerial vehicles using a bank of unknown input observers. In: Australian Control Conference (AuCC); 2015. p. 87–92.
[13] Beard RW, McLain TW. Small Unmanned Aircrafts – Theory and Practice. Princeton, NJ: Princeton University Press; 2012.
[14] Langelaan JW, Alley N, Neidhoefer J. Wind field estimation for small unmanned aerial vehicles. Journal of Guidance, Control, and Dynamics. 2011;34(4): 1016–1030.
[15] Johansen TA, Cristofaro A, Sørensen KL, *et al.* On estimation of wind velocity, angle-of-attack and sideslip angle of small UAVs using standard sensors. In: Intern. Conference on Unmanned Aircraft Systems (ICUAS); 2015. p. 510–519.
[16] Wenz A, Johansen TA, Cristofaro A. Combining model-free and model-based angle of attack estimation for small fixed-wing UAVs using a standard sensor suite. In: Unmanned Aircraft Systems (ICUAS), 2016 International Conference on; 2016. p. 624–632.
[17] Hoblit FM. Gust Loads on Aircraft: Concepts and Applications. Washington, DC: American Institute of Aeronautics and Astronautics; 1988.
[18] Bragg MB, Hutchinson T, Merret J, *et al.* Effect of ice accretion on aircraft flight dynamics. In: Proc 38th AIAA Aerospace Science Meeting and Exhibit; 2000.
[19] Gent RW, Dart NP, Cansdale JT. Aircraft icing. Philosophical Transactions of the Royal Society of London Series A: Mathematical, Physical and Engineering Sciences. 2000;358:2873–2911.
[20] Myers TG. Extension to the Messinger model for aircraft icing. AIAA Journal. 2001;39(2):211–218.
[21] Hajiyev C, Caliskan F. Fault Diagnosis and Reconfiguration in Flight Control Systems. London: Kluwer Academic Publishers; 2003.
[22] Wang D, Lum KY. Adaptive unknown input observer approach for aircraft actuator fault detection and isolation. International Journal of Adaptive Control and Signal Processing. 2007;21(1):31–48.
[23] Massoumnia MA. A geometric approach to the synthesis of failure detection filters. IEEE Transactions on Automatic Control. 1986;31(9):839–846.
[24] White J, Speyer J. Detection filter design: Spectral theory and algorithms. IEEE Transactions on Automatic Control. 1987;32(7):593–603.
[25] Park J, Rizzoni G. An eigenstructure assignment algorithm for the design of fault detection filters. IEEE Transactions on Automatic Control. 1994; 39(7):1521–1524.

[26] Cristofaro A, Johansen TA. Fault-tolerant control allocation: An Unknown Input Observer based approach with constrained output fault directions. In: Proc 52nd IEEE Conf on Decision and Control; 2013. p. 3818–3824.

[27] Cristofaro A, Johansen TA. An unknown input observer based control allocation scheme for icing diagnosis and accommodation in overactuated UAVs. In: 2016 European Control Conference (ECC); 2016. p. 2171–2178.

[28] Zhang Y, Suresh S, Jiang B, *et al.* Reconfigurable control allocation against aircraft control effector failures. In: Proc 16th IEEE Conf on Control Applications; 2007. p. 1197–1202.

[29] Cristofaro A, Johansen TA. Fault-tolerant control allocation with actuator dynamics: Finite-time control reconfiguration. In: Proc 53rd IEEE Conf on Decision and Control; 2014. p. 4971–4976.

[30] Cristofaro A, Polycarpou MM, Johansen TA. Fault diagnosis and fault-tolerant control allocation for a class of nonlinear systems with redundant inputs. In: Proc. 54th IEEE Conf. on Decision and Control; 2015. p. 5117–5123.

[31] Shamma JS. An overview of LPV systems. In: Mohammadpour J, Scherer C, editors. Control of Linear Parameter Varying Systems with Applications. Boston, MA: Springer-Verlag; 2012.

[32] Kwiatkowski A, Boll MT, Werner H. Automated generation and assessment of affine LPV models. In: Proceedings of the 45th IEEE Conference on Decision and Control; 2006. p. 6690–6695.

[33] Rotondo D, Puig V, Nejjari F, *et al.* Automated generation and comparison of Takagi-Sugeno and polytopic quasi-LPV models. Fuzzy Sets and Systems. 2015;277:44–64.

Chapter 5

Actuator fault tolerance for a WAM-V catamaran with azimuth thrusters

Alessandro Baldini[1], Riccardo Felicetti[1], Alessandro Freddi[1], Kazuhiko Hasegawa[2], Andrea Monteriù[1], and Jyotsna Pandey[2]

In this chapter, we present a fault-tolerant control scheme for an over-actuated unmanned surface vehicle (USV) equipped with two azimuth thrusters. The scheme manages the most common actuator faults, i.e., loss of efficiency of the thruster and lock-in-place of the azimuth angle. The scheme is based on a three-layer architecture: a heuristic-based control policy for proper reference generation, a control law for the vehicle dynamics to achieve speed tracking of the generated reference, and a control allocation level for optimally redistributing the control efforts among the thrusters even in presence of actuator faults and failures. The control allocation and the control policies are the main focus of the chapter, since their reconfiguration capabilities allow tolerance with respect to actuator failures. On the contrary, the control law does not depend on the health status of the system. The scheme is then tested in simulation, using a nonlinear model of a wave adaptive modular vessel catamaran.

5.1 Introduction

USVs represent a versatile class of marine vehicles that has many applications in scientific, industrial, and military operations [1]. Such vehicles may also have a high autonomy [2], so reliability has become a major issue, because dependability of such autonomous system is affected by faults. This aspect is even more important in critical scenarios, such as harbour security and mine sweeping [3]. USVs are usually equipped with a single thruster and a rudder, or two thrusters on the stern, which is typical of catamarans with two hulls. Both these configurations cannot work in presence of a thruster failure, and a recovery mission becomes necessary to bring the vehicle back. On the contrary, if a USV is driven by two azimuth thrusters, many failures could be managed, and in many cases the vehicle could also complete its mission without any human intervention.

[1]Department of Information Engineering, Università Politecnica delle Marche, Ancona, Italy
[2]Department of Naval Architecture and Ocean Engineering, Osaka University, Osaka, Suita, Japan

USVs in general comply with the well-known model given by [4]: in particular, we consider the Manoeuvring Mathematical Group (MMG) model [5], which is able to replicate the behaviour of a marine vehicle during manoeuvring motion accurately, even when drift angles are large and forward speed is low. By the way, most of the complexity in modelling comes from the identification of hydrodynamic forces. Many researches in this field have been made for catamaran-like USVs [6–8]: in this chapter, we are considering the representation of hydrodynamic forces for a WAM-V catamaran in calm water proposed by [9]. Also, azimuth thrusters are not so common in the USV literature: some inspiring papers consider the advantages of these actuators for station-keeping in the presence of wind [10] and focus on the nonlinear constrained control allocation problem [11].

Among the open problems in USV control, station-keeping under environmental disturbances [12,13] and stabilization of heading and cross-track error of wind-propelled catamarans [14] have been explored recently, while the problem of fault-tolerant control of catamaran-like USVs has not been investigated exhaustively. Some authors have dealt with the presence of different kind of faults, such as the contact with a submersed obstacle on the path [15]. The presence of faults and failures on the actuators has been tackled in the field of unmanned underwater vehicles (UUVs) and remotely operated vehicles only, in particular as regards failure detection of actuator faults [16] and fault-tolerant control of UUVs [17,18].

In this chapter, we propose a control scheme which exploits active fault tolerance in order to manage faults on the actuators, namely, the azimuth thrusters, together with a control policy to cope with failures as well. The control architecture is composed by three layers, namely, a top level with a set of control policies (control policy layer), a middle level which contains the control laws (control law layer), and a bottom level which accounts for the control effort allocation (control allocation layer). The control policy layer is based on a set of heuristic rules, with the purpose of generating a proper tracking reference to the control law even in presence of faults, without the need of control reconfiguration. In particular, it provides a desired speed and orientation, considering the actuator health status. The line of sight (LOS) manoeuvring [19], which is commonly adopted for under-actuated marine vehicles, is thus considered and extended to take into account possible fault/failure scenario. The control law layer is designed to track the reference speeds in the body frame and consists in a feedback linearisation-like control law. The control allocation layer proposed in this chapter is based on the extended thruster representation [20] which is solved by means of a weighted pseudo-inverse-based method. If the USV becomes under-actuated due to actuator failures, a heuristic switching control is proposed in order to accomplish the mission. The chapter is based on a previous work of the authors [21] and propose a detailed description of the control policy level, together with an extended evaluation of simulated results.

The chapter is structured as follows. In Section 5.2 the catamaran model is shown. Section 5.3 presents the control system architecture in the failure-free case, while the reconfiguration in the case of failures is detailed in Section 5.4. In Section 5.5, simulation results are reported, and, finally, Section 5.6 concludes the chapter.

5.2 Mathematical model

5.2.1 Dynamics

The dynamics of a USV can be generally described, neglecting the roll, pitch, and heave dynamics, as a 3 degree of freedom system (surge, sway, and yaw), as in [4]:

$$\boldsymbol{M}\dot{\boldsymbol{v}} + \boldsymbol{C}(\boldsymbol{v})\boldsymbol{v} + \boldsymbol{D}(\boldsymbol{v})\boldsymbol{v} = \boldsymbol{\tau} \tag{5.1}$$

This well-known equation describes the dynamics of the USV, where the linear velocities $\boldsymbol{v} = [u \ v \ r]^T$ are expressed in the body frame $R_B - \{x_B, y_B, z_B\}$. $\boldsymbol{M}$ is the inertia matrix, $\boldsymbol{C}(\boldsymbol{v})$ is the matrix of Coriolis and centripetal terms, $\boldsymbol{D}(\boldsymbol{v})$ is the drag matrix, and τ represents external forces and moments. Let us introduce the earth frame $R_E - \{x_E, y_E, z_E\}$ and assume that the centre on the body frame is at the mid-ship (Figure 5.1). The body-fixed velocities $\boldsymbol{v}$ are related to the fixed frame velocities $\dot{\eta} = [\dot{x} \ \dot{y} \ \dot{\psi}]^T$ as follows:

$$\dot{\eta} = \boldsymbol{J}(\eta)\boldsymbol{v} = \begin{bmatrix} \cos\psi & -\sin\psi & 0 \\ \sin\psi & \cos\psi & 0 \\ 0 & 0 & 1 \end{bmatrix} \boldsymbol{v} \tag{5.2}$$

In particular, the model can be rewritten in the following form [22]:

$$\begin{cases} m(\dot{u} - vr - x_G r^2) & = X_A + X_S \\ m(\dot{v} + x_G \dot{r} + ur) & = Y_A + Y_S \\ I_{zz}\dot{r} + m x_G(\dot{v} + ur) & = N_A + N_S \end{cases} \tag{5.3}$$

where u, v are the linear velocity components at the mid-ship (along the x_B-axis and y_B-axis, respectively), r is the yaw rate, m is the mass of the ship, I_{zz} is the moment of inertia about z_B, x_G is the position of the centre of mass in the x_B axis and, finally, X_A, Y_A, N_A, X_S, Y_S, and N_S are the external forces and moments.

The terms X_A, Y_A, and N_A represent undesired forces and moments depending on the acceleration of the vehicle, such as added mass. They are modelled by [22]

$$\begin{aligned} X_A &= f_{AX}(\dot{u}) = X_{\dot{u}}\dot{u} \\ Y_A &= f_{AY}(\dot{v}, \dot{r}) = Y_{\dot{v}}\dot{v} + Y_{\dot{r}}\dot{r} \\ N_A &= f_{AN}(\dot{v}, \dot{r}) = N_{\dot{v}}\dot{v} + N_{\dot{r}}\dot{r} \end{aligned} \tag{5.4}$$

The terms X_S, Y_S, and N_S represent forces and moments which depend on the speed components instead. Considering the so-called MMG model [5,22], they can be written as

$$\begin{aligned} X_S &= X_H + X_R + X_P \\ Y_S &= Y_H + Y_R + Y_P \\ N_S &= N_H + N_R + N_P \end{aligned} \tag{5.5}$$

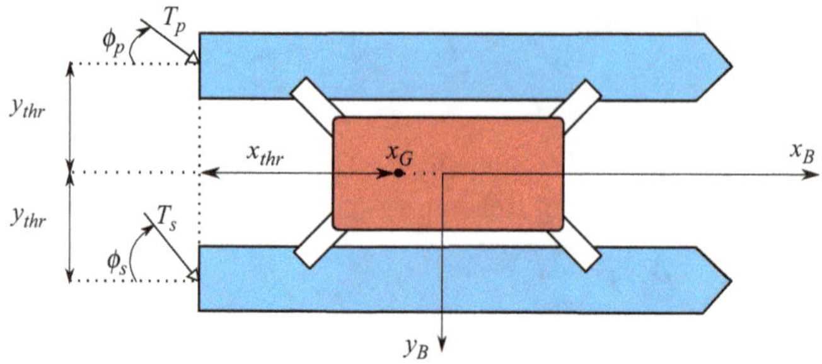

Figure 5.1 Catamaran scheme

The subscript H denotes the hydrodynamic forces acting on the hull, R the ones related to the rudder and, finally, P those related to the thrusters/propellers. As we are consider a catamaran equipped with two azimuth thrusters and no rudder (Figure 5.1), $X_R = Y_R = N_R = 0$ holds. The hydrodynamic forces and moments acting on hull are modelled by the following polynomial approximation:

$$\begin{aligned} X_H &\approx \bar{\rho}\left(-R_0' + X_{vv}'v'^2 + X_{vr}'v'r' + X_{rr}'r'^2 + X_{vvvv}v'^4\right) \\ Y_H &\approx \bar{\rho}\left(Y_v'v' + Y_r'r' + Y_{vvv}'v'^3 + Y_{vvr}'v'^2r' + Y_{vrr}'v'r'^2 + Y_{rrr}'r'^3\right) \\ N_H &\approx \bar{\rho}L_{pp}\left(N_v'v' + N_r'r' + N_{vvv}'v'^3 + N_{vvr}'v'^2r' + N_{vrr}'v'r'^2 + N_{rrr}'r'^3\right) \end{aligned} \tag{5.6}$$

where L_{pp} is the length between perpendiculars, d is the ship draft, $\bar{\rho} = (\rho/2)L_{pp}dU^2$ is a constant expressed in $[N]$, $v' = v/U$ is the non-dimensional lateral velocity, $r' = r(L_{pp}/U)$ is the non-dimensional turning rate and U is the resultant speed at the mid-ship.

The last contribution in (5.5) is given by the thrusters, namely, the port (P) and the starboard (S) azimuth thrusters. Let us denote with T_p and T_s the thrust forces of each actuator, while ϕ_p and ϕ_s represent the thruster orientation, and x_{thr}, y_{thr} are the distances between the thrusters and the USV centre of mass in the body frame (Figure 5.1). Thus, the virtual inputs X_P, Y_P, N_P can be expressed as

$$\begin{aligned} X_P &= T_p\cos(\phi_p) + T_s\cos(\phi_s) \\ Y_P &= T_p\sin(\phi_p) + T_s\sin(\phi_s) \\ N_P &= T_p(y_{thr}\cos(\phi_p) - x_{thr}\sin(\phi_p)) - T_s(y_{thr}\cos(\phi_s) + x_{thr}\sin(\phi_s)) \end{aligned} \tag{5.7}$$

5.2.2 Actuator faults and failures

The actual force produced by the thrusters may be significantly different from the desired one, as an effect of propeller jamming, propeller breakage, motor failure, bus voltage drop, etc. Provided that an appropriate fault detection and isolation is performed (see, e.g., [23]), we assume the loss of effectiveness of each thruster is described by a scalar $w_i \in [0, 1]$, $i \in \{p, s\}$, where

$$\begin{cases} w_i = 1 & \text{if the actuator is healthy (nominal behaviour)} \\ w_i \in (0, 1) & \text{if the actuator experiences loss of effectiveness} \\ w_i = 0 & \text{if the actuator has failed (no thrust)} \end{cases}$$

Hereafter, we call 'fault' a partial loss of thrust and 'failure' a total loss of thrust.

As azimuth thrusters also rely on servo-actuators to control their orientation, we consider an additional case of failure, i.e., servo failure. When a servo failure occurs, we assume the thruster orientation becomes locked-in-place. This is reasonable when the servo operates via a self-locking gear that does not allow back driving.

Summarizing, let us denote with $T_{p,d}$, $T_{s,d}$, $\phi_{p,d}$, and $\phi_{s,d}$ the desired values for T_p, T_s, ϕ_p, and ϕ_s. Fault/failure effects are described by

$$T_p = w_p T_{p,d} \qquad T_s = w_s T_{s,d} \tag{5.8}$$

$$\phi_p = \begin{cases} \phi_{p,d} & \text{if the servo is healthy} \\ \bar{\phi}_p & \text{if the servo is locked-in-place} \end{cases} \qquad \phi_s = \begin{cases} \phi_{s,d} & \text{if the servo is healthy} \\ \bar{\phi}_s & \text{if the servo is locked-in-place} \end{cases} \tag{5.9}$$

where $\bar{\phi}_p$, $\bar{\phi}_s$ denote the (constant) orientation of the azimuth thruster when a lock-in-place occurs, while w_p and w_s are time-varying effectiveness indices. Also note that we have neglected actuator dynamics because their response time is small w.r.t. the vehicle's time constants.

5.3 Control system architecture in the failure-free scenario

The control system architecture of the considered catamaran can be subdivided into different logical layers, as shown in Figure 5.2. The control allocation distributes the control effort among the thrusters; the control law solves the velocity tracking problem, and the control policy provides the reference trajectories for the controller. Above these layers, we consider an external fault estimation module, such as the one in [23], which provides an estimation of w_p and w_s, namely, $\hat{w}_p$ and $\hat{w}_s$.

5.3.1 Control law

We rewrite the system (5.3) in state space form, substituting (5.4), (5.5), and (5.6) into (5.3). We obtain the control affine model:

$$\begin{cases} \dot{u} = f_u(u, v, r) + G_{ux} X_P + G_{uy} Y_P + G_{un} N_P \\ \dot{v} = f_v(u, v, r) + G_{vx} X_P + G_{vy} Y_P + G_{vn} N_P \\ \dot{r} = f_r(u, v, r) + G_{rx} X_P + G_{ry} Y_P + G_{rn} N_P \end{cases} \tag{5.10}$$

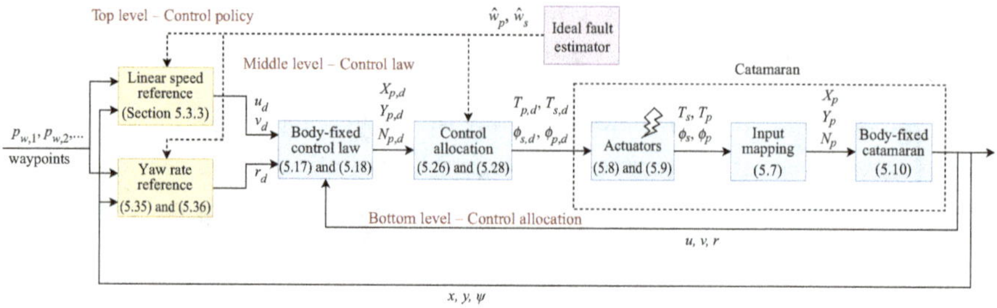

Figure 5.2 Control scheme. The light blue loop represents the speed control loop in the body-fixed reference and includes both the control law and the control allocation layers, while the orange loop generates the speed reference provided the vehicle position and the actual waypoint. The subscript 'd' applied to a variable indicates the desired valued for that variable, which may be different from the actual one due to faults/failures

$$f_u = \frac{rv + \frac{X_H}{m} + r^2 x_G}{1 - \frac{X_{\dot{u}}}{m}} \qquad f_v = \frac{(I_{zz} - N_{\dot{r}})\left(ru - \frac{Y_H}{m}\right) + \left(x_G - \frac{Y_{\dot{r}}}{m}\right)(N_H - mrux_G)}{(I_{zz} - N_{\dot{r}})\left(\frac{Y_{\dot{v}}}{m} - 1\right) - \left(x_G - \frac{Y_{\dot{r}}}{m}\right)(N_{\dot{v}} - mx_G)}$$

$$f_r = \frac{1}{I_{zz} - N_{\dot{r}}}\left(N_H + (N_{\dot{v}} - mx_G)\frac{(I_{zz} - N_{\dot{r}})\left(ru - \frac{Y_H}{m}\right) + \left(x_G - \frac{Y_{\dot{r}}}{m}\right)(N_H - mrux_G)}{(I_{zz} - N_{\dot{r}})\left(\frac{Y_{\dot{v}}}{m} - 1\right) - \left(x_G - \frac{Y_{\dot{r}}}{m}\right)(N_{\dot{v}} - mx_G)} - mrux_G\right) \tag{5.11}$$

where u, v, r are the state variables, i.e., surge, sway, and yaw speed respectively, while the nonlinear function $\boldsymbol{f} = [f_u f_v f_r]^T$ is defined in (5.11). Then, we collect the quantities as follows:

$$\underbrace{\begin{bmatrix} \dot{u} \\ \dot{v} \\ \dot{r} \end{bmatrix}}_{\dot{\boldsymbol{\nu}}} = \underbrace{\begin{bmatrix} f_u(u,v,r) \\ f_v(u,v,r) \\ f_r(u,v,r) \end{bmatrix}}_{\boldsymbol{f}} + \underbrace{\begin{bmatrix} G_{ux} & G_{uy} & G_{un} \\ G_{vx} & G_{vy} & G_{vn} \\ G_{rx} & G_{ry} & G_{rn} \end{bmatrix}}_{\boldsymbol{G}} \underbrace{\begin{bmatrix} X_P \\ Y_P \\ N_P \end{bmatrix}}_{\boldsymbol{\xi}_P} = \tag{5.12}$$

$$= \underbrace{\begin{bmatrix} f_u(u,v,r) \\ f_v(u,v,r) \\ f_r(u,v,r) \end{bmatrix}}_{\boldsymbol{f}} + \underbrace{\begin{bmatrix} \mu_u \\ \mu_v \\ \mu_r \end{bmatrix}}_{\boldsymbol{\mu}} \tag{5.13}$$

We can design a control law which imposes the desired values, namely, $X_{P,d}$, $Y_{P,d}$, and $N_{P,d}$, for the virtual inputs $\boldsymbol{\xi}_P = [X_P, Y_P, N_P]$. Given a speed trajectory reference

$\boldsymbol{\nu}_d = [u_d \; v_d \; r_d]^T$, the control laws are designed, for instance, with a simple feedback linearisation-like approach to asymptotically stabilize the tracking error $\boldsymbol{s} = \boldsymbol{\nu}_d - \boldsymbol{\nu}$. We define the Lyapunov function $V = \boldsymbol{s}^T\boldsymbol{s}/2$ and it follows that

$$\dot{V} = \boldsymbol{s}^T\dot{\boldsymbol{s}} = \boldsymbol{s}^T\left(\dot{\boldsymbol{\nu}}_d - \dot{\boldsymbol{\nu}}\right) = \boldsymbol{s}^T\left(\dot{\boldsymbol{\nu}}_d - \boldsymbol{f} - \boldsymbol{\mu}\right) \tag{5.14}$$

Choosing a symmetric and positive definite $\boldsymbol{K}$ and

$$\boldsymbol{\mu} = \dot{\boldsymbol{\nu}}_d - \boldsymbol{f} + \boldsymbol{K}(\boldsymbol{\nu}_d - \boldsymbol{\nu}) \tag{5.15}$$

then

$$\dot{V} = \boldsymbol{s}^T\dot{\boldsymbol{s}} = -\boldsymbol{s}^T\boldsymbol{K}(\boldsymbol{\nu}_d - \boldsymbol{\nu}) = -\boldsymbol{s}^T\boldsymbol{K}\boldsymbol{s} < 0 \tag{5.16}$$

so the error system is globally exponentially stable. The control law can be also expressed in scalar form:

$$\begin{aligned} \mu_u &= \dot{u}_d - f_u(u, v, r) + k_u(u_d - u) \\ \mu_v &= \dot{v}_d - f_v(u, v, r) + k_v(v_d - v) \\ \mu_r &= \dot{r}_d - f_r(u, v, r) + k_r(r_d - r) \end{aligned} \tag{5.17}$$

Note that the resulting controller is composed of trivial, scalar control laws, and each scalar law is designed on a single equation of the system. Then, the virtual control inputs $\boldsymbol{\xi}_P$ are given by

$$\boldsymbol{\xi}_P = \boldsymbol{G}^{-1}\boldsymbol{\mu} \tag{5.18}$$

where $\boldsymbol{G}$ is the non-singular and its inverse is

$$\boldsymbol{G}^{-1} = \begin{bmatrix} m - X_{\dot{u}} & 0 & 0 \\ 0 & m - Y_{\dot{v}} & mx_G - Y_{\dot{r}} \\ 0 & mx_G - N_{\dot{v}} & I_{zz} - N_{\dot{r}} \end{bmatrix} \tag{5.19}$$

5.3.2 *Control allocation*

Once the control law has determined the virtual inputs $X_{P,d}, Y_{P,d}, N_{P,d}$ to track the desired trajectory, it is necessary to find $T_{p,d}$, $T_{s,d}$, and ϕ_p, ϕ_s that satisfy (5.7), (5.8), and (5.9). In fact, the actual inputs for the system (5.3) are the desired thrust forces and azimuth angles. Note that (5.7) shows a nonlinear relation between the actual inputs and the virtual ones. This problem can be solved by means of linear methods using the extended thruster representation, which proposes a simple method to tackle unconstrained problem characterized by thrusters with varying angle [20]. It is sufficient to define

$$T_{px} = T_p\cos(\phi_p) \quad T_{py} = T_p\sin(\phi_p) \tag{5.20}$$

$$T_{sx} = T_s\cos(\phi_s) \quad T_{sy} = T_s\sin(\phi_s) \tag{5.21}$$

which are the coordinates of the vectors T_p, T_s in the body frame, and $\boldsymbol{f_e} = [T_{px}\ T_{py}\ T_{sx}\ T_{sy}]^T$, so that (5.7) can be represented by

$$\boldsymbol{\xi_P} = \begin{bmatrix} 1 & 0 & 1 & 0 \\ 0 & 1 & 0 & 1 \\ y_{thr} & -x_{thr} & -y_{thr} & -x_{thr} \end{bmatrix} \boldsymbol{f_e} = \boldsymbol{T}\boldsymbol{f_e} \tag{5.22}$$

Analogously, we define the components of the desired inputs:

$$T_{px,d} = T_{p,d}\cos(\phi_p) \quad T_{py,d} = T_{p,d}\sin(\phi_p) \tag{5.23}$$

$$T_{sx,d} = T_{s,d}\cos(\phi_s) \quad T_{sy,d} = T_{s,d}\sin(\phi_s) \tag{5.24}$$

and $\boldsymbol{f_{e,d}} = [T_{px,d}\ T_{py,d}\ T_{sx,d}\ T_{sy,d}]^T$. To consider loss off effectiveness, we note that $\boldsymbol{f_e} = \boldsymbol{W}\boldsymbol{f_{e,d}}$ holds, where $\boldsymbol{W} = diag\{w_p, w_p, w_s, w_s\}$ and $w_p, w_s \in [0, 1]$. As the extended thruster allocation does not consider saturation constraints, we totally remove a thruster from the control allocation when $w_i \in [0, 0.25]$, i.e., once a thruster loses more than 75% of its thrust capabilities, it is considered to be failed. Since $\boldsymbol{W}$ is unknown, it is replaced by its estimation $\hat{\boldsymbol{W}} = diag\{\hat{w}_p, \hat{w}_p, \hat{w}_s, \hat{w}_s\}$ provided by an external module, as shown in Figure 5.2, which is not in the scope of the present chapter and it is assumed to be already designed according to the one of the many techniques available in the literature [24].

The optimization problem is expressed by

$$\begin{aligned} &\min_{f_e}\{\boldsymbol{f_{e,d}}^T\boldsymbol{Q}\boldsymbol{f_{e,d}}\} \\ &s.t. \quad \boldsymbol{\xi_P} - \boldsymbol{T}\hat{\boldsymbol{W}}\boldsymbol{f_{e,d}} = \boldsymbol{0} \end{aligned} \tag{5.25}$$

and a direct solution is given by

$$\boldsymbol{f_{e,d}} = \boldsymbol{Q}^{-1}\hat{\boldsymbol{W}}\boldsymbol{T}^T\left(\boldsymbol{T}\hat{\boldsymbol{W}}\boldsymbol{Q}^{-1}\hat{\boldsymbol{W}}\boldsymbol{T}^T\right)^{-1}\boldsymbol{\xi_P} \tag{5.26}$$

Finally, the desired thrust forces and angles can be calculated from $\boldsymbol{f_{e,d}}$:

$$T_{p,d} = \sqrt{T_{px,d}^2 + T_{py,d}^2} \quad \phi_p = \tan^{-1}\left(\frac{T_{py,d}}{T_{px,d}}\right) \tag{5.27}$$

$$T_{s,d} = \sqrt{T_{sx,d}^2 + T_{sy,d}^2} \quad \phi_s = \tan^{-1}\left(\frac{T_{sy,d}}{T_{sx,d}}\right) \tag{5.28}$$

The solution expressed by (5.27) and (5.28) is characterized by a positive thrust and a $[-\pi, +\pi]$ azimuth angle. Assuming that each azimuth thruster is bound in the set $[-\pi/2, +\pi/2]$, while the thruster blade can rotate in both directions, we can deal with infeasible angles. When the azimuth angle is not in the interval $[-\pi/2, +\pi/2]$, it is sufficient to add (or subtract) π and to provide an opposite thrust. In this way, both components of the thrust remain the same, but the azimuth angle becomes feasible. Also note that the control allocation module introduces errors which depend on the fault estimation errors. Anyway, since the nominal inner loop system is globally exponentially stable, assuming bounded fault estimation error, we expect locally bounded error dynamics [25].

This solution tackles the loss of effectiveness problem, but it is not sufficient in the case of actuator failures. When a failure occurs, the control architecture is reconfigured, as described in Section 5.4.

5.3.3 Control policy

The reference is defined as a set of waypoints $\mathbb{X} = \{[x_{w,i}\ y_{w,i}]^T,\ i = 1, \ldots, m\}$ in the earth frame that the catamaran has to reach. The vehicle has to reach each waypoint $\boldsymbol{p}_{w,i} = [x_{w,i}\ y_{w,i}]^T$ with a tolerance defined by the acceptance radius r_{th}. The control law needs a continuous time reference, so it is generated low-pass filtering the sequence of waypoints, so that the output $\boldsymbol{p}_w$ of this filter is sufficiently smooth. We define the position error:

$$\boldsymbol{e} = \boldsymbol{p}_w - \boldsymbol{x} = [x_w - x \quad y_w - y]^T \tag{5.29}$$

which is the vector that starts from the current position of the catamaran and ends at the desired position. The magnitude of the desired velocity influences the time of arrival to the desired waypoints, nevertheless we can impose a heuristic speed profile for the catamaran which takes into account the distance from the current waypoint, the saturation limits, the fault information about the thrusters and the position of the next waypoint. The desired speed is also chosen as a function of the track configuration, in order to allow tight manoeuvres. The WAM-V velocity is set to have the same orientation of $\boldsymbol{e}$:

$$\dot{\boldsymbol{x}}_d = [\dot{x}_d\ \dot{y}_d]^T = \frac{\boldsymbol{e}}{\|\boldsymbol{e}\|} V_f(\|\boldsymbol{e}\|, \alpha, \beta) \cdot \gamma_{sat} \tag{5.30}$$

while $V_f(\|\boldsymbol{e}\|, \alpha, \beta)$ is the desired speed when no saturations occur and γ_{sat} considers the thruster saturations.

In particular, the term γ_{sat} is a correction factor to adjust the speed and is expressed by

$$\gamma_{sat} = 1 - 0.9 \cdot e^{-\lambda_\eta (1-\eta)^2} \tag{5.31}$$

where

$$\eta = \max\left(\frac{|T_p|}{T_{\max} \cdot w_p}, \frac{|T_s|}{T_{\max} \cdot w_s}\right) \tag{5.32}$$

and $\lambda_\eta > 0$ is a design parameter. In Figure 5.3, a plot of γ_{sat} is reported; the desired speed V_f in the saturation-free case is kept until actuators' usage are lower than 80% and then rapidly decreases for higher value. This mitigates the effects of unmodelled saturations, which are not considered in the control scheme otherwise.

To discuss the term $V_f(\|\boldsymbol{e}\|, \alpha, \beta)$, we first identify the following quantities that are useful to determine the desired speed: β is the current orientation error, while α is an estimation of the manoeuvring error angle once the current point is reached. The orientation error β is defined by

$$\beta = \psi - \tan^{-1}\left(\frac{y_{w,i} - y}{x_{w,i} - x}\right) \tag{5.33}$$

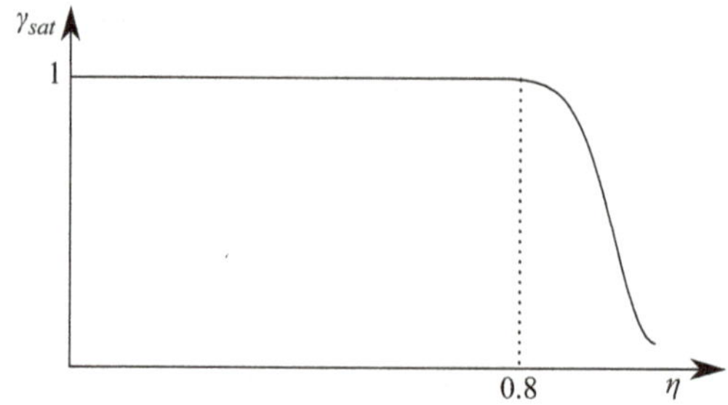

Figure 5.3 Correction factor of speed profile due to saturations

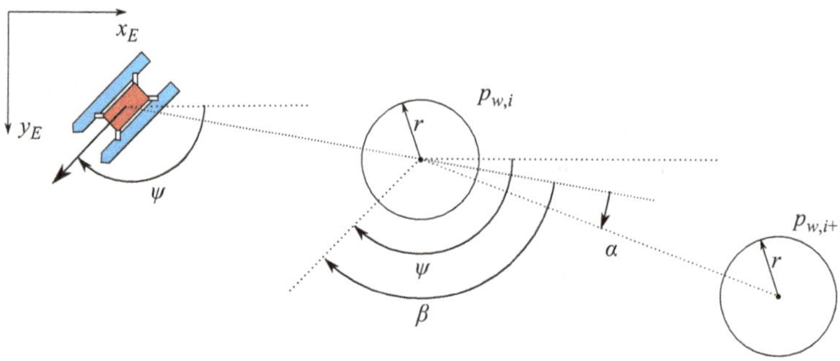

Figure 5.4 Representation of α and β in the most general case

where $\boldsymbol{p_{w,i}} = [x_{w,i}\ y_{w,i}]^T$ is the current waypoint. Note that $\beta = 0$ when the vehicle is oriented towards $\boldsymbol{p_{w,i}}$, so the vehicle's speed should be inversely proportional to β, in order to allow the steering manoeuvre. The angle α is calculated by

$$\alpha = \tan^{-1}\left(\frac{y_{w,i+1} - y_{w,i}}{x_{w,i+1} - x_{w,i}}\right) - \tan^{-1}\left(\frac{y_{w,i} - y}{x_{w,i} - x}\right) \tag{5.34}$$

where $\boldsymbol{p_{w,i+1}} = [x_{w,i+1}\ y_{w,i+1}]^T$ is the position of the next waypoint, as shown in Figure 5.4. Note that $\alpha = 0$ when the position of the vehicle, the waypoint we are approaching and the following waypoint are collinear, so the idea is to reduce the speed only in the neighbourhood of the approaching waypoint.

The desired heading is defined using the components of the vector $\boldsymbol{e}$:

$$\psi_d = \tan^{-1}\left(\frac{y_d - y}{x_d - x}\right) \tag{5.35}$$

and the desired yaw rate is given by

$$r_d = \frac{d\psi_d}{dt} + K_\psi(\psi_d - \psi) \tag{5.36}$$

The term $K_\psi(\psi_d - \psi)$, where $K_\psi > 0$ is a design parameter, forces the yaw rate to converge to the desired one.

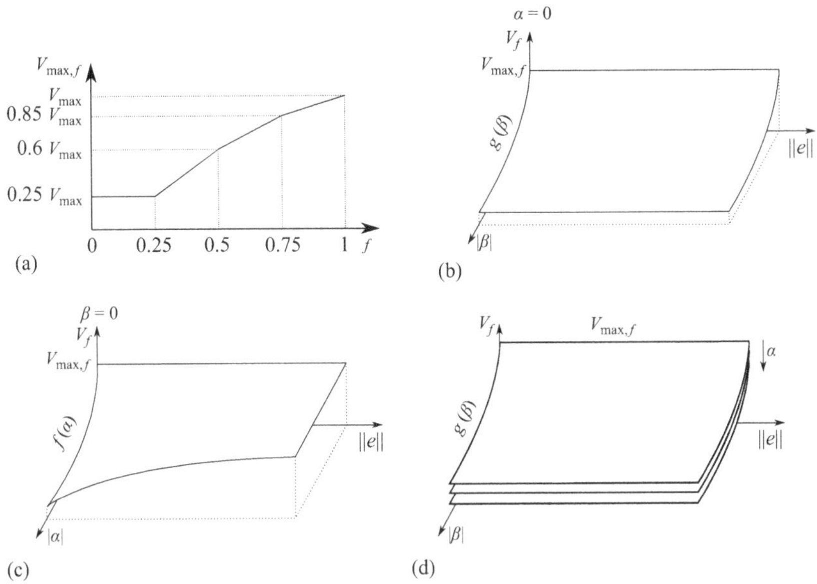

Figure 5.5 A parametric representation of (a) maximum speed for linear motion, (b) desired speed with $\alpha = 0$, (c) desired speed with $\beta = 0$, and (d) desired speed V_f

The speed references are then rotated into the body-fixed frame in order to be fed to the control law:

$$\boldsymbol{v_d} = \begin{bmatrix} u_d \\ v_d \\ r_d \end{bmatrix} = \begin{bmatrix} \cos(\psi) & \sin(\psi) & 0 \\ -\sin(\psi) & \cos(\psi) & 0 \\ 0 & 0 & 1 \end{bmatrix} \begin{bmatrix} \dot{x}_d \\ \dot{y}_d \\ r_d \end{bmatrix} \tag{5.37}$$

The speed V_f takes into account several parameters: let us start from the definition of $V_{\max}$, namely, the maximum linear speed in the absence of faults and failures. The maximum linear speed in the presence of faults is expressed by $V_{\max,f}$, which can be arbitrarily defined as a safe speed, based on experiments and/or simulations. In Figure 5.5(a) a representation for the proposed $V_{\max,f}$ is shown, which is the piecewise continuous function:

$$V_{\max,f} = \begin{cases} V_{\max}\left[\frac{-0.15}{0.25}(1-f)+1\right] & 0.75 \le f \le 1 \\ V_{\max}\left[\frac{-0.25}{0.25}(1-f-0.25)+0.85\right] & 0.5 \le f < 0.75 \\ V_{\max}\left[\frac{-0.35}{0.25}(1-f-0.5)+0.6\right] & 0.25 \le f < 0.5 \\ 0.25V_{\max} & 0 \le f < 0.25 \end{cases} \tag{5.38}$$

where $f = \min(w_p, w_s)$.

In Figure 5.5(b) a representation for the speed profile $V_f^{\alpha=0} = V_f(\|\boldsymbol{e}\|, 0, \beta)$ in function of $\|\boldsymbol{e}\|$ and β is shown. Regardless of the distance $\|\boldsymbol{e}\|$, when $\beta = 0$ the system can follow its maximum speed $V_{\max,f}$, but when β increases, the desired speed should decrease, so the desired speed $V_f^{\alpha=0}$ is chosen as

$$V_f^{\alpha=0}(\|\boldsymbol{e}\|, \beta) = V_{\max,f} \cdot g(\beta) \tag{5.39}$$

where

$$g(\beta) = e^{-\lambda_\beta |\beta|} \tag{5.40}$$

with $\lambda_\beta > 0$. As a consequence of (5.39), there is no actual dependance on $\|\boldsymbol{e}\|$, because the distance from the waypoint is not important when $\alpha = 0$, i.e., the vehicle's position and the next waypoints are collinear.

In Figure 5.5(c), a representation of the speed profile $V_f^{\beta=0} = V_f(\|\boldsymbol{e}\|, \alpha, 0)$ in function of $\|\boldsymbol{e}\|$ and α is shown, which is defined as

$$V_f^{\beta=0}(\|\boldsymbol{e}\|, \alpha) = V_{\max,f}\left[1 - (1 - f(\alpha))e^{-\lambda_e \|\boldsymbol{e}\|}\right] \tag{5.41}$$

where

$$f(\alpha) = e^{-\lambda_\alpha \cdot |\alpha|} \tag{5.42}$$

and $\lambda_e > 0, \lambda_\alpha > 0$ are the design parameters.

The overall desired speed profile $V_f(\|\boldsymbol{e}\|, \alpha, \beta)$ reduces to $V_f^{\beta=0}$ when $\beta = 0$ and to $V_f^{\alpha=0}$ when $\alpha = 0$. We propose the desired speed V_f to be

$$V_f = \frac{V_f^{\alpha=0} V_f^{\beta=0}}{V_{\max,f}} = V_{\max,f} \cdot g(\beta) \cdot \left[1 - (1 - f(\alpha))e^{-\lambda_e \|\boldsymbol{e}\|}\right] \tag{5.43}$$

A parametric representation for V_f is reported in Figure 5.5(d).

5.4 Control reconfiguration in the presence of failures

In this section, we show the control reconfiguration to cope with actuator failures. For an azimuth thruster, the two main types of failures are on the thrust driver and on the angle driver. A failure on the thrust driver implies a complete loss of propulsion; hence, the thruster cannot be used. A failure on the angle driver can arise in two different ways: the azimuth angle is moving uncontrollably, or the azimuth angle is locked-in-place and cannot move from the current position. In the first case, we choose to treat this failure as a failure on the thrust driver and set the corresponding thruster output to zero. In other words, whenever this failure occurs, the system becomes under-actuated and it is not possible to control both heading and position (or, analogously, linear speed, and angular velocity). In the second case, we can use the azimuth thruster as a simple fixed thruster; hence; we can control the generated thrust only. This case is more favourable than the previous: a servo failure produces a lock-in-place whenever a self-locking gear is employed to control the orientation. For each scenario, we propose a reconfiguration of control policy and control allocation layers only, allowing one to continue the navigation with reduced capabilities. Failures are shown on the S azimuth thruster, since failures in P have analogous solutions.

5.4.1 Failure on S azimuth thruster

When $0 \leq w_s \leq 0.25$, the S azimuth thruster is switched off, so that it provides no thrust. The control allocation problem (5.7) becomes

$$\begin{aligned} X_P &= T_p \cos(\phi_p) \\ Y_P &= T_p \sin(\phi_p) \\ N_P &= T_p\left(y_{thr} \cos(\phi_p) - x_{thr} \sin(\phi_p)\right) \end{aligned} \tag{5.44}$$

The system is under-actuated as we dispose of two inputs $[T_p\ \phi_p]^T$, while the control allocation has to produce three required efforts $[X_{P,d}\ Y_{P,d}\ N_{P,d}]^T$, so, in general, (5.45) has no feasible solutions and the least squares solution leads to unacceptable performances. We propose to split the control into two phases, and to switch between them with a hysteresis threshold, whose bounds are $\Delta\psi_{inf}$ and $\Delta\psi_{sup}$ ($\Delta\psi_{inf} < \Delta\psi_{sup}$). Such bounds define the range of acceptable heading errors. The control policy layer and the control allocation layer together contribute to the following solution:

- $|\psi_d - \psi| \geq \Delta\psi_{sup}$: in this case the heading error is relevant, so the priority is to adjust the WAM-V orientation to accomplish the LOS guidance. In particular, the system tries to allocate the surge force $X_{P,d}$, as well as the yaw moment $N_{P,d}$, while neglecting the sway force $Y_{P,d}$, so (5.45) is intentionally reduced to

$$\begin{bmatrix} X_P \\ N_P \end{bmatrix} = \begin{bmatrix} 1 & 0 \\ y_{thr} & -x_{thr} \end{bmatrix} \begin{bmatrix} w_p & 0 \\ 0 & w_p \end{bmatrix} \begin{bmatrix} T_{px} \\ T_{py} \end{bmatrix} \tag{5.45}$$

  where $0.25 < w_p \leq 1$ is the efficiency of the thruster P. The matrix is always invertible if $w_p > 0$, so both T_{px} and T_{py} can be obtained and (5.27) determines T_p and ϕ_p.

- $|\psi_d - \psi| \leq \Delta\psi_{inf}$: in this case, the heading error is sufficiently small, so the priority is to reduce the position error. The system tries to allocate the surge force $X_{P,d}$ and the sway force $Y_{P,d}$, leaving out the yaw moment $N_{P,d}$; thus (5.45) is reduced to

$$\begin{bmatrix} X_P \\ Y_P \end{bmatrix} = \begin{bmatrix} 1 & 0 \\ 0 & 1 \end{bmatrix} \begin{bmatrix} w_p & 0 \\ 0 & w_p \end{bmatrix} \begin{bmatrix} T_{px} \\ T_{py} \end{bmatrix} \tag{5.46}$$

  where $0.25 < w_p \leq 1$ measures the efficiency of thruster P, which can be faulty as well. This problem can always be solved by trivial inversion of a square matrix.

As an additional precaution, in the case of actuator failure, the azimuth angle of the working thruster is artificially saturated in the range $[-\pi/4, \pi/4]$ to avoid inefficient thruster orientations that could limit even more the performances. To reduce the control effort, the parameter K_ψ is also replaced by $K_{\psi,fail}$ (see Table 5.1).

Table 5.1 Design parameters

K	Q	K_ψ	$K_{\psi,fail}$	r_{th}
$0.6I_3$	$0.6I_4$	2	0.2	4 m

5.4.2 Blocked angle on S azimuth thruster

When the angle of the azimuth thruster S is locked-in-place (i.e., ϕ_s becomes constant and with a value equal to the stuck value $\bar{\phi}_s$, namely, $\phi_s = \bar{\phi}_s$) the system is no more over-actuated. Equation (5.7) becomes

$$\begin{aligned} X_P &= T_p\cos(\phi_p) + T_s\cos(\bar{\phi}_s) \\ Y_P &= T_p\sin(\phi_p) + T_s\sin(\bar{\phi}_s) \\ N_P &= T_p\left(y_{thr}\cos(\phi_p) - x_{thr}\sin(\phi_p)\right) - T_s\left(y_{thr}\cos(\bar{\phi}_s) + x_{thr}\sin(\bar{\phi}_s)\right) \end{aligned} \tag{5.47}$$

The extended thruster representation becomes

$$\begin{bmatrix} X_P \\ Y_P \\ N_P \end{bmatrix} = \begin{bmatrix} 1 & 0 & \cos\bar{\phi}_s \\ 0 & 1 & \sin\bar{\phi}_s \\ y_{thr} & -x_{thr} & l_s \end{bmatrix} \begin{bmatrix} w_p & 0 & 0 \\ 0 & w_p & 0 \\ 0 & 0 & w_s \end{bmatrix} \begin{bmatrix} T_{px} \\ T_{py} \\ T_s \end{bmatrix} \tag{5.48}$$

where $l_s = -y_{thr}\cos(\bar{\phi}_s) - x_{thr}\sin(\bar{\phi}_s)$. The solution can be obtained by matrix inversion, because it is nonsingular $\forall\bar{\phi}_s \neq \pi/2 + k\pi, k \in \mathbb{N}$. T_p, ϕ_p are given by (5.27).

5.4.3 Other cases

When two (or more) of the previous failures happen at the same time on two different thrusters (e.g., P is blocked and S is completely failed), the vehicle cannot be generally controlled in order to achieve the task (i.e., reach the waypoint).

The only exception is when both thrusters have a blocked angle: in this circumstance, the system is under-actuated but still partially controllable. However, this system can be seen as a traditional USV with two fixed-angle thrusters; hence, the solution of this problem is not analysed further. Note that fault/failure reconfiguration is performed by the control allocation, while the speed references are adjusted by the control policy.

5.5 Simulation results

In this section, we present the simulation results that have been obtained in four scenarios with different fault/failure conditions. In scenario I, no faults and no failures occur on the actuators. In scenario II, a faulty situation is shown. In the third scenario,

Table 5.2 WAM-V and actuator parameters

Variable name	Symbol	Value	Unit
Mass	m	181	kg
Inertia	I_{zz}	239.14	kg m^2
Centre of mass (CM)	x_G	0	m
Length	L	4.88	m
Beam	B	2.44	m
Thruster x distance w.r.t. CM	x_{thr}	2.44	m
Thruster y distance w.r.t. CM	y_{thr}	0.915	m
Added inertia	J_z	−0.089	kg m^2
Added mass along x axis	m_x	−0.830	kg
Added mass along y axis	m_y	−1.120	kg
Draft	d	0.15	m
Length between perpendiculars	L_{pp}	3.20	m
Maximum thruster force	$T_{\max}$	500	N
Maximum thruster angle	–	$\pi/2$	rad
Thrust time constant	–	0.05	s
Azimuth angle time constant	–	0.1	s

Table 5.3 Hydrodynamic parameters [9]

R'_0	X'_{vv}	Y'_v	Y'_r	
0.0896	−0.880	−1.155	−0.500	
Y'_{vvv}	N'_v	N'_r	N'_{vvv}	
−0.0008	−0.265	−0.177	1.659	
$X_{\dot{u}}$	$Y_{\dot{v}}$	$Y_{\dot{r}}$	$N_{\dot{v}}$	$N_{\dot{r}}$
−0.830	−1.120	−0.435	0.025	−0.089

we investigate the performance of the system with a total failure on an azimuth thruster. In the last scenario, we consider the WAM-V control when an azimuth thruster is locked on a fixed angle. In order to compare the performances of the system, the same trajectory is considered.

A 16′ WAM-V USV catamaran is considered for the simulation: parameters and external forces and moments are experimentally obtained in calm water in [9]. The vehicle's parameters are reported in Tables 5.2 and 5.3. The maximum speed $V_{\max}$ for thc WAM-V is sct to 11 knots [26]. Finally, the actuators have been simulated as first-order low-pass filters, subject to saturation constraints; moreover, the ideal estimation of the fault weights has been corrupted by additive coloured noise (standard deviation ≈ 0.01).

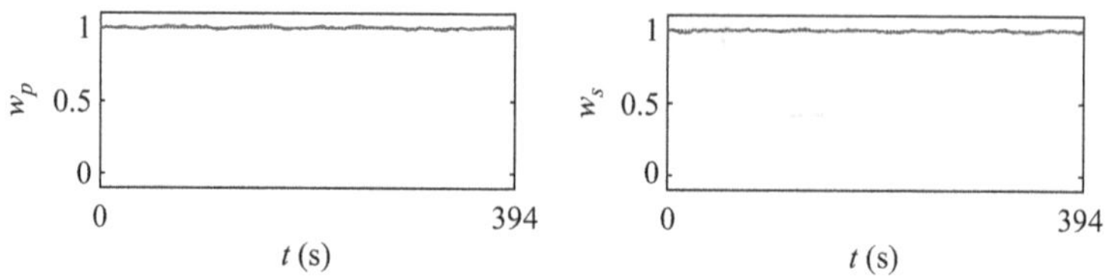

Figure 5.6 Thruster efficiency in scenario I (actual value in dotted blue, noisy estimation in solid red)

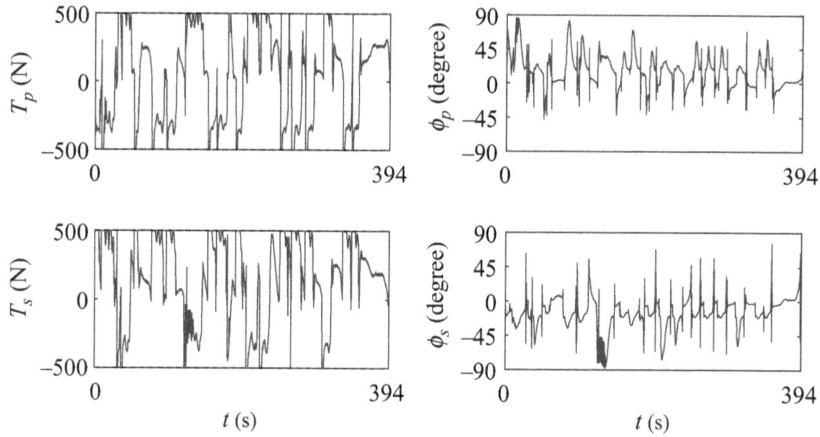

Figure 5.7 Control efforts in scenario I

5.5.1 Scenario I – fault-free actuators

In this scenario, we study the WAM-V in the absence of actuator faults/failures (i.e., $w_p = w_s = 1$, see Figure 5.6), to validate the controller and test the feasibility of the reference. Control efforts are reported in Figure 5.7: the control system exploits both variable thrusts and orientable azimuth angles, although the thrusters reach their saturation limit in some samples. The WAM-V completes the route (solid line in Figure 5.14(a)) in about 394 s.

5.5.2 Scenario II – double thruster faults

In this scenario, we consider the WAM-V system with faults on both thrusters, but without failures. In Figure 5.8 the value of the weights w_i is reported. Although the system is still over-actuated, the main issues are related to the reduced saturation limits, which cannot be avoided by the control allocation. The vehicle's desired speed is reduced by the control policy layer, allowing one to track the desired path and thus increasing the total time to about 513 s. We notice in Figure 5.9 a similar number of saturation events with respect to the previous scenario.

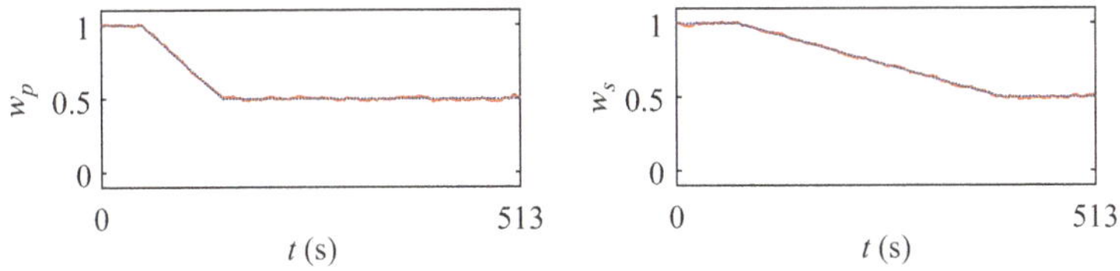

Figure 5.8 Thruster efficiency in scenario II (actual value in dotted blue, noisy estimation in solid red)

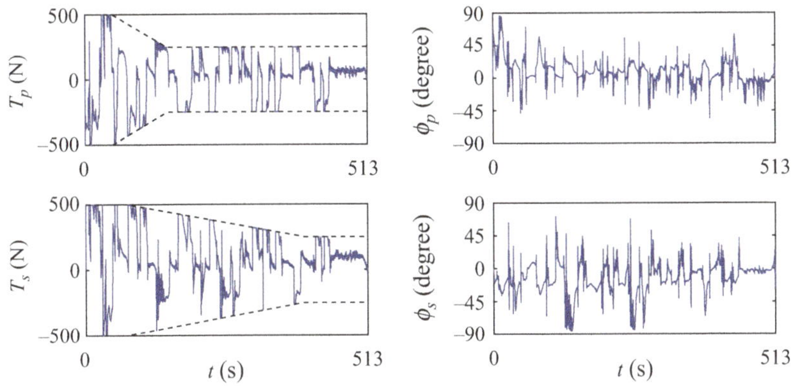

Figure 5.9 Control efforts in scenario II

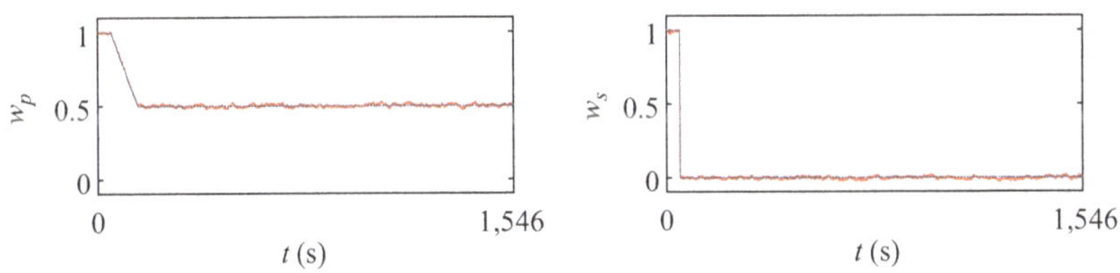

Figure 5.10 Thruster efficiency in scenario III (actual value in dotted blue, noisy estimation in solid red)

5.5.3 Scenario III – fault and failure on thrusters

In this scenario, the system performances with a total failure on the starboard thruster, along with a fault on the port thruster, are shown. The vehicle becomes under-actuated as soon as the thruster fails, so the control policy and the control allocation have a key role. In Figure 5.10 thruster efficiencies w_i are shown. The starboard thruster is subject to a failure after 50 s, while the port thruster efficiency decreases up to $w_s = 0.5$. Control efforts are shown in Figure 5.11. As expected, the starboard thruster is switched off after the failure occurs. The port thruster supplies a variable but positive thrust in almost every instant, to compensate for the longitudinal hydrodynamic friction. The

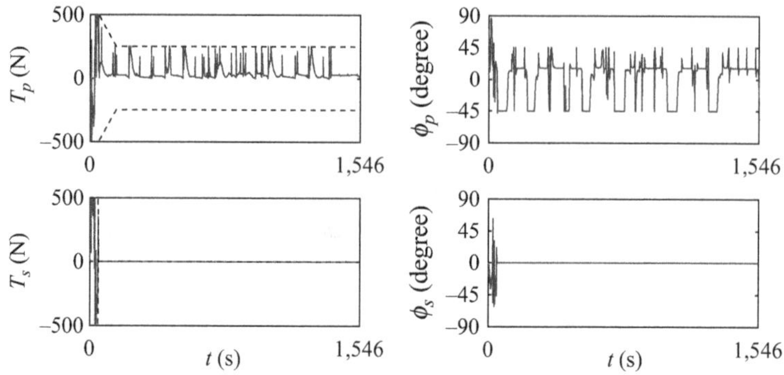

Figure 5.11 Control efforts in scenario III

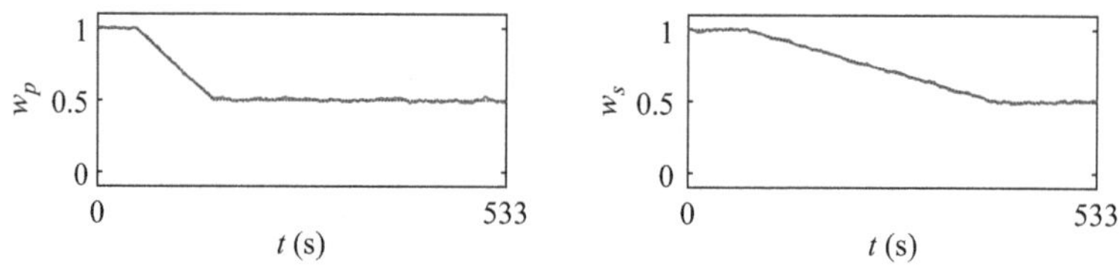

Figure 5.12 Thruster efficiency in scenario IV (actual value in dotted blue, noisy estimation in solid red)

azimuth angle of the starboard thruster is heavily exploited in order to control the direction of the USV. We remark that the azimuth angle in this case is imposed to be less than $\pi/4$. The system completes the route (dash-dotted line in Figure 5.14(c)) in about 1,546 s, so its travel time is more than three times the one of the first scenario.

5.5.4 Scenario IV – stuck and faulty thruster

In this scenario, we consider the tracking problem when an azimuth thruster is locked-in-place. The system becomes fully actuated, so a reduction of performances due to the saturations is expected. In this simulation, the starboard thruster's azimuth angle becomes locked-in-place after 45 s at a position of approximately 23 degrees. In Figure 5.12 the thrust efficiencies w_i are shown. Note that they are the same of scenario II, for the purpose of comparison. We notice in Figure 5.13 that the control efforts change faster than in scenario II: in fact, the presence of a considerable lateral force provided by the starboard thruster obliges the control system to compensate it with the port thruster; hence, leading to frequent saturation of the thrusters. The system completes the route (solid line in Figure 5.14(d)) in about 533 s, so its travel time is slightly higher than in scenario II.

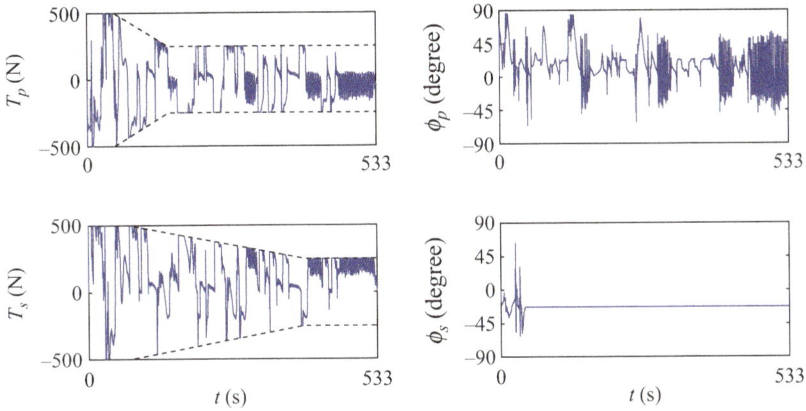

Figure 5.13 Control efforts in scenario IV

5.5.5 Discussion of results

The path of the WAM-V in the four scenarios is reported in Figure 5.14. At first glance, it can be noticed that the best tracking performances have been obtained in the first scenario. This is not a trivial consideration, because in this scenario the highest speed is requested; thus, manoeuvring complications can easily occur. Scenarios II and IV show comparable performances: in both cases, some narrow curves are unattainable due to the presence of faults; hence, in this control scheme the lock-in-place is less critical than a failure. Scenario III, as expected, is the most critical because of the under-actuated nature of the system. Many curves cannot be achieved; hence, a complete turn often occurs in order to recover the direction of the vehicle. We also notice that the WAM-V cannot proceed in a straight line, as can be seen in the longer sections of the track. This is a consequence of the switching behaviour of the control allocation, with the purpose of controlling linear velocity and angular speed.

We compare the performances with three indices: travel time (T_f), length of the actual path (Δs) and control effort. The indices are reported in Table 5.4. Since the thrusters' power can be estimated by [27]

$$P_{p,s} = kT_p^{3/2} + kT_s^{3/2} \tag{5.49}$$

we define an adimensional index of consumption:

$$\Phi_{p,s} = \int_0^{T_f} \left(T_p^{3/2} + T_s^{3/2}\right) dt \tag{5.50}$$

where the constant k is not important because we are interested only on the ratios between the scenarios.

Coherently with the previous considerations, the path length is minimal in the first scenario and maximal in the third scenario. The length of the path in scenarios II and

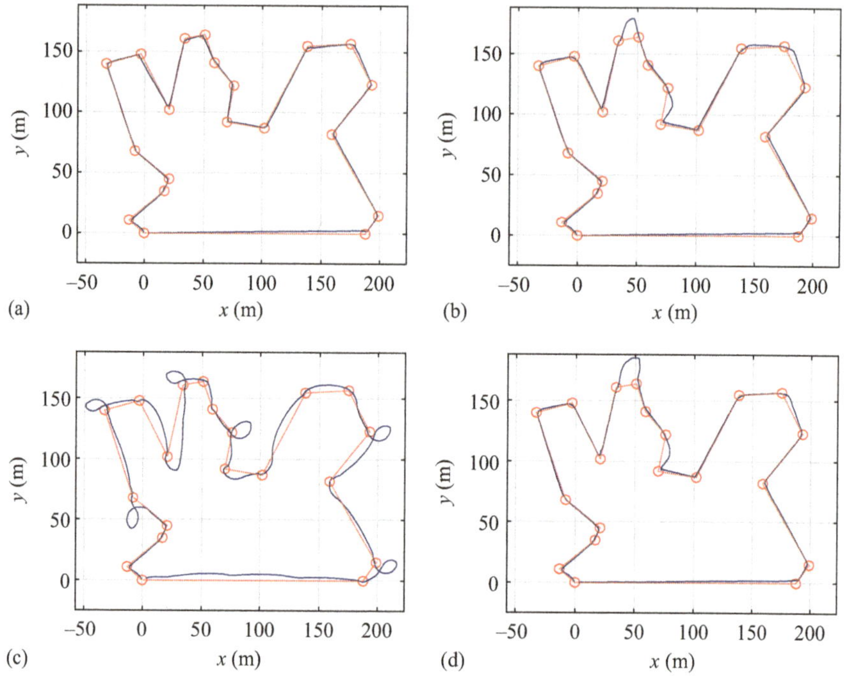

Figure 5.14 Track performances for the four scenarios: (a) scenario I, (b) scenario II, (c) scenario III, and (d) scenario IV

Table 5.4 Simulation results

Scenario	Travel time T_f (s)	Path length Δs (m)	Consumption index $\Phi_{p,s}$
I	394	917	4.629
II	513	945	2.943
III	1,546	1,264	1.246
IV	533	960	3.297

IV is intermediate, with a slightly better behaviour in the former one. The path length in scenarios I, II, and IV differ by less than 5%, while the length in scenario III is 38% higher than in scenario I, highlighting a major tracking performance deterioration. The reason is clear in Figure 5.14 that shows the path in the different scenarios.

As already revealed in the previous section, the shortest travel time is obtained in the first scenario, as a direct consequence of the higher speed imposed by the control policy. In scenario IV, the timing is slightly higher than in scenario II: this is a consequence of the fixed angle of the port thruster that has a reduced capability to generate a force in the longitudinal direction, in addition to the lateral force that must be compensated by the other thruster. Note that the same speed reference has been imposed by the control policy in these two scenarios, because the control policy does not take into account the presence of failures on the azimuth angles. Scenario III shows the greatest travel time, due to both low speed reference and increased path length.

As regards the consumption index, we notice that it is strongly related to the speed of the vehicle: the greatest consumption is obtained in the first scenario and the lowest in the third scenario, which involves, respectively, the highest and the lowest speeds. Scenarios II and IV have the same desired speed and comparable path length and timings: nevertheless, the consumption in the last scenario is higher due to the fixed angle of the port thruster, which leads to a lower efficiency.

5.6 Conclusion

In this chapter, a fault/failure tolerant control scheme for the 16′ WAM-V system has been proposed, which is based on a three-layer control architecture. The approach relies on control allocation and control policy layers to perform fault/failure tolerance, without varying the control law. In particular, given the estimation of the efficiencies, the control allocation layer provides a straightforward method to manage such fault. When a failure occurs, the control allocation is reconfigured to end the task with reduced performances. The control policy layer provides a reference for the control law, taking into account saturation and faults. Simulation results show that the system is able to complete a complex track even under severe actuator faults and failures. Three failures are considered: lock-in-place for the azimuth angle is tackled with comparable performances, while a failure of the thrust driver (or, equivalently, an uncontrollable azimuth angle) involves a considerable reduction of performances.

References

[1] Manley JE. Unmanned surface vehicles, 15 years of development. In: OCEANS 2008. Quebec City, QC: IEEE; 2008. p. 1–4.

[2] Rynne PF, von Ellenrieder KD. Development and preliminary experimental validation of a wind-and solar-powered autonomous surface vehicle. IEEE Journal of Oceanic Engineering. 2010;35(4):971–983.

[3] Schnoor RT. Modularized unmanned vehicle packages for the littoral combat ship mine countermeasures missions. In: Oceans 2003. Celebrating the Past... Teaming Toward the Future (IEEE Cat. No. 03CH37492). vol. 3. IEEE; 2003. p. 1437–1439.

[4] Fossen TI. Guidance and control of ocean vehicles. New York: John Wiley & Sons Inc; 1994.

[5] Yasukawa H, Yoshimura Y. Introduction of MMG standard method for ship maneuvering predictions. Journal of Marine Science and Technology. 2015;20(1):37–52.

[6] Zlatev Z, Milanov E, Chotukova V, *et al.* Combined model-scale EFD-CFD investigation of the maneuvering characteristics of a high speed catamaran. In: Proceedings of FAST09, Athens, Greece; 2009.
[7] Sutulo S, Guedes Soares C. Simulation of a fast catamaran's manoeuvring motion based on a 6DOF regression model. In: Proceedings of International Conference on Fast Sea Transportation (FAST'05), St. Petersburg, Russia; 2005. p. 1–8.
[8] Zhang X, Lyu Z, Yin Y, *et al.* Mathematical model of small water-plane area twin-hull and application in marine simulator. Journal of Marine Science and Application. 2013;12(3):286–292.
[9] Pandey J, Hasegawa K. Study on turning manoeuvre of catamaran surface vessel with a combined experimental and simulation method. IFAC-PapersOnLine. 2016;49(23):446–451.
[10] Qu H, Sarda EI, Bertaska IR, *et al.* Wind feed-forward control of a USV. In: OCEANS 2015, Genova, IEEE; 2015. p. 1–10.
[11] Oswald M, Blaich M, Wirtensohn S, *et al.* Optimal control allocation for an ASC with two azimuth-like thrusters with limited panning range. In: Control and Automation (MED), 2014 22nd Mediterranean Conference of. IEEE; 2014. p. 580–586.
[12] Sarda EI, Bertaska IR, Qu A, *et al.* Development of a USV station-keeping controller. In: OCEANS 2015, Genova, IEEE; 2015. p. 1–10.
[13] Sarda EI, Qu H, Bertaska IR, *et al.* Station-keeping control of an unmanned surface vehicle exposed to current and wind disturbances. Ocean Engineering. 2016;127:305–324.
[14] Elkaim G, Kelbley R. Station keeping and segmented trajectory control of a wind-propelled autonomous catamaran. In: Decision and Control, 2006 45th IEEE Conference on. IEEE; 2006. p. 2424–2429.
[15] Zanoli S, Astolfi G, Bruzzone G, *et al.* Application of fault detection and isolation techniques on an unmanned surface vehicle (USV). IFAC Proceedings Volumes. 2012;45(27):287–292.
[16] Alessandri A, Caccia M, Veruggio G. Fault detection of actuator faults in unmanned underwater vehicles. Control Engineering Practice. 1999;7(3): 357–368.
[17] Antonelli G. A survey of fault detection/tolerance strategies for AUVs and ROVs. In: Fault diagnosis and fault tolerance for mechatronic systems: Recent advances. Berlin, Heidelberg: Springer; 2003. p. 109–127.
[18] Baldini A, Ciabattoni L, Felicetti R, *et al.* Dynamic surface fault tolerant control for underwater remotely operated vehicles. ISA Transactions. 2018;78:10–20.
[19] Fossen TI, Breivik M, Skjetne R. Line-of-sight path following of underactuated marine craft. In: Proceedings of the 6th IFAC MCMC, Girona, Spain. 2003;p. 244–249.
[20] Sørdalen O. Optimal thrust allocation for marine vessels. Control Engineering Practice. 1997;5(9):1223–1231.
[21] Baldini A, Felicetti R, Freddi A, *et al.* Fault tolerant control for an over-actuated WAM-V catamaran. In: 12th IFAC Conference on Control Applications in

Marine Systems, Robotics, and Vehicles (CAMS), Daejeon, South Korea; 2019.

[22] Yoshimura Y. Mathematical model for manoeuvring ship motion (MMG Model). In: Workshop on Mathematical Models for Operations involving Ship-Ship Interaction. Tokyo, Japan; 2005. p. 1–6.

[23] Capocci R, Omerdic E, Dooly G, *et al.* Fault-tolerant control for ROVs using control reallocation and power isolation. Journal of Marine Science and Engineering. 2018;6(2):40.

[24] Blanke M, Kinnaert M, Lunze J, *et al.* Diagnosis and fault-tolerant control. vol. 691. Berlin, Heidelberg: Springer; 2006.

[25] Khalil HK. Nonlinear systems. Upper Saddle River, NJ: Prentice Hall; 2002.

[26] Marine Advanced Research, inc. WAM-V 16' datasheet. Accessed: 2017-04-10. https://static1.squarespace.com/static/53cb3ecbe4b03cd266be8fa7/t/5672108569a91a5538e35b45/1450315909468/WAM-V+BOSS+Brochure.pdf.

[27] Veksler A, Johansen TA, Skjetne R. Thrust allocation with power management functionality on dynamically positioned vessels. In: American Control Conference (ACC), 2012. IEEE; 2012. p. 1468–1475.

Chapter 6
Fault-tolerant control of a service robot

Alberto San Miguel[1], *Vicenç Puig*[1], *and Guillem Alenyà*[1]

In this chapter, the problem of fault-tolerant control of a service robot is addressed. The proposed approach is based on using a fault estimation scheme based on a robust unknown-input observer (RUIO) that allows one to estimate the fault as well as the robot state. This fault estimation scheme is integrated with the control algorithm that is based on observer-based state-feedback control. After the fault occurrence, from the fault estimation, a feedforward control action is added to the feedback control action to compensate the fault effect. To cope with the robot non-linearity, its non-linear model is transformed into a Takagi–Sugeno (TS) model. Then, the state feedback and RUIO are designed using a linear matrix inequality (LMI)-based approach considering a gain-scheduling scheme. To illustrate the proposed fault-tolerant scheme a mobile service robot TIAGo, developed by PAL Robotics, is used.

6.1 Introduction

Over the last few-year service, robots have been increasingly introduced in our daily lives (see e.g. Figure 6.1). According to the International Federation of Robotics, since 2016 there has been a yearly increase of 15% on its sales [1].

Although service robots have been designed to successfully perform tasks on highly dynamic and unpredictable anthropic domains some faults can appear. A wide range of faults can be considered regarding interaction, such as misleading interpretations of the human actions or unexpected scenarios beyond nominal operation. Also, their inherent complexity makes them prone to failures at all their levels, from the low-level actuators and sensors to the high-level decision layers. All these factors can lead to a degradation on the performance of the robot or imply critical damage to it, which might even jeopardise its safety. Thus, their ability to autonomously overcome most of these situations in a safe and efficient manner must play a fundamental role in their implementation.

[1]Institut de Robòtica i Informàtica Industrial (CSIC-UPC), Llorens i Artigas, Barcelona, Spain

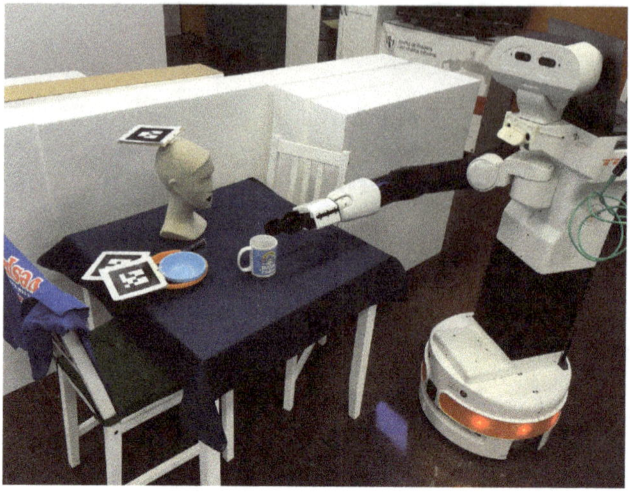

Figure 6.1 TIAGo robot in a domestic environment

6.1.1 State of the art

Fault detection and diagnosis (FDD) field has been widely studied for many years on classic control problems [2]. Only on certain generic robotic platforms, some of these approaches have been successfully applied, like for wheeled mobile robots [3], being still considered as a relative new field of study for robotic systems.

Current FDD techniques for robots can be classified into three different categories according to their common key characteristics [4]:

Data-driven: rely on the extraction and processing of data from different parts of the system, in order to detect and determine a faulty behaviour. One example is the work presented in [5], where a neural network is trained and integrated within a sliding mode control structure to enhance its robustness.

Model-based: depend on *a priori* analytical model that depicts the behaviour of a certain fault in the system or the nominal behaviour of the system itself. By comparison of the expected performance given by the model against the current one, faults can be detected and isolated. In [6], a residual is computed using a dynamical model of a 7-DOF robotic arm to detect and obtain information on unexpected collisions to determine suitable reaction strategies.

Knowledge-based: mimic the behaviour of a human expert, directly associating certain evidences with their corresponding faults. Some of these methods are seen as hybrid techniques that combine data-driven and model-based approaches. On this line, in [7], a two-layered structure is used, where a model aims at detection and a decision tree (data-driven) at isolation of different actuator faults.

On the suggested classification, key advantages and weaknesses of the different approaches are worth to be pointed out. It should be mentioned that this classification does not draw clear boundaries between the different groups but determine certain general characteristics that are usually present, which have been considered on this remark.

The main drawback of model-based techniques is the use of a model itself. For some robotic platforms, it is extremely complex to determine analytic expressions that describe their behaviour or establish relationship between their components. This issue is even more significant when trying to describe interactions with the environment or the effect of a fault in the unaffected parts, for example. Here data-driven approaches present their main advantage, as no knowledge about the robot is assumed and relies on data of the particular robot where the method is applied to. But most data-driven methods require a high computational expense which might make them unfeasible for an on-board and on-line implementation that some robots could require (e.g. a spatial exploration robot). When a learning phase (supervised or unsupervised) is involved, part (or all) of it is carried out offline, as in [8] where high-dimensional data of a robotic arm is recorded to obtain dimensional reduction transformations and train a binary support vector machine model to be used online. It should be considered that online learning allows one to obtain a dynamic method able to capture unexpected behaviours. Computational cost issue can also appear on some model-based methods but is usually overcome by simplification or reformulation of the model, as in [9], where a modification of the classical Newton–Euler is introduced in order to reduce its computation time for on-line execution purposes. Some knowledge-based approaches that consist in combining data- and model-driven methods are able to establish a trade-off between the discussed issues and suggest feasible solutions for FDD on robotic systems.

Regarding the characteristics that are required for service robots, autonomy is one of the most relevant ones. Service robots have to usually perform without the intervention of a human that supervises its actions or include a human in its operation loop that is not a robotic expert [10]. Thereby, being able to detect faults, identifying them and overcoming their effects in the execution of a task are crucial to achieve autonomy. As FDD methods should operate on a supervision level concurrently with all the techniques used for the desired performance, the computational demand on the method plays an important role on its implementation. High-demanding processes might interfere with others and be the source of faults themselves. Thus, according to the discussion above, model-based approaches are preferred, although data-driven methods can be also applied if they have a low-computational burden or some of its implementation is carried out offline.

6.1.2 Objectives

In this chapter, the problem of fault-tolerant control of the TIAGo humanoid robot by PAL Robotics [11] is addressed (Figure 6.2). Specifically, and as a proof of concept, the focus has been put into its 2-DOF head subsystem, to tackle the scenario where

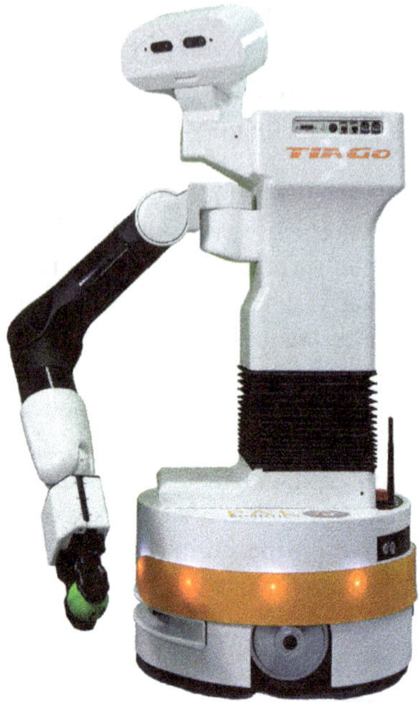

Figure 6.2 TIAGo robotic platform by PAL Robotics

external forces (e.g. a mass, a contact with a human) generate torques on its joints such that desired configurations are not reached.

The proposed model-based approach relies on a TS model of the robot, in order to cope with its non-linearity. From it, a state-feedback plus feed-forward control strategy will be designed using a fault estimation scheme based on a RUIO. To obtain the desired performance of the implemented method, a combination of LMIs and linear matrix equalities (LME), respectively, is used for the design. Additionally, the affecting fault is estimated using a reference model and its compensation will be included within the control loop, in order to overcome possible errors derived from the design process. All the procedures will be developed in the discrete-time domain, in order to bring the implementation on the real platform together with the proposed approach.

6.2 Takagi–Sugeno model

6.2.1 Robot model

The method presented in this chapter considers the implementation of a fault detection and isolation scheme by means of a model-based approach. Thereby, an analytic model which describes the behaviour of our system has to be determined. As aforementioned,

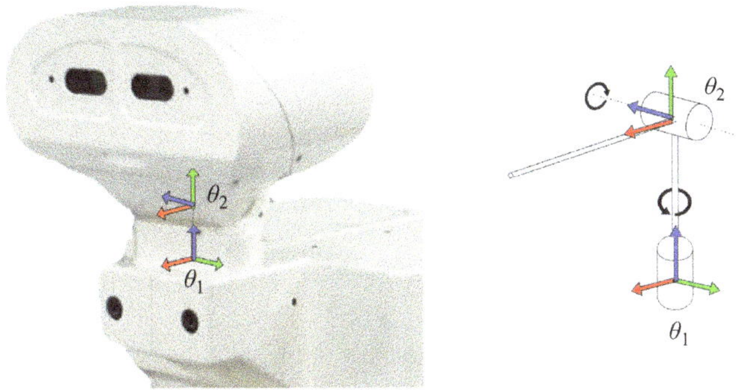

Figure 6.3 TIAGo's head subsystem representation as a two-manipulator link

the target system on this chapter is the 2-DOF head subsystem of the TIAGo robotic platform, presented in Figure 6.3. This type of systems is usually named Pan and Tilt structures, where the Pan movement corresponds to the rotation around an axis and the Tilt to the rotation around its perpendicular one.

To obtain this analytic model, the Newton–Euler method [12] has been applied considering the system as a two-link manipulator with two rotational joints θ_i ($i = 1, 2$), represented on the left part of Figure 6.3. For the sake of brevity this process is omitted, presenting only on this chapter the final expression for the model's dynamics.

The model can be stated in the so-called configuration-space form, which gives the joints torque vector τ as a function of $\ddot{\theta}, \dot{\theta}$ and θ, which are the joint acceleration, velocity and position vectors:

$$\tau = M(\theta)\ddot{\theta} + B(\theta)[\dot{\theta}\dot{\theta}] + C(\theta)[\dot{\theta}^2] + G(\theta), \tag{6.1}$$

where $M(\theta)_{n\times n}$ describes the mass matrix of the manipulator, $B(\theta)_{n\times n(n-1)/2}$ the Coriolis terms, $C(\theta)_{n\times n}$ the centrifugal coefficients and $G(\theta)_{n\times 1}$ the gravity effects; being the number of joints $n = 2$.

Applying (6.1) on TIAGo's head model:

$$\begin{bmatrix} \tau_1 \\ \tau_2 \end{bmatrix} = \begin{bmatrix} I_{zz_1} + m_2 d_4^2 + I_{xx_2} c_2^2 + I_{yy_2} s_2^2 - m_2 d_4^2 c(2\theta_2)^2 & 0 \\ 0 & I_{zz_2} + m_2 d_4^2 \end{bmatrix} \begin{bmatrix} \ddot{\theta}_1 \\ \ddot{\theta}_2 \end{bmatrix}$$
$$+ \begin{bmatrix} -2c_2 s_2 (I_{xx_2} - I_{yy_2}) + m_2 d_4^2 s(4\theta_2) \\ 0 \end{bmatrix} \dot{\theta}_1 \dot{\theta}_2$$
$$+ \begin{bmatrix} 0 & 0 \\ c_2 s_2 (I_{xx_2} - I_{yy_2}) - \frac{1}{2} m_2 d_4^2 s(4\theta_2) & 0 \end{bmatrix} \begin{bmatrix} \dot{\theta}_1^2 \\ \dot{\theta}_2^2 \end{bmatrix} + g \begin{bmatrix} 0 \\ -m_2 d_4 s(2\theta_2) \end{bmatrix}. \tag{6.2}$$

Terms in the form I_{a_i}, where $a = xx, yy, zz$ and $i = 1, 2$, correspond to the inertial tensor diagonal values of the links.

In order to simplify the expression into a shorter more intuitive manner, all the expressions have been arranged into constant and variable-dependant terms, obtaining

$$\begin{bmatrix}\tau_1\\ \tau_2\end{bmatrix} = \begin{bmatrix}\alpha+\beta(\theta_2) & 0\\ 0 & \xi\end{bmatrix}\begin{bmatrix}\ddot{\theta}_1\\ \ddot{\theta}_2\end{bmatrix} + \begin{bmatrix}\delta(\theta_2)\\ 0\end{bmatrix}\dot{\theta}_1\dot{\theta}_2 + \begin{bmatrix}0 & 0\\ \eta(\theta_2) & 0\end{bmatrix}\begin{bmatrix}\dot{\theta}_1^2\\ \dot{\theta}_2^2\end{bmatrix} + g\begin{bmatrix}0\\ \lambda(\theta_2)\end{bmatrix}. \tag{6.3}$$

From this form, the model equations of the head subsystem can be derived for the joint accelerations and velocities:

$$\begin{cases}\ddot{\theta}_1 = -\dfrac{\delta(\theta_2)}{\alpha+\beta(\theta_2)}\dot{\theta}_1\dot{\theta}_2 + \dfrac{1}{\alpha+\beta(\theta_2)}\tau_1,\\ \ddot{\theta}_2 = -\dfrac{\eta(\theta_2)}{\xi}\dot{\theta}_1^2 + \dfrac{1}{\xi}\tau_2 - \dfrac{\lambda(\theta_2)}{\xi},\\ \dot{\theta}_1 = \dfrac{d}{dt}\theta_1,\\ \dot{\theta}_2 = \dfrac{d}{dt}\theta_2.\end{cases} \tag{6.4}$$

Considering the complete equations and arrangements from (6.2) and (6.3), terms can be further arranged as follows:

$$\delta(\theta_2) = -2\eta(\theta_2), \tag{6.5}$$

$$\gamma(\theta_2) = \alpha+\beta(\theta_2), \tag{6.6}$$

$$\varphi(\theta_2,\dot{\theta}_1) = \delta(\theta_2)\dot{\theta}_1. \tag{6.7}$$

The non-linear model from (6.4) can be expressed in the state-space linear-parameter varying formulation (considering the states as scheduling parameters), according to its general expression:

$$\begin{cases}x(t+1) = A(x)x(t) + B(x)u(t) + d(x,t),\\ y(t) = C(x)x(t) + D(x)u(t).\end{cases} \tag{6.8}$$

Regarding TIAGo's physical system deployment, it is considered the input action u as the joints torque τ and the system output y as the joint position θ given by internal sensor measurements. The state vector x has been defined as $[\dot{\theta}_1\ \dot{\theta}_2\ \theta_1\ \theta_2]^T$, so (6.4) and term arrangements from (6.5) to (6.7) can be stated according to (6.8), obtaining the following model representation:

Table 6.1 Value limits for the TIAGo head subsystem variables

State	Min	Max heads
$\dot{\theta}_1$	0 rad/s	3 rad/s
$\dot{\theta}_2$	0 rad/s	3 rad/s
θ_1	$-75°$	$75°$
θ_2	$-60°$	$45°$

$$\begin{bmatrix} \ddot{\theta}_1 \\ \ddot{\theta}_2 \\ \dot{\theta}_1 \\ \dot{\theta}_2 \end{bmatrix} = \begin{bmatrix} 0 & -\frac{\varphi(\theta_2,\dot{\theta}_1)}{\gamma(\theta_2)} & 0 & 0 \\ \frac{\varphi(\theta_2,\dot{\theta}_1)}{2\xi} & 0 & 0 & 0 \\ 1 & 0 & 0 & 0 \\ 0 & 1 & 0 & 0 \end{bmatrix} \cdot \begin{bmatrix} \dot{\theta}_1 \\ \dot{\theta}_2 \\ \theta_1 \\ \theta_2 \end{bmatrix} + \begin{bmatrix} \frac{1}{\gamma(\theta_2)} & 0 \\ 0 & \frac{1}{\xi} \\ 0 & 0 \\ 0 & 0 \end{bmatrix} \cdot \begin{bmatrix} \tau_1 \\ \tau_2 \end{bmatrix}$$
$$+ \begin{bmatrix} 0 \\ -\frac{\lambda(\theta_2)}{\xi} \\ 0 \\ 0 \end{bmatrix},$$
$$\begin{bmatrix} \theta_1 \\ \theta_2 \end{bmatrix} = \begin{bmatrix} 0 & 0 & 1 & 0 \\ 0 & 0 & 0 & 1 \end{bmatrix} \cdot \begin{bmatrix} \dot{\theta}_1 \\ \dot{\theta}_2 \\ \theta_1 \\ \theta_2 \end{bmatrix}. \tag{6.9}$$

In order to ease the development of the system model, matrix $d(x,t)$ will be neglected until Section 6.4, where its effect will be further analysed.

It should be also pointed out that the physical model imposes limits on the joint variables, summarised in Table 6.1, on which the TS model will be developed on the next subsection.

6.2.2 Takagi–Sugeno formulation

Fuzzy logic, introduced by Zadeh [13], in contrast with classical Boolean logic does not define the output of a decision as binary (*Yes*/*No*, 1/0) but gives a degree of belonging of the output to the extreme values of the decision, according to a set of rules. This definition has been widely used in the artificial intelligence field as an approximation of the human decision-making process, where in most cases it admits a range of possibilities between the limits.

Takagi–Sugeno (TS) models [14] (henceforth TS models), named after their designers, apply the concept of fuzzy logic to the description of non-linear dynamics

Table 6.2 Upper and lower bounds of the premise variables for the TS formulation

Premise variables	**Min**	**Max heads**
$z_1 \equiv \gamma(\theta_2)$	0.0055	0.0091
$z_2 \equiv \varphi(\theta_2, \dot{\theta}_1)$	−0.0110	0.0110

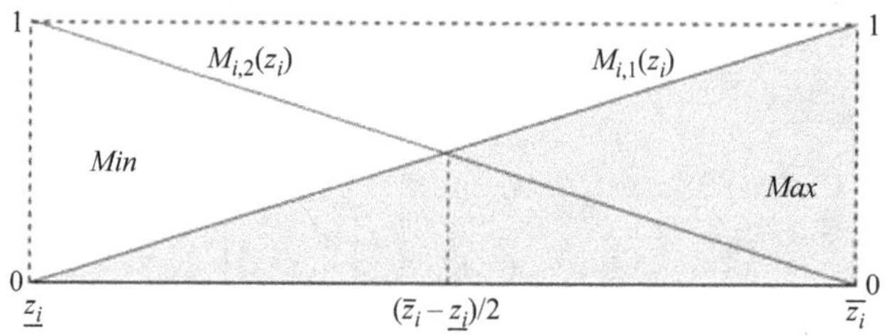

Figure 6.4 Graphical representation of the linear membership function

of the systems by a set of rules, associated to linear descriptions. Overall model is obtained by *blending* this linear system according to their rules.

In our case, the TS model starts with the obtained system representation from (6.9), which includes non-linearities that make the direct application of classical control strategies unfeasible. First, non-linear terms are embedded on the so-called premise variables z_i. For our particular case, these terms have been already arranged in (6.6) and (6.7), being $z_1 \equiv \gamma(\theta_2)$ and $z_2 \equiv \varphi(\theta_2, \dot{\theta}_1)$.

The concept of sector non-linearity has been used in the construction of the TS model to assure its exact representation. Its objective is to find sectors in the system state space where the non-linear behaviour lies. As these terms include parameters and variables which are bounded in the physical system, sectors can be defined in local regions, delimited by these bounds. From the limits of $\dot{\theta}_1$,$\dot{\theta}_2$,θ_1 and θ_2 presented in Table 6.1, our premise variables bounds can be found, included in Table 6.2.

Thereby, (local) sector non-linearity approach [16] is applied to reformulate the premise variables according to their limits, by means of the membership functions. These functions represent the degree of belonging to the upper or lower bounds according to a certain trend, defined in the fuzzy logic field as *fuzzy sets*. For this problem, linear membership functions have been considered, formulated for the *i*th premise variable as follows:

$$z_i = M_{i,1}(z_i)\bar{z}_i + M_{i,2}(z_i)\underline{z_i} \text{ where } \quad M_{i,1}(z_i) = \frac{\bar{z}_i - z_i}{\bar{z}_i - \underline{z_i}}, \quad M_{i,2}(z_i) = \frac{z_i - \underline{z_i}}{\bar{z}_i - \underline{z_i}}. \tag{6.10}$$

Figure 6.4 presents the graphical representation of (6.10).

Aforementioned fuzzy rules can be stated using the defined membership functions. These rules have the form of *IF–THEN* structures where premise variables z_i are evaluated w.r.t. the membership functions. The number of rules N is equal to 2^p, being p the number of chosen premise variables, as they consider all the possible permutations between the limits of z_i. Thereby, each rule is associated with a linear system that describes the behaviour of the *freezed* original system on the corresponding limits. For the addressed system, its fuzzy rules and associated linear systems have been stated next.

Model rule 1

IF z_1 is "*Max*" and z_2 is "*Max*" THEN $x(t+1) = A_1x(t) + B_1u(t)$

$$A_1 = \begin{bmatrix} 0 & -\bar{\varphi}/\bar{\gamma} & 0 & 0 \\ \bar{\varphi}/2\xi & 0 & 0 & 0 \\ 1 & 0 & 0 & 0 \\ 0 & 1 & 0 & 0 \end{bmatrix} \qquad B_1 = \begin{bmatrix} 1/\bar{\gamma} & 0 \\ 0 & 1/\xi \\ 0 & 0 \\ 0 & 0 \end{bmatrix}.$$

Model rule 2

IF z_1 is "*Min*" and z_2 is "*Max*" THEN $x(t+1) = A_2x(t) + B_2u(t)$

$$A_2 = \begin{bmatrix} 0 & -\underline{\varphi}/\bar{\gamma} & 0 & 0 \\ \underline{\varphi}/2\xi & 0 & 0 & 0 \\ 1 & 0 & 0 & 0 \\ 0 & 1 & 0 & 0 \end{bmatrix} \qquad B_2 = \begin{bmatrix} 1/\bar{\gamma} & 0 \\ 0 & 1/\xi \\ 0 & 0 \\ 0 & 0 \end{bmatrix}.$$

Model rule 3

IF z_1 is "*Max*" and z_2 is "*Min*" THEN $x(t+1) = A_3x(t) + B_3u(t)$

$$A_3 = \begin{bmatrix} 0 & -\bar{\varphi}/\underline{\gamma} & 0 & 0 \\ \bar{\varphi}/2\xi & 0 & 0 & 0 \\ 1 & 0 & 0 & 0 \\ 0 & 1 & 0 & 0 \end{bmatrix} \qquad B_3 = \begin{bmatrix} 1/\underline{\gamma} & 0 \\ 0 & 1/\xi \\ 0 & 0 \\ 0 & 0 \end{bmatrix}.$$

Model rule 4

IF z_1 is "*Min*" and z_2 is "*Min*" THEN $x(t+1) = A_4x(t) + B_4u(t)$

$$A_4 = \begin{bmatrix} 0 & -\underline{\varphi}/\underline{\gamma} & 0 & 0 \\ \underline{\varphi}/2\xi & 0 & 0 & 0 \\ 1 & 0 & 0 & 0 \\ 0 & 1 & 0 & 0 \end{bmatrix} \qquad B_4 = \begin{bmatrix} 1/\underline{\gamma} & 0 \\ 0 & 1/\xi \\ 0 & 0 \\ 0 & 0 \end{bmatrix}.$$

Finally, the defuzzification step has to be carried out to completely represent the system according to the defined fuzzy rules and sets. This process gives a complete representation of the system behaviour according to its fuzzy rules. Thereby, the system is described by a weighted sum of all the limit systems considered on the fuzzy rules. See [16] for more details.

For the considered TIAGo's head subsystem, this defuzzification process has to be applied only for system's A and B matrices, as they both depend on premise variables:

$$x(t+1) = \sum_{n=1}^{N=4} h_n(z_p(t))[A_n \cdot x(t) + B_n u(t)], \tag{6.11}$$

$$x(t+1) = A(z_p(t))x(t) + B(z_p(t))u(t), \tag{6.12}$$

where

$$h_1(z_p(t)) = M_{1,1}(z_1) \cdot M_{2,1}(z_2), \qquad h_2(z_p(t)) = M_{1,2}(z_1) \cdot M_{2,1}(z_2),$$

$$h_3(z_p(t)) = M_{1,1}(z_1) \cdot M_{2,2}(z_2), \qquad h_4(z_p(t)) = M_{1,2}(z_1) \cdot M_{2,2}(z_2).$$

6.3 Control design

6.3.1 Parallel distributed control

As has been presented, TS models are based upon a set of rules which enclose the behaviour of a non-linear system using linear descriptions on its *bounds*. Thereby, its performance can be described at a certain instant by a combination of the membership to these *bounds*. Following this concept, a control strategy for a system can be defined as a set of linear control laws defined at its limit operation points, being the system control at a certain point defined as a combination of these limit controllers. This concept was initially presented in [15] by Kang and Sugeno under the name of parallel distributed compensation (PDC from now on).

The PDC offers a procedure to design a control strategy from a given TS model using linear techniques. For each one of the fuzzy rules defined for the model, a control rule can be stated sharing the same premise variables and their corresponding fuzzy sets (membership functions). For this chapter, a state-feedback control has been used as the linear control strategy. From the already presented TS model for the considered system in this chapter, control rules have been stated as follows:

Control rule 1

IF z_1 is "*Max*" and z_2 is "*Max*" THEN $u(t) = -K_1 x(t)$.

Control rule 2

IF z_1 is "*Min*" and z_2 is "*Max*" THEN $u(t) = -K_2 x(t)$.

Control rule 3

IF z_1 is "*Max*" and z_2 is "*Min*" THEN $u(t) = -K_3x(t)$.

Control rule 4

IF z_1 is "*Min*" and z_2 is "*Min*" THEN $u(t) = -K_4x(t)$.

As for the TS models, the defuzzification step is applied on the control action vector $u(t)$, using the same procedure and h_n functions from the TS model:

$$u(t) = -\sum_{i=1}^{N=4} h_i(z_p(t))[K_ix(t)] = -K(z_p(t))x(t). \tag{6.13}$$

The key point of PDC is to design the feedback control gains K_n assuring stability and a certain number of performance specifications. Although this strategy only implies the definition of the system in the limit operation points (bounds of our premise variables), the design has to consider global design conditions.

According to Lyapunov's theory, global asymptotically stability exists for a set of subsystems if there exists a common positive definite matrix P for all the subsystems such that the following condition holds [16]:

$$A_i^T PA_i - P < 0 \qquad \forall i = 1, 2, \ldots, N. \tag{6.14}$$

Considering the *defuzzified* system expression from (6.11) and the state-feedback control formulation, global asymptotically stable condition is assured if the following expressions hold:

$$(A_i - B_iK_i)^T P(A_i - B_iK_i) - P < 0, \qquad \forall i = 1, 2, \ldots, N, \tag{6.15}$$

$$G_{ij}^T PG_{ij} - P < 0, \qquad \forall i < j \le N \quad \text{s.t.} \quad h_i \cap h_j \ne \phi, \tag{6.16}$$

where

$$G_{ij} = \frac{(A_i - B_iK_j) + (A_j - B_jK_i)}{2}.$$

The solution for stated inequalities is found by means of the LMIs-based approach. This method implies the formulation of the problem in a set of inequality constraints that define convex sets. Henceforth, all the design conditions that imply inequalities will be directly formulated as LMIs. The reader is referred to [17] for further details on LMIs, as they are out of the scope of this chapter.

6.3.2 *Optimal control design*

Robotic applications quite often require optimal performance in order to minimise costs in terms of resources and time consumption [12] as well as present some characteristics on their output(s) related with their application. This issue has been tackled on this chapter through the formulation of an optimal design oriented to achieve certain performance indexes.

Regarding optimality, the design is stated as a linear quadratic control (LQC from now on) problem, stated in the desired discrete time domain. The gains K_i of the state-feedback control (6.13) are found such that the following quadratic performance criterion is minimised:

$$J = \sum_{0}^{\infty} \left[x_k^T Q x_k + u_k^T R u_k \right], \tag{6.17}$$

where matrix Q allows one to control convergence speed of the states towards their references while matrix R is selected to limit the control effort required.

For the considered PDC, the minimisation problem is subjected to a set of LMIs included in [16]. In these optimality conditions and on the previously stated ones, there exist a significant number of LMIs to be fulfilled, due to the inclusion of all the possible A_i and B_i permutations. This might arise some tractability and feasibility problems on finding a solution, especially when applied to high-order systems [16]. To avoid these issues, the system has been re-stated such that B_i remains constant, by means of the Apkarian filter [18].

Apkarian filtering consists in pre/post applying a fast enough dynamic filter, such that it does not interfere system own dynamics, with the following form:

$$\begin{aligned} \dot{x}_f &= A_f x_f + B_f u_f, \\ \begin{bmatrix} \dot{\tau}_1 \\ \dot{\tau}_2 \end{bmatrix} &= \begin{bmatrix} -\psi & 0 \\ 0 & -\psi \end{bmatrix} \cdot \begin{bmatrix} \tau_1 \\ \tau_2 \end{bmatrix} + \begin{bmatrix} \psi & 0 \\ 0 & \psi \end{bmatrix} \cdot \begin{bmatrix} u_{\tau_1} \\ u_{\tau_2} \end{bmatrix}, \end{aligned} \tag{6.18}$$

where ψ represents the filter gain and u_f the new control variable vector. Pre-appliance of this filter has been chosen in the considered case, increasing the system matrices order, as follows:

$$\tilde{A}_i = \begin{bmatrix} A_i & B_i \\ 0 & A_f \end{bmatrix}, \qquad \tilde{B}_i = \begin{bmatrix} 0 \\ B_f \end{bmatrix}, \qquad \tilde{C}_i = \begin{bmatrix} C & 0 \end{bmatrix}. \tag{6.19}$$

As can be seen, B_i term is allocated on the $\tilde{A}_i$, remaining $\tilde{B}_i$ constant according to ψ for all TS model rules.

Applied on the considered system in this chapter, the $\tilde{A}_i$ are now formulated in such form. Fuzzy model (and rules) maintain the same structure, as they have been defined only regarding the premise variables definitions.

Considering this new formulation, the aforementioned LQC optimisation problem can be stated into the following LMIs [19]:

$$\begin{bmatrix} -Y & YA_i^T - W_i B_i^T & Y(Q^{1/2})^T & W_i^T \\ A_i Y - B_i W_i & -Y & 0 & 0 \\ Q^{1/2} Y & 0 & -I & 0 \\ W & 0 & 0 & -R^{-}1 \end{bmatrix} < 0, \tag{6.20}$$

$$\begin{bmatrix} \gamma I & I \\ I & Y \end{bmatrix} < 0, \tag{6.21}$$

where $Y \equiv P^{-1}$ and $W_i = K_i Y$, stating the problem to minimise γ, which represents the upper bound for the LQC criterion from (6.17).

6.4 Fault and state estimation

6.4.1 Robust unknown input observer

The main purpose of the work presented in this chapter is to apply a systematic approach for the design of a control strategy for non-linear systems, focusing on those related with robotic platforms. In these cases, besides stability and optimality conditions, some specific performance characteristics are desired. Furthermore, in those environments where robots have to perform along with human beings or even collaborate with them. Then, the robustness and fault-tolerance of the chosen control strategy is essential to avoid behaviours that do not take into account the entropy of the unknown surroundings or might even harm a person. An initial step has been taken to confront this challenge by means of a robust observer for unknown inputs (RUIO from now on) for TS models, presented by Chadli and Karimi [20].

The RUIO presents an observer structure for TS models which allows decoupling its state estimation from the effects of possible unknown disturbances (faults) or inputs that might affect the system. Its design is based on sufficient conditions that assure this behaviour, stated in terms of LMEs and the already presented LMIs. For this observer, the considered TS model includes the effects of unknown disturbances and inputs as follows:

$$\begin{cases} x(t+1) &= \sum_{i=1}^{N} h_i(z_i)(A_i x(t) + B_i u(t) + R_i d(t) + H_i w(t)), \\ y(t) &= Cx(t) + Fd(t) + Jw(t), \end{cases} \tag{6.22}$$

where $d(t)$ stands for the unknown input vector and $w(t)$ for the external disturbance vector. R_i and F represent the influence of $d(t)$ in the system behaviour, and H_i and J the influence of $w(t)$.

The existence of a solution for the problem is assured if the following necessary and sufficient condition is fulfilled:

$$rank \begin{bmatrix} CR_v & CH_v & [F\ J] \\ -F_v & -J_v & 0 \end{bmatrix} = rank \begin{bmatrix} F_v & J_v \\ R_v & H \end{bmatrix} + rank[F\ J], \tag{6.23}$$

where F_v and J_v are diagonal matrices with F and J as diagonal terms, and H_v and R_v are the horizontal concatenation of all the H_i and R_i column matrices.

One of the main improvements of the RUIO with respect to the previous similar techniques, for example the one presented in [21], is the relaxation on the necessary and sufficient condition(s). As the aforementioned matrices which model the influence of the unknown terms are given from design, they can be adjusted so (6.23) holds considering the C matrix of the system. The RUIO structure has the following form:

$$\begin{cases} r(t+1) = \sum_{i=1}^{N} h_i(z_i)(N_i r(t) + G_i u(t) + L_i y(t)), \\ \hat{x}(t) = r(t) - Ey(t), \end{cases} \tag{6.24}$$

where $r(t)$ corresponds to the RUIO internal state vector and the same set of defuzzification functions $h_i(z(t))$ of the TS model are considered. Matrices N_i, G_i, L_i and E are the observer gains to be determined. The design problem is based on assuring an asymptotic stability of the dynamics of the observer, so the estimation error converges to zero as time tends to infinite, disregarding the magnitude of the unknown inputs and disturbances. According to [21], this behaviour is satisfied if there exist matrices $X_i > 0$, S, V and W_i such that the LMI and LME conditions below hold:

$$\begin{bmatrix} \varphi_i & -V - A_i^T(V+SC)^T - C^T W_i^T \\ -V^T - (V+SC)A_i - W_i C & X_j - V - V^T \end{bmatrix} < 0, \quad (6.25)$$

where $\varphi = -X_i + (V+SC)A_i - W_iC + A_i^T(V+SC)^T - C^TW_i^T$,

$$(V+SC)R_i = W_iF, \quad (6.26)$$

$$(V+SC)H_i = W_iJ, \quad (6.27)$$

$$S[F\ J] = 0. \quad (6.28)$$

Observer gains from (6.24) are determined from the found solution according to the following expressions:

$$E = V^{-1}S, \quad (6.29)$$

$$G_i = (I + V^{-1}SC)B_i, \quad (6.30)$$

$$N_i = (I + V^{-1}SC)A_i - V^{-1}W_iC, \quad (6.31)$$

$$L_i = V^{-1}W_i - N_iE. \quad (6.32)$$

It should be noted from (6.22) that the system has not been formulated using the Apkarian filter for the RUIO design. In this case LMI constraints regard only about the output model, defined by constant matrix C. Thus, problem complexity does not increase as in control gain synthesis due to variant input matrix B_i.

6.4.2 Fault concept and design implications

Regarding this robotic platform and the literature insights into faults classification and sources [22], two main types of faults that could affect the TIAGo head subsystem and that could be overcome using the presented scheme have been distinguished:

1. **Sensor measurement error,** existing a difference between the given measures for θ_1 and θ_2 and their real values.
2. **External forces affecting the system** in both rotation axes that will appear in terms of torques in the rotational joints.

Recalling the obtained expression of the system behaviour from (6.9), there exists a term included in $d(x,t)$ that is described as the function of θ_2. From (6.1), it can be noticed that it is derived from the effect of the mass of the second body on the second rotational joint, which produces a torque variable with the distance between its centre of gravity (COG) and the joint axis. In Figure 6.5, this phenomenon is graphically depicted.

Thereby, this term can be understood as a second type fault, and only this definition has been considered in the presented work of this chapter. Future work could be carried out on the first one using a similar scheme.

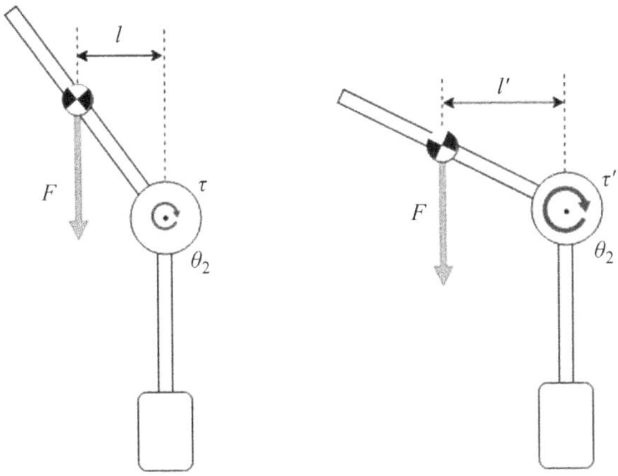

Figure 6.5 Induced torque on θ_2 by the effect of the distance between the COG and the rotation axis

Using this definition, aforementioned R_i, H_i, F and J matrices from the RUIO structure can be designed such that they agree with the effects of the faults. From stated concepts in the RUIO, the fault will affect the system as an unknown input (torque) $d(t)$, being $w(t)$ null (and consequently H_i and J). Matrix F will be also null as the fault does not occur on the system output.

This last consideration does not imply that joint angles will not be affected by faults. By construction, our state variables and its derivatives are joint accelerations, velocities and positions. If $d(t)$ is a torque, affects in the computation of $\ddot{\theta}$, so $\dot{\theta}$ and θ will *perceive* this effect even if it has not been injected directly on their calculation. Thereby, R_i variables have been set as unitary column vectors for all the N subsystems, as the fault will present the same behaviour disregarding the operational point of the system.

6.4.3 Fault estimation and compensation

The presented RUIO scheme is able to decouple the estimation of the system's states from unknown inputs or disturbances that might degrade the performance of the system, i.e. faults in this work. Using this decoupled estimation, faults are overcome internally by the classical state-feedback mechanism.

This behaviour is highly desired for autonomous robotic platforms as it has been discussed, but also information about fault characteristics could be useful in order to evaluate further measures to be taken according to it. Taking as an illustrative example a robotic arm affected by a sensor fault, information about its effects could lead to an isolation of the faulty sensor or maybe other further measures that allow the robot achieve its goal disregarding the fault (if it is not critical). But also if the robotic arm is collaborating along with a human and a direct contact, undesired or desired, happens unexpectedly for the robot, information about the detected effects could be crucial to avoid harming the person or to gather data that specifies the task to be performed together.

This latter example where unknown forces might occur agrees with the aforementioned concept of fault considered in this chapter and justifies the importance of estimating its magnitude. Thus, in this work the estimation of its value and isolation has been studied and included in the developed scheme.

For this purpose a reference control structure is introduced within the presented approach, which consists of the aforementioned state-feedback scheme where the head subsystem has been substituted by a continuous time model implemented according to the system description described in (6.9), but without the $d(x, t)$ term nor other faults. Desired positions of θ_1 and θ_2 for the real system are also set to the reference structure, so the corresponding states for a non-faulty behaviour are obtained. Using the estimation of the states $\hat{x}_{sys}$ from the RUIO and the computation of the system matrices, the value of the disturbance can be obtained as the difference between the reference and the real system:

$$\hat{d}(t) = O_d[x^a_{ref}(t+1) - \hat{x}^a_{sys}(t+1)], \tag{6.33}$$

where O_d matrix is defined so $\hat{d}$ has the components of the disturbance in both axes $\hat{d}_{\theta_1}$ and $\hat{d}_{\theta_2}$, corresponding to the differences on the first and second components of $x(t+1)$, and

$$\begin{aligned} \hat{x}^a_{sys}(t+1) &= A(\hat{x}_{sys})\,\hat{x}_{sys}(t) + B(\hat{x}_{sys})u_{sys}(t) + \hat{d}, \\ x^a_{ref}(t+1) &= A(x_{ref})x_{ref}(t) + B(x_{ref})u_{ref}(t), \end{aligned} \tag{6.34}$$

standing *sys* and *ref* subindexes for variables of the current system and of the reference structure, respectively. The superindex a points out that the values at $t+1$ are obtained by means of the analytic model of the TIAGo head subsystem, which is considered as a close enough approximation of the evolution of the states.

The active compensation mechanism consists of the direct injection of the opposite value of $\hat{d}$ into the control action transferred to the head subsystem. Presumably, with this procedure, the state-feedback plus RUIO control scheme will regard only for the part of control actions related with the nominal operation (i.e. movement between different set points), as the disturbance will be compensated by the injection of its estimation.

6.5 Fault-tolerant scheme

In this section, all the different components of the presented approach in this chapter are integrated into the general fault-tolerant scheme presented in Figure 6.6.

At each time instant, the desired positions for both axes θ_{des} are externally given (e.g. by a task planner layer) to both the reference and the main control structures. A corresponding control action is computed for the desired positions by means of a feedforward scaling matrix M [19], which depends on the current states (of the respective structure) according to the following expression:

$$M = -[\tilde{C}(-\tilde{B}K(z_p(t)) + \tilde{A}(z_p(t)))^{-1}\tilde{B}]^{-1}, \tag{6.35}$$

recalling that $z_p(t)$ corresponds to the set of defined premise variables (computed from $x(t)$) and $K(z_p(t))$ to the state-feedback gain from PDC.

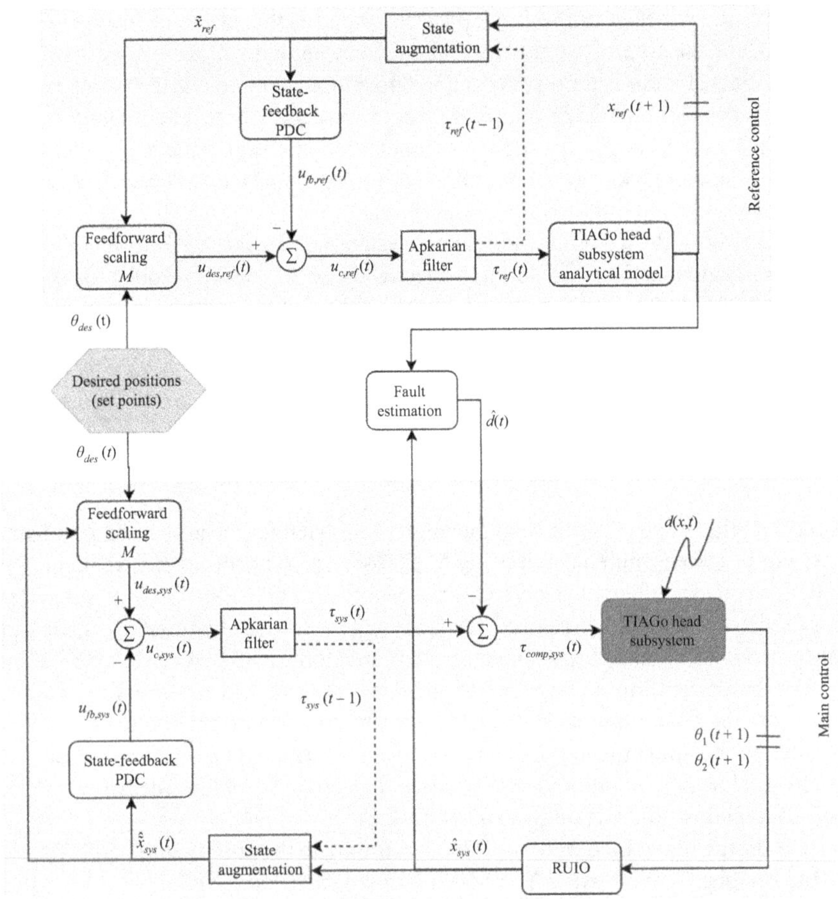

Figure 6.6 Schematic representation of the complete fault-tolerant control approach on discrete time

State-feedback PDC computes the feedback control action $u_{fb}(t)$ from the augmented state space taking into account the introduction of the Apkarian filter:

$$\tilde{x}(t) = \begin{bmatrix} x(t) \\ x_f(t-1) \end{bmatrix} = \begin{bmatrix} \dot{\theta}_1(t) \\ \dot{\theta}_2(t) \\ \theta_1(t) \\ \theta_2(t) \\ \tau_1(t-1) \\ \tau_2(t-1) \end{bmatrix}. \tag{6.36}$$

It should be noted that for the main control structure, a decoupled estimation of the states $\hat{x}(t)$ is given by the RUIO, which is the one to be used in the augmented states, as the TIAGo head subsystem only provides as output the angular positions.

The value of the control action $u_c(t)$ is computed as the difference between $u_{des}(t)$ and $u_{fb}(t)$, and the control torque $\tau(t)$ is obtained by applying the discrete time formulation of the Apkarian filter. In parallel, the magnitudes of the faults in both axes are estimated according to (6.33) and (6.34) from $\hat{x}_{sys}(t)$ and $x_{ref}(t)$. Their counter values are injected into $\tau_{sys}(t)$, obtaining the compensated control torque $\tau_{comp,sys}(t)$ that is sent to the TIAGo head subsystem, affected by $d(x,t)$ (which is assumed to include all possible faults in both axes in this scheme).

Finally, both the TIAGo head real subsystem and its model from the reference structure update their outputs on the respective injected control torques for the next time instant $t+1$. As already mentioned earlier, the real subsystem only provides the values of $\theta_1(t+1)$ and $\theta_2(t+1)$, but for the reference structure all states are known as its part of the implementation.

6.6 Application results

To evaluate the proposed fault-tolerant control approach for the addressed problem, a simulator has been implemented using MATLAB® programming environment [23]. Analytic expressions and physical limits and parameters of the TIAGo head subsystem have used to implement a continuous-time system into the simulation. The Dormand–Prince method [24] has been applied to solve these differential analytic expressions.

For the discrete-time implementation, classical Euler discretisation method has been used, for a sampling time $T_s = 10$ ms. The Apkarian filter has been designed considering this implementation, and it has been determined that its gain ψ has to be less or equal to $1/T_s$ in order to avoid consistency related problems between samples from consecutive time instants on its application.

The design of the PDC and the RUIO, according to the aforementioned LMIs and LMEs, has been solved using YALMIP toolbox [25,26] and SeDuMi optimisation software [27]. Additional LMIs regarding pole placement have been included on the design stage in order to achieve fast response time and zero overshooting on the PDC and to avoid couplings between RUIO and the subsystem dynamics (ten times faster). The reader is referred to [17] for more details of them.

According to the given description of the faults considered in this chapter, a single scenario is contemplated for all the results to be shown. On the Pan (θ_1) axis, a constant positive torque of $d_{\theta_1} = 4$ N m is injected at $t = 20$ s (half of the complete simulation time). For the Tilt (θ_2) one, only the aforementioned variable-dependant effect of the mass of the second link in θ_2 is present during all the simulation. The desired positions of θ_1 and θ_2 are their upper and lower limits, respectively (included in Table 6.1). At $t = 20$ s, position references change to half of these values.

To demonstrate the effectiveness of the proposed solution, results have been obtained for methods that incrementally incorporate all the presented techniques towards the final fault-tolerant control scheme. Thus, the approach is proved against the results of partial solutions in an *incremental improvement* fashion. Initially, from the basic control structure, only including the state-feedback PDC, the feedforward scaling matrix and the Apkarian filter, a classical Luenberger observer [17] is used to estimate the system states. Then, it is substituted by the RUIO, and finally the reference control structure and the fault estimation are included. In this latter case, the fault estimation mechanism is evaluated with and without its compensation.

6.6.1 Basic control structure with the Luenberger observer

The design of the Luenberger observer has been performed applying the LQC problem as in the state-feedback PDC. In addition, the same aforementioned additional pole placement LMI conditions for the RUIO have been applied. In Figure 6.7, the evolution of the system states along the simulation on an scenario without faults (but maintaining the position references) are included, to be used for determining differences with respect to it.

Figure 6.8 shows the same results for the common fault scenario. As expected, the classical Luenberger observer does not decouple the effects of the fault(s) affecting the system in any of the axes, producing a noticeable error in the estimation of the state variables. Also, oscillatory behaviours during the transitory appear in both axes with respect to results in Figure 6.7, due to the injection of unseen faults within the control structure. It should be pointed out that as a result from the coupling effects between

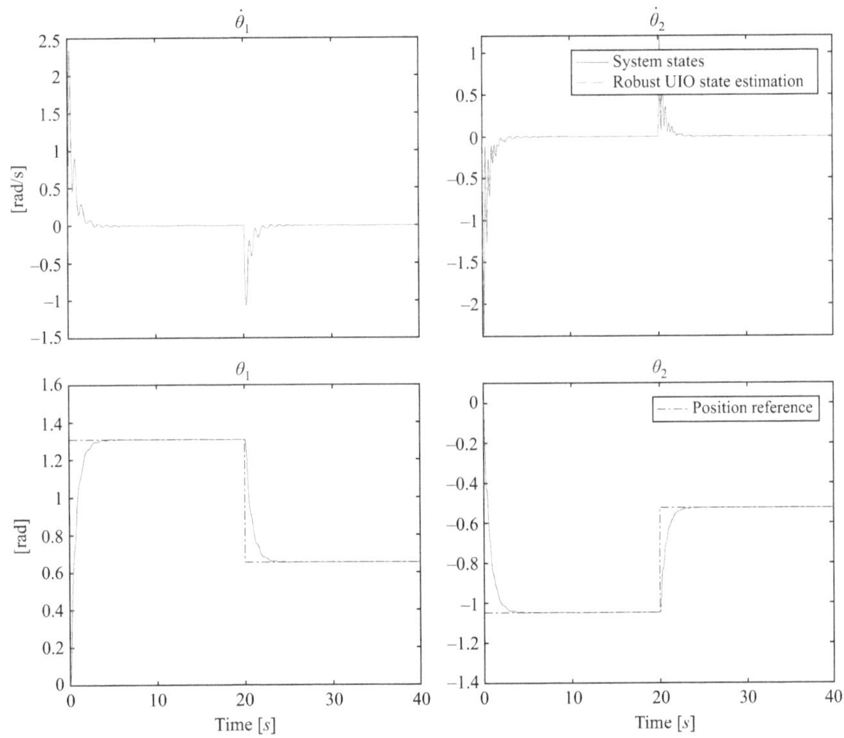

Figure 6.7 System states evolution on a non-fault scenario using the basic control structure with the Luenberger observer

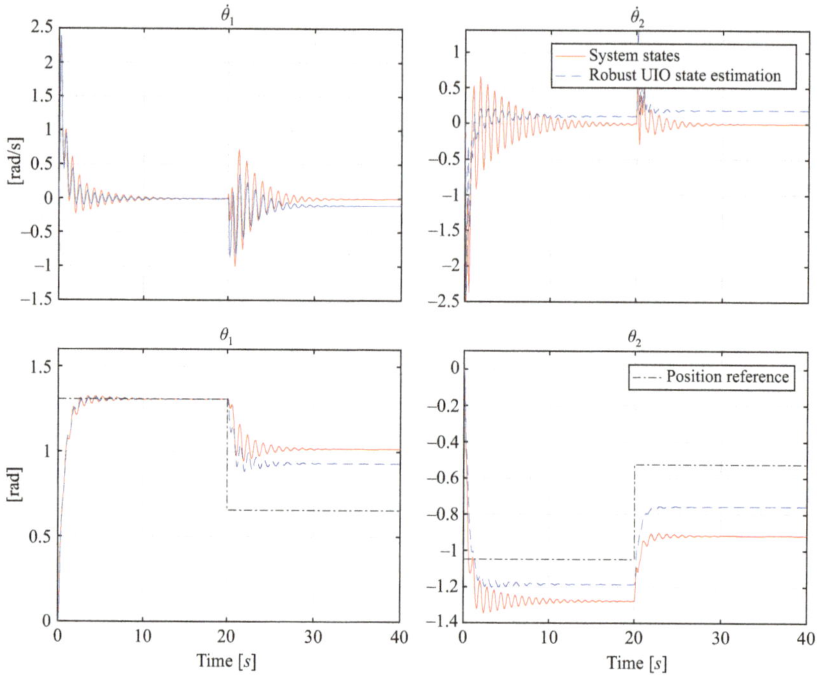

Figure 6.8 System states evolution on fault scenario using the basic control structure with the Luenberger observer

both axes, even when there is no fault torque on θ_1, the oscillatory behaviour still remains. Regarding the angular velocities, besides the oscillatory behaviour, there exists a difference between their estimated and real values, also produced due to the effect of the fault.

6.6.2 Basic control structure with RUIO

From the previous control structure, the substitution of the classical Luenberger observer by the RUIO leads to an overall reduction of the estimation errors along with the cancellation of the oscillatory phenomena in both axes, as is shown in Figure 6.9.

Although there is a remarkable improvement with respect to the previous control structure, the presented behaviour does not correspond to the expected one as the position errors are not null. The RUIO was supposed to completely decouple the effect of the disturbance from the system states by design, as the LMI conditions assure that the estimation error converges to zero. As in the results on Figure 6.8, a nearly equal estimation error appears in both angular velocities. Further analysis has been performed on this issue, finding that it might be related with the design, as SeDuMi presents limited capabilities to solve strict LME.

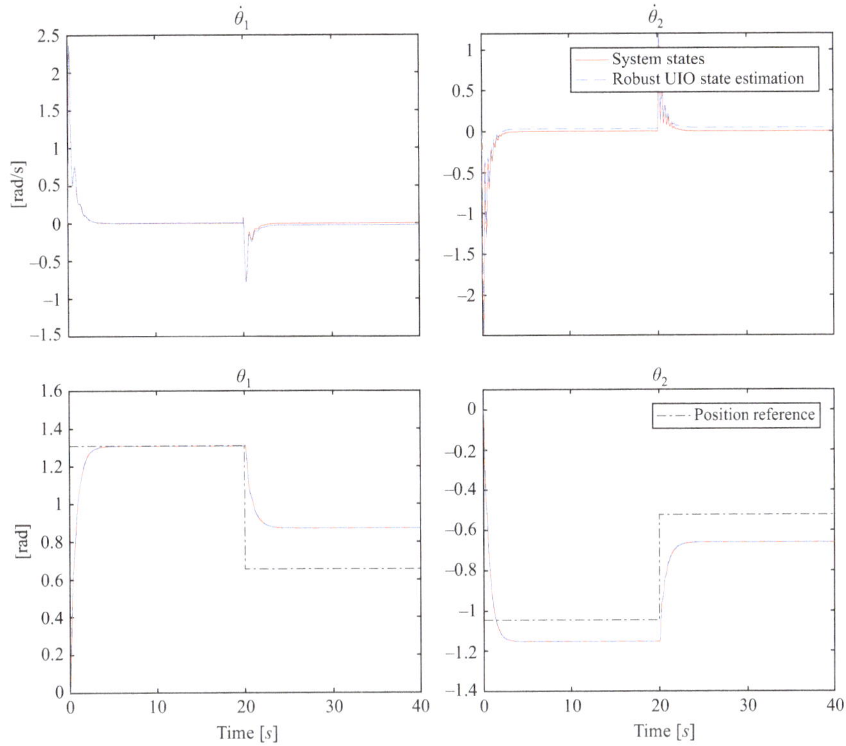

Figure 6.9 System's states evolution on fault scenario using the basic control structure with RUIO

6.6.3 Complete fault-tolerant control scheme

With the complete fault-tolerant control scheme, the magnitude of faults can be estimated using the reference control structure. In Figure 6.10, the magnitudes of the estimated injected torques are presented without including its compensation mechanism. They converge to the real values of the injected faults for both axes, achieving their complete isolation disregarding the existing coupling effects between the axes. Oscillatory effects appear during the transitory phase in both estimations, being more significant on the Tilt axis due to the dependency of the affecting fault in θ_2.

Results in Figure 6.11 correspond to the evolution of the system states for the final control scheme compensating the estimated faults. The evolution of the estimations for this case are included in Figure 6.12.

The obtained results present an improvement with respect to the previous control structure, but still the position error is not null as expected, and there exists a (slightly reduced) difference between estimated and real position velocities. The cause behind this issue for this case is related with the error in the faults' estimation: magnitudes converge to approximately half the values of the current faults being injected. Further

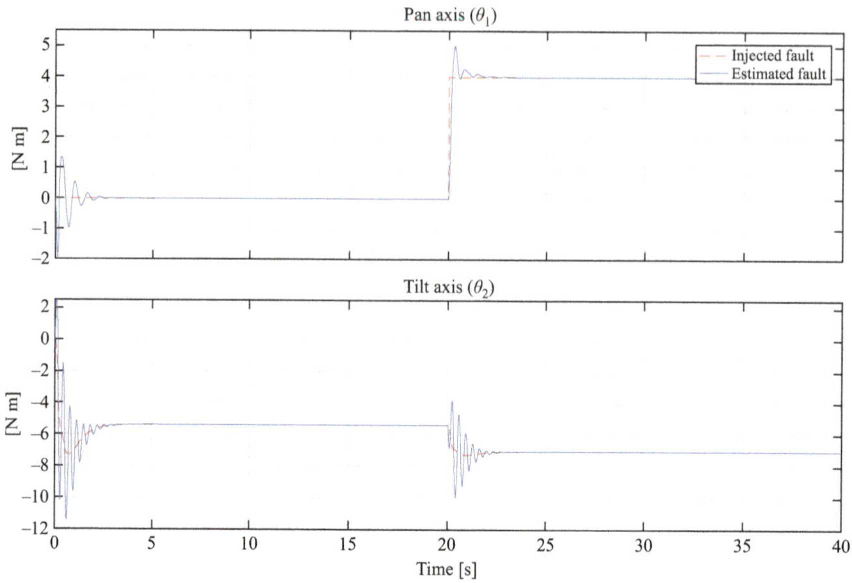

Figure 6.10 Fault estimation in the Pan and Tilt axes without including its compensation mechanism

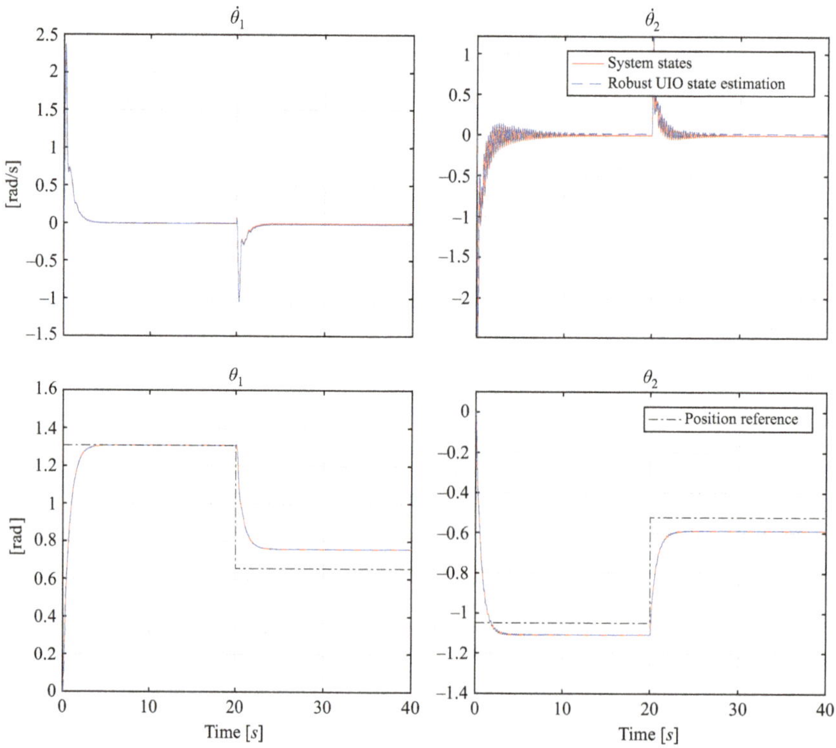

Figure 6.11 System states evolution in fault scenario using the complete fault-tolerant control scheme

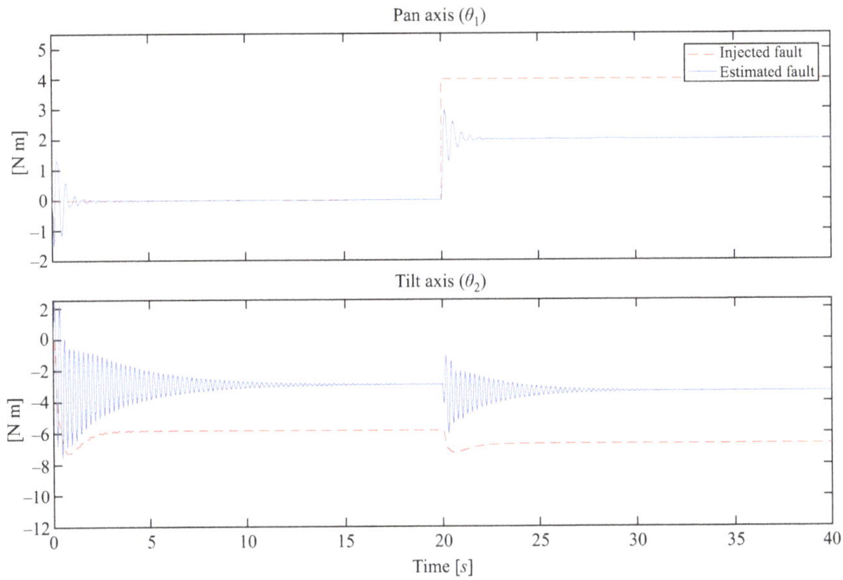

Figure 6.12 Fault estimation in the Pan and Tilt axes, including its compensation mechanism

analysis has been performed, revealing that the estimation within the compensation mechanism has not fast enough dynamics to avoid that part of the disturbance is assumed by the RUIO. Thus, the aforementioned problem derived from the design is induced into the solution, and errors do not converge to zero.

Finally, it should be pointed out that some strategies have been evaluated to overcome this problem, for example weighting the injected estimations by gains greater than the unit in order to initially over-compensate it as in [28]. This strategy led to an improvement in the results, but oscillatory effects (as those present in the Tilt axis estimation in Figure 6.12) compromised the stability of the control scheme, so it had to be discarded.

6.7 Conclusions

In this chapter, the proposed fault-tolerant model-based control approach for the head subsystem of a TIAGo humanoid robot has been proved to tackle the stated problem. Although results show that the problem is not completely answered, used methodology presents a first line of action towards a final fulfilling solution, upon which improvements can be made.

As future work, the proposed scheme will be implemented in the real system available at our labs. Other research opportunities regarding this problem can be related with gaining more information on the fault itself (e.g. in the case of an unexpected collisions, its point of contact) and its integration within a complete autonomous architecture.

Acknowledgement

This work is supported by the Spanish State Research Agency through the María de Maeztu Seal of Excellence to IRI MDM-2016-0656.

References

[1] International Federation of Robotics (IFR). *Executive Summary World Robotics. 2016 Service Robot.*

[2] Gertler J. *Fault Detection and Diagnosis in Engineering Systems*. New York: CRC Press; 1998.

[3] Zhuo-hua D., Zi-xing C., Jin-xia Y. 'Fault diagnosis and fault tolerant control for wheeled mobile robots under unknown environments: A survey'. *Proceedings of the International Conference on Robotics and Automation (ICRA)*. Barcelona: IEEE; 2005.

[4] Khalastchi E., Kalech M. 'On Fault Detection and Diagnosis in Robotic Systems'. *ACM Computing Surveys*. 2018, vol. 51(1), pp. 1–24.

[5] Van M., Kang H. 'Robust Fault-Tolerant Control for Uncertain Robot Manipulators Based on Adaptive Quasi-Continuous High-Order Sliding Mode and Neural Network'. *Proceedings of the Institution of Mechanical Engineers, Part C: Journal of Mechanical Engineering Science*. 2015, vol. 229(8), pp. 1425–1446.

[6] Haddadin S., Albu-Schaffer A., De Luca A., Hirzinger G. 'Collision detection and reaction: A contribution to safe physical human-robot interaction'. *Proceedings of the International Conference on In Intelligent Robots and Systems (IROS)*. Nice: IEEE; 2008; pp. 3356–3363.

[7] Stavrou D., Eliades D. G., Panayiotou C. G., Polycarpou M. M. 'Fault Detection for Service Mobile Robots Using Model-Based Method'. *Autonomous Robots*. 2016, vol. 40, pp. 383–394.

[8] Hornung R., Urbanek H., Klodmann J., Osendorfer C., van der Smagt P. 'Model-free robot anomaly detection'. *Proceedings of the International Conference on Intelligent Robots and Systems (IROS)*. Chicago, IL: IEEE; 2014; pp. 3676–3683.

[9] De Luca A., Ferrajoli L. 'A modified Newton-Euler method for dynamic computations in robot fault detection and control'. *Proceedings of the IEEE International Conference on Robotics and Automation (ICRA)*. Kobe: IEEE; 2009; pp. 3359–3364.

[10] Prassler E., Kazuhiro K. 'Domestic Robotics'. *Springer Handbook of Robotics*. Berlin, Heilderberg: Springer; 2008; pp. 1253–1281.

[11] PAL Robotics. TIAGo Robotic Platform. Available from: http://tiago.pal-robotics.com [Accessed 31 January 2019]

[12] Craig J. J. *Introduction to Robotics. Mechanics and Control.* 3rd edn. London: Pearson Education International; 2005; p. 173.

[13] Zadeh L. A. 'Fuzzy Sets'. *Information and Control*. 1965, vol. 8(3), pp. 338–353.

[14] Takagi T., Sugeno M. 'Fuzzy Identification of Systems and Its Applications to Modelling and Control'. *IEEE Transactions on Systems, Man and cybernetics.* 1985, vol. 1, pp. 116–132.

[15] Sugeno M., Kang G. T. 'Fuzzy Modeling and Control of Multilayer Incinerator' *Fuzzy Sets and Systems.* 1986, vol. 18(3), pp. 329–364.

[16] Tanaka K., Wang H. *Fuzzy Control Systems Design and Analysis.* New York: Wiley; 2010.

[17] Duan G. R., Yu H. H. *LMIs in Control Systems: Analysis, Design and Applications.* Boca Raton, FL: CRC Press; 2013.

[18] Apkarian P., Gahinet P., Becker, G. 'Self-Scheduled H_∞ Control of Linear Parameter-Varying Systems: A Design Example'. *Automatica.* 1995, vol. 31(9), pp. 1251–1261.

[19] Ostertag E. *Mono- and Multivariable Control and Estimation: Linear, Quadratic and LMI Methods.* Berlin, Heilderberg: Springer Science and Business Media; 2011.

[20] Chadli M., Karimi H.R. 'Robust Observer Design for Unknown Inputs Takagi–Sugeno Models'. *IEEE Transactions on Fuzzy Systems.* 2013, vol. 21(1), pp. 158–164.

[21] Chadli M. 'An LMI Approach to Design Observer for Unknown Inputs Takagi-Sugeno Fuzzy Models'. *Asian Journal of Control.* 2010, vol. 12(4), pp. 524–530.

[22] Crestani D., Godary-Dejean K. 'Fault tolerance in control architectures for mobile robots: Fantasy or reality?'. *Presented in CAR: Control Architectures of Robots.* Nancy, France; 2012.

[23] MATLAB. MathWorks. Available from: https://www.mathworks.com/ [Accessed 31 January 2019].

[24] Dormand J. R., Prince P. J. 'A Family of Embedded Runge-Kutta Formulae'. *Journal of Computational and Applied Mathematics.* 1980, vol. 6(1), pp. 19–26.

[25] Lofberg J., 'YALMIP: A toolbox for modeling and optimization in MATLAB', *IEEE International Conference on Robotics and Automation.* Taipei, Taiwan; 2004.

[26] YALMIP toolbox. Available from: https://yalmip.github.io/ [Accessed 31 January 2019].

[27] SeDuMi: Optimization over symmetric cones. Available from: https://github.com/sqlp/sedumi [Accessed 31 January 2019].

[28] Witczak M. *Fault Diagnosis and Fault-Tolerant Control Strategies for Nonlinear Systems.* Lecture Notes in Electrical Engineering (LNEE), no. 266. Switzerland: Springer; 2014; p. 119–142.

Chapter 7

Distributed fault detection and isolation strategy for a team of cooperative mobile manipulators

Giuseppe Gillini[1], Martina Lippi[2], Filippo Arrichiello[1], Alessandro Marino[1], and Francesco Pierri[3]

Applications involving multi-robot systems have been increasing day by day, since they allow one to accomplish complex tasks otherwise impossible for a single unit. Common control approaches for these robot systems are based on distributed architecture, where each robot computes its own control input, only based on local information from onboard sensors or received from its neighbor robots. This means that the failure of one or more agents might jeopardize the task execution. For this reason, fault detection and isolation (FDI) strategies become crucial to accomplish the assigned task in the aforementioned case as well.

This chapter presents a distributed fault diagnosis architecture aimed at detecting failures in a team of robots working in tight cooperation. The proposed approach relies on a distributed observer–controller scheme, where each robot estimates the overall system state by means of a local observer, and it uses such an estimate to compute the local control input to achieve a specific task. The local observer is also used to define a set of residual vectors aimed at detecting and isolating faults occurring on any robot of the team, even if there is no direct communication. The approach is validated through experiments where a heterogeneous team of three robots perform a cooperative task.

7.1 Introduction

The need to accomplish complex and/or heavy tasks in a high-performance manner has led to the massive use of teams of robots in different fields of application, ranging, for example, from exploration, navigation [1], and surveillance [2] to cooperative manipulation and load transportation tasks [3]. More in detail, the adoption of multiple cooperative robots rather than individual robots improves both efficiency and

[1]Department of Electrical and Information Engineering (DIEI), University of Cassino and Southern Lazio, Cassino, Italy
[2]Department of Information and Electrical Engineering and Applied Mathematics (DIEM), University of Salerno, Fisciano, Italy
[3]School of Engineering, University of Basilicata, Potenza, Italy

robustness to faults of the overall system and it allows one to perform tasks otherwise not feasible. In addition, in the case of distributed control architectures, characterized by the lack of a central control unit coordinating the robots of the team, the scalability and the flexibility of the system enhance as well. However, in both centralized and distributed architectures, fault robustness is only potential if not properly handled, i.e., if the coordination strategy relies on the operativity of all robots, the entire mission may be compromised in case a failure occurs. For this reason, appropriate FDI methodologies should be generally included in the control architecture in order to (i) detect the occurrence of a fault and (ii) identify the faulty robot, thus allowing one to online monitor the health of the system and possibly ensure its reliability.

Among the FDI strategies dealing with single-unit systems, observer-based approaches have been widely investigated. For example, the study in [4] presents a centralized FDI scheme in which a bank of nonlinear adaptive estimators is responsible for detecting a fault while a second bank of estimators is responsible for isolating it. In addition, a centralized fault-tolerant control scheme is devised in [5] where both FDI and accommodation schemes for recovering from possible faults are introduced. A reinforcement learning approach is presented in [6] where a fault-tolerant control strategy with discrete-time systems is proposed.

FDI strategies are much more challenging in the case of distributed control architectures where each robot is required to recognize a fault of a teammate exclusively on the basis of information gathered from local sensors and neighboring robots, thus lacking knowledge of global information associated with the overall multi-robot system. In this regard, observer-based approaches are traditionally adopted, which involve the definition of residual signals, obtained by comparing measured and estimated quantities, that are then monitored to recognize and isolate faults. According to this approach, an FDI strategy for nonlinear uncertain systems is designed in [7] which also relies on adaptive thresholds to detect faults. Large-scale systems composed of nonlinear agents are analyzed in [8], where a decomposition into smaller overlapping subsystems is proposed, and local observers are considered for non-overlapping parts whereas a consensus-like strategy is designed for the overlapping parts. An extension of this work dealing with possible delays and packets loss in the intraagent communication can be found in [9]. As a different approach, a bank of unknown input observers is adopted in [10] with second-order linear time-invariant systems, where the existence of these observers is provided for two classes of distributed control laws. Building on this architecture, model uncertainty caused by the removal of communication links or nodes in the system has been then included in [11]. In addition, heterogeneous linear agents with possible simultaneous faults are considered in [12], but the designed strategy only allows each robot to detect its own fault or the one of a direct neighbor. Analogously, the same limitation is present in [13] where an H_∞/H_- formulation is proposed to solve a consensus task with linear systems while detecting possible faults. Furthermore, a leader-tracking task with nonlinear systems is addressed in [14] and both fault detection and accommodation strategies are envisaged; however, similarly to the previous works, the former relies on local observers that exclusively allow one to detect fault of the robot itself. Recently, a particular class of nonlinear systems subject to actuator faults has been analyzed in [15] and a hierarchical architecture for fault estimation and fault-tolerant control with bidirectional information exchange has been devised.

Although much effort has been devoted to distributed FDI schemes, few contributions deal with multi-manipulator systems that are characterized by nonlinear continuous-time dynamics. In this regard, the work in [16] focuses on multiple Euler–Lagrange systems involved in a synchronized set-point regulation problem; to this aim, H_∞ theory is leveraged and an FDI module for actuator faults is introduced in the control architecture. The latter is assumed to be available and only the consequences of its possible imperfections or uncertainties are analyzed. A leader-tracking task with possible communication link faults and actuator faults is discussed in [17] where a fault-tolerant distributed control protocol is proposed, which does not rely on an FDI scheme. Furthermore, a distributed formation control problem with multiple Euler–Lagrange systems is addressed in [18] where a fault-diagnosis strategy is devised as well. Nevertheless, the latter is based on the fact that each robot is able to detect its own possible fault through an observer system and, in the event, it occurs, an alarm signal is broadcasted to the entire network.

Motivated by the above reasons and differently from the existing solutions, this chapter investigates a general distributed framework for multiple mobile-manipulator systems that allows each robot to detect and isolate possible faults of any other robot in the team without the need of direct communication with it.

It is worth remarking that, especially in the case of tight cooperation, the health awareness of the entire team by each robot is a key feature without which the fault of one may even cause damage to others. For this purpose, an observer-controller scheme is devised based on which each robot estimates the overall system state by means of a local observer and, then, it exploits this estimate to compute the local control input in order to achieve the desired cooperative task. Based on the same local observer, residual signals are also defined, which enable one to detect and identify possible faults in the network. The proposed solution builds on the schemes presented in [19,20], where the first focuses on continuous-time single integrator dynamic systems, whereas the latter deals with discrete-time linear systems. Here, these contributions are extended with the following novel contributions:

- The observer-controller scheme is extended to continuous-time Euler–Lagrange systems for which a formal analysis is provided.
- Experiments on a real setup composed of three cooperative mobile robots and manipulators are performed to corroborate the proposed approach.

The remainder of this chapter is organized as follows. Section 7.2 introduces the mathematical background and discusses the problem setting; Section 7.3 presents the distributed controller-observer architecture, while Section 7.4 presents the fault diagnosis and isolation scheme. Section 7.5 shows the experimental results using a team of mobile robots and manipulators. Finally, Section 7.6 draws conclusions and future works.

7.2 Mathematical background and problem setting

7.2.1 Robot modeling

Let us consider a work-cell with N mobile manipulators equipped with a wrist-mounted force/torque sensor and involved in a cooperative task. The dynamic model of the ith manipulator can be expressed as follows:

$$\boldsymbol{M}_i(\boldsymbol{q}_i)\ddot{\boldsymbol{q}}_i + \boldsymbol{C}_i(\boldsymbol{q}_i, \dot{\boldsymbol{q}}_i)\dot{\boldsymbol{q}}_i + \boldsymbol{F}_i\dot{\boldsymbol{q}}_i + \boldsymbol{g}_i(\boldsymbol{q}_i) + \boldsymbol{d}_i(\boldsymbol{q}_i, \dot{\boldsymbol{q}}_i) = \boldsymbol{\tau}_i - \boldsymbol{J}_i^T(\boldsymbol{q}_i)\boldsymbol{h}_i, \tag{7.1}$$

where $\boldsymbol{q}_i(\dot{\boldsymbol{q}}_i, \ddot{\boldsymbol{q}}_i) \in \mathbb{R}^{n_i}$ is the joint position (velocity, acceleration) vector, $\boldsymbol{M}_i \in \mathbb{R}^{n_i \times n_i}$ is the symmetric and positive definite inertia matrix, $\boldsymbol{C}_i \in \mathbb{R}^{n_i \times n_i}$ is the matrix collecting the Coriolis and centrifugal terms, $\boldsymbol{F}_i \in \mathbb{R}^{n_i \times n_i}$ is the matrix modeling the viscous friction, $\boldsymbol{g}_i \in \mathbb{R}^{n_i}$ is the vector of generalized gravitational terms, $\boldsymbol{d}_i \in \mathbb{R}^{n_i}$ is a vector collecting the modeling uncertainties and disturbances (e.g., low-velocity friction, motor electromagnetic disturbances, noise), $\boldsymbol{\tau}_i \in \mathbb{R}^{n_i}$ is the vector of joint torques, $\boldsymbol{J}_i \in \mathbb{R}^{6 \times n_i}$ is the Jacobian matrix and, finally, $\boldsymbol{h}_i \in \mathbb{R}^6$ is the interaction wrench exerted by the robot end effector on the environment. The dynamics in (7.1) can be rewritten in compact form as

$$\boldsymbol{M}_i(\boldsymbol{q}_i)\ddot{\boldsymbol{q}}_i + \boldsymbol{n}_i(\boldsymbol{q}_i, \dot{\boldsymbol{q}}_i) + \boldsymbol{d}_i(\boldsymbol{q}_i, \dot{\boldsymbol{q}}_i) = \boldsymbol{\tau}_i - \boldsymbol{J}_i^T(\boldsymbol{q}_i)\boldsymbol{h}_i, \tag{7.2}$$

where the vector $\boldsymbol{n}_i(\boldsymbol{q}_i, \dot{\boldsymbol{q}}_i) \in \mathbb{R}^{n_i}$ contains Coriolis, centrifugal, friction, and gravitational terms. Let us consider a fault affecting the joint actuators of the ith robot, $\boldsymbol{\phi}_i \in \mathbb{R}^{n_i}$, and thus, in the presence of the fault, (7.2) becomes

$$\boldsymbol{M}_i(\boldsymbol{q}_i)\ddot{\boldsymbol{q}}_i + \boldsymbol{n}_i(\boldsymbol{q}_i, \dot{\boldsymbol{q}}_i) + \boldsymbol{d}_i(\boldsymbol{q}_i, \dot{\boldsymbol{q}}_i) = \boldsymbol{\tau}_i - \boldsymbol{J}_i^T(\boldsymbol{q}_i)\boldsymbol{h}_i + \boldsymbol{\phi}_i. \tag{7.3}$$

Based on (7.3), the equation of motion of the end effector in the Cartesian space can be derived [21], i.e.,

$$\boldsymbol{\Lambda}_i(\boldsymbol{q}_i)\ddot{\boldsymbol{x}}_i + \boldsymbol{\mu}_i(\boldsymbol{x}_i, \dot{\boldsymbol{x}}_i) + \boldsymbol{\xi}_i(\boldsymbol{x}_i, \dot{\boldsymbol{x}}_i) = \boldsymbol{\gamma}_i - \boldsymbol{h}_i + \boldsymbol{J}_{M,i}^T\boldsymbol{\phi}_i, \tag{7.4}$$

where $\boldsymbol{x}_i$ $(\dot{\boldsymbol{x}}_i, \ddot{\boldsymbol{x}}_i)$ is the (6×1) end effector pose (velocity and acceleration) vector of the ith robot, and

$$\boldsymbol{\Lambda}_i = \left[\boldsymbol{J}_i(\boldsymbol{q}_i)\boldsymbol{M}_i^{-1}(\boldsymbol{q}_i)\boldsymbol{J}^T(\boldsymbol{q}_i)\right]^{-1}, \qquad \boldsymbol{\gamma}_i = \boldsymbol{J}_{M,i}^T\boldsymbol{\tau}_i,$$

$$\boldsymbol{\mu}_i = \boldsymbol{J}_{M,i}^T\boldsymbol{n}_i(\boldsymbol{q}_i, \dot{\boldsymbol{q}}_i) - \boldsymbol{\Lambda}_i\dot{\boldsymbol{J}}_i(\boldsymbol{q})\dot{\boldsymbol{q}}_i, \quad \boldsymbol{\xi}_i = \boldsymbol{J}_{M,i}^T\boldsymbol{d}_i(\boldsymbol{q}_i, \dot{\boldsymbol{q}}_i),$$

with $\boldsymbol{\Lambda}_i$ the inertia matrix in the operational space and $\boldsymbol{J}_{M,i}$ a dynamically consistent generalized inverse of $\boldsymbol{J}_i$ given by

$$\boldsymbol{J}_{M,i} = \boldsymbol{M}_i^{-1}(\boldsymbol{q}_i)\boldsymbol{J}_i^T(\boldsymbol{q}_i)\boldsymbol{\Lambda}_i.$$

Remark 7.1. *It is worth noticing that the fault $\boldsymbol{\phi}_i$ does not affect the end effector dynamics if it belongs to the null space of the matrix $\boldsymbol{J}_{M,i}^T$. Thus, such faults cannot be detected by the proposed method since they affect the internal dynamics of a given manipulator and cannot be visible to the other ones through the observation of the dynamics at the end effector.*

Remark 7.2. *The devised diagnosis method can be applied also for detecting and isolating faults affecting the force/torque sensor, $\boldsymbol{\phi}_h \in \mathbb{R}^6$. In this chapter, they are not considered for the sake of brevity.*

By considering the following state vector

$$z_i = \begin{bmatrix} \boldsymbol{x}_i \\ \dot{\boldsymbol{x}}_i \end{bmatrix} \in \mathbb{R}^{12}, \tag{7.5}$$

the end effector equation of motion (7.4) can be written in the state space form as

$$\begin{aligned} \dot{z}_{i,1} &= z_{i,2} \\ \dot{z}_{i,2} &= \boldsymbol{\Lambda}_i^{-1}(z_i)\left(\boldsymbol{\gamma}_i - \boldsymbol{h}_i + \boldsymbol{J}_{M,i}^T\boldsymbol{\phi}_i - \boldsymbol{\mu}_i(z_i) - \boldsymbol{\xi}_i(z_i)\right), \end{aligned} \tag{7.6}$$

which can be then rearranged in matrix form as follows:

$$\dot{z}_i = \boldsymbol{A}z_i + \boldsymbol{B}_i(z_i)(\boldsymbol{\gamma}_i - \boldsymbol{h}_i + \boldsymbol{J}_{M,i}^T\boldsymbol{\phi}_i - \boldsymbol{\mu}_i(z_i) - \boldsymbol{\xi}_i(z_i)), \tag{7.7}$$

with

$$\boldsymbol{A} = \begin{bmatrix} \boldsymbol{O}_6 & \boldsymbol{I}_6 \\ \boldsymbol{O}_6 & \boldsymbol{O}_6 \end{bmatrix}, \qquad \boldsymbol{B}_i(z_i) = \begin{bmatrix} \boldsymbol{O}_6 \\ \boldsymbol{\Lambda}_i^{-1}(z_i) \end{bmatrix},$$

being $\boldsymbol{O}_m$ and $\boldsymbol{I}_m$ the $(m \times m)$ zero and identity matrices, respectively.

The following assumption on the mobile manipulators is made in this chapter.

Assumption 7.1. *For each robot, the vector collecting the modeling uncertainties and disturbances in the operational space, $\boldsymbol{\xi}_i$, is norm-bounded by a positive scalar $\bar{\xi}$, i.e., it holds $\|\boldsymbol{\xi}_i\| \leq \bar{\xi} \quad \forall i = 1, \ldots, N$.*

7.2.2 Communication

As usual in multi-robot systems, the robot information exchange is modeled via a connectivity graph [22,23], $\mathcal{G}(\mathcal{E}, \mathcal{V})$, where $\mathcal{V}$ is a set of indexes labeling the N vertices (nodes) representing the robots, and $\mathcal{E} = \mathcal{V} \times \mathcal{V}$ is the set of edges (arcs) representing the communications between robots. In detail, robot i and robot j can communicate only if there is an arc between node i and node j. From the mathematical point of view, the graph topology is represented by the $(N \times N)$ adjacency matrix

$$\boldsymbol{W} = \{w_{ij}\} : \quad w_{ii} = 0, \quad w_{ij} = \begin{cases} 1 & \text{if}(j, i) \in \mathcal{E} \\ 0 & \text{otherwise,} \end{cases}$$

where $w_{ij} = 1$ if there exists an arc from vertex j to vertex i; such matrix is symmetric in the case of undirected graphs, i.e., in case all the communication links among the robots are bidirectional. The communication topology can be also characterized by the $(N \times N)$ Laplacian matrix [24,25] defined as

$$\boldsymbol{L} = \{l_{ij}\} : \quad l_{ii} = \sum_{j=1, j \neq i}^{N} w_{ij}, \quad l_{ij} = -w_{ij}, \; i \neq j.$$

The Laplacian matrix is often preferred in the field of multi-agent systems due to the following property.

Property 7.1. *All eigenvalues of* $\boldsymbol{L}$ *have real part equal to or greater than zero; moreover* $\boldsymbol{L}$ *exhibits at least a zero eigenvalue with corresponding right eigenvector the* $N \times 1$ *vector of all ones* $\boldsymbol{1}_N$*. Hence, rank* $(\boldsymbol{L}) \leq N - 1$*, with rank* $(\boldsymbol{L}) = N - 1$ *if and only if the graph is strongly connected, i.e., if any two distinct nodes of the graph can be connected via a directed path, and* $\boldsymbol{L}\boldsymbol{1}_N = \boldsymbol{0}_N$*, where* $\boldsymbol{0}_N$ *is the* $(N \times 1)$ *null vector.*

The following assumptions about the communication hold in the rest of this chapter:

- The ith robot receives information only from its neighbors $\mathcal{N}_i = \{j \in \mathcal{V} : (j, i) \in \mathcal{E}\}$. The cardinality of $\mathcal{N}_i$ is the in-degree of node i, i.e., $d_i = |\mathcal{N}_i| = \sum_{j=1}^{N} w_{ij}$. Moreover, the cardinality of the set of nodes receiving information from node i represents the out-degree of node i, i.e., $D_i = \sum_{k=1}^{N} w_{ki}$.
- The graph topology is assumed fixed, i.e., there are no communication links that can appear or disappear over the time.

7.2.3 Problem description

We can now formally state the problem addressed in this work.

Problem 7.1. *Let us consider a team of N mobile manipulators with dynamics in* (7.1) *for which a cooperative task is assigned. Let us assume a central control unit is not present and robot i has a fault at a certain time* t_f*, i.e.,* $\|\boldsymbol{\phi}_i(t_f)\| > 0$*. Our objective is to enable each robot in the team (i) to detect that a fault has occurred and (ii) to identify which robot is in fault.*

7.3 Observer and controller scheme

Let us assume that the robots in the work-cell are in charge of achieving a global task depending on the overall state $\boldsymbol{z} = [\boldsymbol{z}_1^T \quad \boldsymbol{z}_2^T \quad \cdots \quad \boldsymbol{z}_N^T]^T \in \mathbb{R}^{12N}$ of the cell. The observer-controller scheme proposed for the solution of Problem 7.1 considers that each robot first estimates the overall state of the cell through an observer that is based on the information exchange with the neighboring robots; then, the state estimate is used to calculate the input that allows one to achieve a global tasks as in a centralized configuration. The overall distributed control architecture of the ith robot, combining both the observer-controller and the fault detection schemes, is reported in Figure 7.1.

It is considered that a local fault signal is generated by robot i in case a fault is detected in such a way that it can be exploited to determine the local control input, e.g., the robot may stop when a fault of a teammate is detected. In the section, we first present the centralized solution to achieve cooperative tasks and then we extend

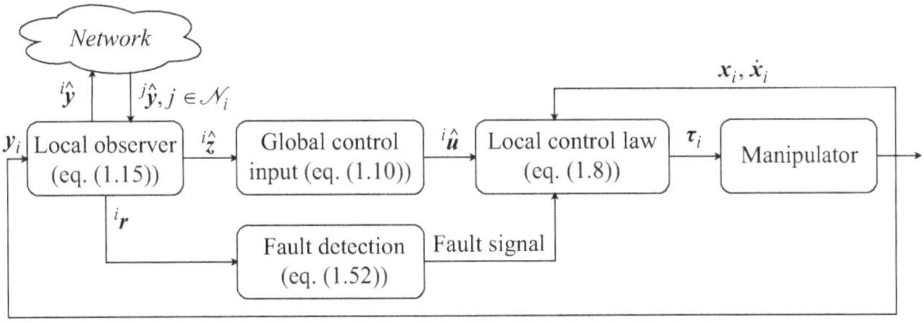

Figure 7.1 Distributed control architecture of the ith robot

it to a decentralized framework through the proposed observer-controller scheme. For each robot, let us assume that the following control law is adopted

$$\boldsymbol{\gamma}_i = \boldsymbol{\Lambda}_i(\boldsymbol{z}_i)\,(\boldsymbol{u}_i + \boldsymbol{\mu}_i(\boldsymbol{z}_i) + \boldsymbol{h}_i)\,, \tag{7.8}$$

where $\boldsymbol{u}_i$ is an auxiliary input depending on the overall state $\boldsymbol{z}$ which is not available in a distributed setting. In the case of a centralized system, in which a central unit, capable of computing the control input for each robot, is present, the collective auxiliary input

$$\boldsymbol{u} = [\boldsymbol{u}_1^T \quad \boldsymbol{u}_2^T \quad \cdots \quad \boldsymbol{u}_N^T] \in \mathbb{R}^{6N}$$

for achieving a global task would assume the following form

$$\boldsymbol{u} = \boldsymbol{K}\boldsymbol{z} + \boldsymbol{u}_f, \tag{7.9}$$

where $\boldsymbol{K} \in \mathbb{R}^{6N \times 12N}$ is a constant gain matrix and $\boldsymbol{u}_f \in \mathbb{R}^{6N}$ represents a feedforward term [26].

In the absence of the central unit, the computation of the control law (7.9) requires that the ith robot estimates the overall cell state in order to use such an estimate, denoted as ${}^i\hat{\boldsymbol{z}}$, in lieu of $\boldsymbol{z}$ (see, e.g., [27]). More specifically, based on the estimate ${}^i\hat{\boldsymbol{z}}$, the ith robot computes an estimate ${}^i\hat{\boldsymbol{u}}_g$ of the collective control input (7.9), i.e.,

$${}^i\hat{\boldsymbol{u}}_g = \boldsymbol{K}\,{}^i\hat{\boldsymbol{z}} + \boldsymbol{u}_f. \tag{7.10}$$

Moreover, in the decentralized solution, a local stabilizing term must be added in order to ensure the convergence of the state estimation to the actual value. Therefore, the effective auxiliary control input is obtained by considering the following selection matrix:

$$\boldsymbol{\Gamma}_{u_i} = \{\boldsymbol{O}_6 \ \cdots \ \underbrace{\boldsymbol{I}_6}_{i\text{th node}} \ \cdots \ \boldsymbol{O}_6\} \in \mathbb{R}^{6\times 6N} \tag{7.11}$$

such that

$$\boldsymbol{u}_i = \boldsymbol{u}_{g,i} + \boldsymbol{u}_{s,i} = \boldsymbol{\Gamma}_{u_i}{}^i\hat{\boldsymbol{u}}_g + \boldsymbol{K}_s\boldsymbol{z}_i, \tag{7.12}$$

where $\boldsymbol{K}_s \in \mathbb{R}^{6\times 12}$ is a constant gain matrix. Based on (7.8), (7.7) can be rearranged as

$$\begin{aligned}\dot{\boldsymbol{z}}_i &= \boldsymbol{A}\boldsymbol{z}_i + \boldsymbol{C}(\boldsymbol{u}_{g,i} + \boldsymbol{u}_{s,i}) + \boldsymbol{B}_i(-\boldsymbol{\xi}_i(\boldsymbol{z}_i) + \boldsymbol{J}_{M,i}^T\boldsymbol{\phi}_i)\\ &= \boldsymbol{F}\boldsymbol{z}_i + \boldsymbol{C}\boldsymbol{u}_{g,i} + \boldsymbol{B}_i(-\boldsymbol{\xi}_i(\boldsymbol{z}_i) + \boldsymbol{J}_{M,i}^T\boldsymbol{\phi}_i)\end{aligned} \tag{7.13}$$

with $\boldsymbol{C} = [\boldsymbol{O}_6 \quad \boldsymbol{I}_6]^T$ and $\boldsymbol{F} = \boldsymbol{A} + \boldsymbol{C}\boldsymbol{K}_s$, which leads to the following dynamics of the collective state:

$$\dot{\boldsymbol{z}} = (\boldsymbol{I}_N \otimes \boldsymbol{F})\boldsymbol{z} + (\boldsymbol{I}_N \otimes \boldsymbol{C})\boldsymbol{u}_g + \bar{\boldsymbol{B}}(\boldsymbol{\phi}_M - \boldsymbol{\xi}) = \bar{\boldsymbol{F}}\boldsymbol{z} + \bar{\boldsymbol{C}}\boldsymbol{u}_g + \bar{\boldsymbol{B}}(\boldsymbol{\phi}_M - \boldsymbol{\xi}), \tag{7.14}$$

where the symbol $\otimes$ denotes the Kronecker product and

$$\boldsymbol{u}_g = \begin{bmatrix}\boldsymbol{u}_{g,1}\\ \vdots \\ \boldsymbol{u}_{g,N}\end{bmatrix}, \bar{\boldsymbol{B}} = \begin{bmatrix}\boldsymbol{B}_1(\boldsymbol{z}_1)\\ \vdots \\ \boldsymbol{B}_N(\boldsymbol{z}_N)\end{bmatrix}, \boldsymbol{\phi}_M = \begin{bmatrix}\boldsymbol{J}_{M,1}^T\boldsymbol{\phi}_1\\ \vdots \\ \boldsymbol{J}_{M,N}^T\boldsymbol{\phi}_N\end{bmatrix}, \boldsymbol{\xi} = \begin{bmatrix}\boldsymbol{\xi}_1\\ \vdots \\ \boldsymbol{\xi}_N\end{bmatrix}.$$

7.3.1 Collective state estimation

In order to estimate the collective state, $\boldsymbol{z}$, each robot runs an observer requiring only local information provided by robot sensors and information received from neighbor robots. Moreover, the same observer system is exploited for both control purposes and FDI strategy, without increasing the information exchange burden.

The observer of robot i has the following dynamics:

$${}^i\dot{\hat{\boldsymbol{z}}} = k_o\left(\sum_{j\in\mathcal{N}_i}({}^j\hat{\boldsymbol{y}} - {}^i\hat{\boldsymbol{y}}) + \boldsymbol{\Pi}_i(\boldsymbol{y} - {}^i\hat{\boldsymbol{y}})\right) + \bar{\boldsymbol{C}}\,{}^i\hat{\boldsymbol{u}}_g + \bar{\boldsymbol{F}}\,{}^i\hat{\boldsymbol{z}}, \tag{7.15}$$

where ${}^i\hat{\boldsymbol{z}}$ is the estimate of the collective state made by robot i and $\boldsymbol{\Pi}_i \in \mathbb{R}^{12N\times 12N}$ is a selection matrix defined as

$$\boldsymbol{\Pi}_i = \mathrm{diag}\{\boldsymbol{O}_{12}, \ldots, \underbrace{\boldsymbol{I}_{12}}_{i\text{th robot}}, \ldots, \boldsymbol{O}_{12}\}, \tag{7.16}$$

which selects the components of $\boldsymbol{y}$ and ${}^i\hat{\boldsymbol{y}}$ referred to robot i. The variable $\boldsymbol{y}$ is an auxiliary state defined as

$$\boldsymbol{y} = \boldsymbol{z} - \int_{t_0}^{t}\left(\bar{\boldsymbol{F}}\boldsymbol{z} + \bar{\boldsymbol{C}}\boldsymbol{u}_g\right)d\sigma, \tag{7.17}$$

where t_0 is the initial time instant. Its estimate ${}^i\hat{\boldsymbol{y}}$ in (7.15) is then obtained as

$${}^i\hat{\boldsymbol{y}} = {}^i\hat{\boldsymbol{z}} - \int_{t_0}^{t}\left(\bar{\boldsymbol{F}}\,{}^i\hat{\boldsymbol{z}} + \bar{\boldsymbol{C}}\,{}^i\hat{\boldsymbol{u}}_g\right)d\sigma, \tag{7.18}$$

which depends only on local information available to robot i. It is worth noticing that each observer is updated using only the estimates ${}^j\hat{\boldsymbol{y}}$ received from direct neighbors, which is, thus, the only information exchanged among neighbors.

The collective estimation dynamics is

$$\dot{\hat{\boldsymbol{z}}}^\star = -k_o\boldsymbol{L}^\star\hat{\boldsymbol{y}}^\star + k_o\boldsymbol{\Pi}^\star\tilde{\boldsymbol{y}}^\star + \boldsymbol{I}_N \otimes \bar{\boldsymbol{C}}\hat{\boldsymbol{u}}_g^\star + \boldsymbol{I}_N \otimes \bar{\boldsymbol{F}}\hat{\boldsymbol{z}}^\star, \tag{7.19}$$

where

$$\begin{aligned}
\boldsymbol{L}^\star &= \boldsymbol{L} \otimes \boldsymbol{I}_{12N} \in \mathbb{R}^{12N^2 \times 12N^2},\\
\boldsymbol{\Pi}^\star &= \mathrm{diag}\{\boldsymbol{\Pi}_1, \ldots, \boldsymbol{\Pi}_N\} \in \mathbb{R}^{12N^2 \times 12N^2},\\
\hat{\boldsymbol{z}}^\star &= \left[{}^1\hat{\boldsymbol{z}}^T \ \cdots \ {}^N\hat{\boldsymbol{z}}^T\right]^T \in \mathbb{R}^{12N^2},\\
\hat{\boldsymbol{y}}^\star &= \left[{}^1\hat{\boldsymbol{y}}^T \ \cdots \ {}^N\hat{\boldsymbol{y}}^T\right]^T \in \mathbb{R}^{12N^2},\\
\hat{\boldsymbol{u}}_g^\star &= \left[{}^1\hat{\boldsymbol{u}}_g^T \ \cdots \ {}^N\hat{\boldsymbol{u}}_g^T\right]^T \in \mathbb{R}^{6N^2},\\
\tilde{\boldsymbol{y}}^\star &= \boldsymbol{1}_N \otimes \boldsymbol{y} - \hat{\boldsymbol{y}}^\star.
\end{aligned}$$

7.3.2 Observer convergence

Based on the properties of the Lagrangian matrix and by using the same arguments adopted in [19], the following lemma can be proven.

Lemma 7.1. *In the presence of a strongly connected directed communication graph and in the absence of faults and model uncertainties (i.e.,* $\|\boldsymbol{\phi}_M\| = 0$ *and* $\|\boldsymbol{\xi}\| = 0$*), the term* $\tilde{\boldsymbol{y}}^\star$ *is globally exponentially convergent to zero, given the observer* (7.15) *and for any* $k_o > 0$.

Proof. By differentiating (7.18) and after some straightforward steps, the following holds

$$\dot{\hat{\boldsymbol{y}}}^\star = \dot{\hat{\boldsymbol{z}}}^\star - \boldsymbol{I}_N \otimes \bar{\boldsymbol{C}}\hat{\boldsymbol{u}}_g^\star - \boldsymbol{I}_N \otimes \bar{\boldsymbol{F}}\hat{\boldsymbol{z}}^\star, \tag{7.20}$$

thus, from (7.19) it follows

$$\dot{\hat{\boldsymbol{y}}}^\star = -k_o\boldsymbol{L}^\star\hat{\boldsymbol{y}}^\star + k_o\boldsymbol{\Pi}^\star\tilde{\boldsymbol{y}}^\star. \tag{7.21}$$

By exploiting the following property of the Laplacian matrix [24],

$$\boldsymbol{L}^\star(\boldsymbol{1}_N \otimes \boldsymbol{y}) = (\boldsymbol{L}^\star \otimes \boldsymbol{I}_{12N})(\boldsymbol{1}_N \otimes \boldsymbol{y}) = (\boldsymbol{L}\boldsymbol{1}_N) \otimes (\boldsymbol{I}_{12N}\boldsymbol{y}) = \boldsymbol{0}_{12N^2}, \tag{7.22}$$

(7.21) can be rearranged as

$$\dot{\hat{\boldsymbol{y}}}^\star = -k_o\left(\boldsymbol{L}^\star + \boldsymbol{\Pi}^\star\right)\tilde{\boldsymbol{y}}^\star = -k_o\tilde{\boldsymbol{L}}^\star\tilde{\boldsymbol{y}}^\star. \tag{7.23}$$

By considering from (7.14) that $\dot{\boldsymbol{y}} = \bar{\boldsymbol{B}}(\boldsymbol{z})\left(\boldsymbol{\phi}_M - \boldsymbol{\xi}\right)$, it holds

$$\dot{\tilde{\boldsymbol{y}}}^\star = \boldsymbol{1}_N \otimes \dot{\boldsymbol{y}} - \dot{\hat{\boldsymbol{y}}}^\star = -k_o\tilde{\boldsymbol{L}}^\star\tilde{\boldsymbol{y}}^\star + \boldsymbol{\phi}^\star - \boldsymbol{\xi}^\star, \tag{7.24}$$

where $\boldsymbol{\phi}^\star = \mathbf{1}_N \otimes \bar{\boldsymbol{B}}(\boldsymbol{z})\boldsymbol{\phi}_M$ and $\boldsymbol{\xi}^\star = \mathbf{1}_N \otimes \bar{\boldsymbol{B}}(\boldsymbol{z})\boldsymbol{\xi}$. Finally, in the absence of faults and uncertainties, the dynamics of $\tilde{\boldsymbol{y}}^\star$ is

$$\dot{\tilde{\boldsymbol{y}}}^\star = \mathbf{1}_N \otimes \dot{\boldsymbol{y}} - \dot{\hat{\boldsymbol{y}}}^\star = -k_o \tilde{\boldsymbol{L}}^\star \tilde{\boldsymbol{y}}^\star. \tag{7.25}$$

By leveraging the results in [28], it can be stated that $-\tilde{\boldsymbol{L}}^\star$ is Hurwitz provided that the communication graph is strongly connected. Thus, in the absence of faults and model uncertainties, (7.25) proves the lemma $\forall k_o > 0$. □

In the presence of bounded uncertainties $\boldsymbol{\xi}$ and on the basis of the results in [29], it can be easily proven that the system (7.19) is globally uniformly ultimately bounded (see [19]).

Based on Lemma 7.1, under the same assumptions, the following theorem states the convergence properties of the collective state estimation error $\tilde{\boldsymbol{z}}^\star = \mathbf{1}_N \otimes \boldsymbol{z} - \hat{\boldsymbol{z}}^\star$.

Theorem 7.1. *In the presence of a strongly connected directed communication graph and in the absence of faults and model uncertainties (i.e., $\|\boldsymbol{\phi}_M\| = 0$ and $\|\boldsymbol{\xi}\| = 0$), the term $\tilde{\boldsymbol{z}}^\star$ is convergent to zero, given the observer* (7.15) *for any $k_o > 0$ and for a proper choice of the controller gain matrices $\boldsymbol{K}$ and $\boldsymbol{K}_s$ in* (7.12).

Proof. By considering the dynamics of $\boldsymbol{z}$ in (7.14) and of $\hat{\boldsymbol{z}}^\star$ in (7.19), and the Laplacian matrix property (7.22), in the absence of faults and uncertainties, the dynamics of $\tilde{\boldsymbol{z}}^\star$ can be written as

$$\dot{\tilde{\boldsymbol{z}}}^\star = -k_o \tilde{\boldsymbol{L}}^\star \tilde{\boldsymbol{y}}^\star + \boldsymbol{I}_N \otimes \bar{\boldsymbol{C}} \tilde{\boldsymbol{u}}_g^\star + \boldsymbol{I}_N \otimes \bar{\boldsymbol{F}} \tilde{\boldsymbol{z}}^\star, \tag{7.26}$$

where $\tilde{\boldsymbol{u}}_g^\star = \mathbf{1}_N \otimes \boldsymbol{u}_g - \hat{\boldsymbol{u}}_g^\star$.

By virtue of Lemma 7.1, the term $-k_o \tilde{\boldsymbol{L}}^\star \tilde{\boldsymbol{y}}^\star$ is a disturbance converging to zero at steady state; thus the convergence properties of $\tilde{\boldsymbol{z}}^\star$ are those of the non-perturbed system

$$\dot{\tilde{\boldsymbol{z}}}^\star = \boldsymbol{I}_N \otimes \bar{\boldsymbol{C}} \tilde{\boldsymbol{u}}_g^\star + \boldsymbol{I}_N \otimes \bar{\boldsymbol{F}} \tilde{\boldsymbol{z}}^\star. \tag{7.27}$$

Based on (7.12), the auxiliary global input $\boldsymbol{u}_g$ can be written as

$$\boldsymbol{u}_g = \sum_{l=1}^{N} \boldsymbol{\Gamma}_{u_l}^T \boldsymbol{\Gamma}_{u_l} {}^l\hat{\boldsymbol{u}}_g = \sum_{l=1}^{N} \boldsymbol{\Pi}_{u_l} {}^l\hat{\boldsymbol{u}}_g, \tag{7.28}$$

where $\boldsymbol{\Pi}_{u_l} \in \mathbb{R}^{6N \times 6N}$ is a matrix that nullifies all the elements of ${}^l\hat{\boldsymbol{u}}_g$ but $\boldsymbol{u}_{g,l}$. Therefore, the ith element of $\tilde{\boldsymbol{u}}_g^\star$ can be written as

$$\begin{aligned} {}^i\tilde{\boldsymbol{u}}_g &= \sum_{l=1}^{N} \boldsymbol{\Pi}_{u_l}(\boldsymbol{K}\,{}^l\hat{\boldsymbol{z}} + \boldsymbol{u}_f) - \boldsymbol{K}\,{}^i\hat{\boldsymbol{z}} - \boldsymbol{u}_f = \sum_{l=1}^{N} \boldsymbol{\Pi}_{u_l} \boldsymbol{K}({}^l\hat{\boldsymbol{z}} \pm \boldsymbol{z}) - \boldsymbol{K}\,{}^i\hat{\boldsymbol{z}} \\ &= -\sum_{l=1}^{N} \boldsymbol{\Pi}_{u_l} \boldsymbol{K}\,{}^l\tilde{\boldsymbol{z}} + \boldsymbol{K}\,{}^i\tilde{\boldsymbol{z}}, \end{aligned} \tag{7.29}$$

where the property $\sum_{l=1}^{N} \boldsymbol{\Pi}_{u_l} \boldsymbol{v} = \boldsymbol{v}$ for any vector $\boldsymbol{v}$ of proper dimensions has been exploited. From (7.29), $\tilde{\boldsymbol{u}}_g^\star$ is given by

$$\tilde{\boldsymbol{u}}_g^\star = (\boldsymbol{I}_N \otimes \boldsymbol{K})\tilde{\boldsymbol{z}}^\star - \boldsymbol{\Pi}_u^\star(\boldsymbol{I}_N \otimes \boldsymbol{K})\tilde{\boldsymbol{z}} = (\boldsymbol{I}_{6N^2} - \boldsymbol{\Pi}_u^\star)(\boldsymbol{I}_N \otimes \boldsymbol{K})\tilde{\boldsymbol{z}}^\star, \tag{7.30}$$

where $\boldsymbol{\Pi}_u^\star = \boldsymbol{1}_N \otimes \left[\boldsymbol{\Pi}_{u_1}\boldsymbol{\Pi}_{u_2}\cdots\boldsymbol{\Pi}_{u_N}\right] \in \mathbb{R}^{6N^2\times 6N^2}$. It can be trivially verified that the following property holds for the matrix $\boldsymbol{I}_{6N^2} - \boldsymbol{\Pi}_u^\star$.

Property 7.2. *Matrix* $\boldsymbol{I}_{6N^2} - \boldsymbol{\Pi}_u^\star$ *is idempotent, it is characterized by* $6N$ *null rows and* $\operatorname{rank}(\boldsymbol{I}_{6N^2} - \boldsymbol{\Pi}_u^\star) = 6N(N-1)$.

Based on the Property 7.2, there exists a nonsingular permutation matrix $\boldsymbol{P} \in \mathbb{R}^{N^2\times N^2}$ such as

$$(\boldsymbol{P} \otimes \boldsymbol{I}_6)(\boldsymbol{I}_{6N^2} - \boldsymbol{\Pi}_u^\star) = \left[\frac{\boldsymbol{0}_{6N\times 6N^2}}{\boldsymbol{S}}\right], \tag{7.31}$$

where $\boldsymbol{S} \in \mathbb{R}^{6N(N-1)\times 6N^2}$ is a full row rank matrix, i.e., $\operatorname{rank}(\boldsymbol{S}) = 6N(N-1)$. From (7.27) and (7.30), the dynamics of $\tilde{\boldsymbol{z}}^\star$ is

$$\dot{\tilde{\boldsymbol{z}}}^\star = (\boldsymbol{I}_N \otimes \bar{\boldsymbol{F}})\tilde{\boldsymbol{z}}^\star + (\boldsymbol{I}_N \otimes \bar{\boldsymbol{C}})(\boldsymbol{I}_{6N^2} - \boldsymbol{\Pi}_u^\star)(\boldsymbol{I}_N \otimes \boldsymbol{K})\tilde{\boldsymbol{z}}^\star. \tag{7.32}$$

Then, by considering the state transformation $\tilde{\boldsymbol{\omega}}^\star = (\boldsymbol{P} \otimes \boldsymbol{I}_{12})\tilde{\boldsymbol{z}}^\star$, the system (7.32) can be rewritten as

$$\begin{aligned}\dot{\tilde{\boldsymbol{\omega}}}^\star &= (\boldsymbol{P} \otimes \boldsymbol{I}_{12})(\boldsymbol{I}_N \otimes \bar{\boldsymbol{F}})(\boldsymbol{P} \otimes \boldsymbol{I}_{12})^{-1}\tilde{\boldsymbol{\omega}}^\star \\ &\quad + (\boldsymbol{P} \otimes \boldsymbol{I}_{12})(\boldsymbol{I}_N \otimes \bar{\boldsymbol{C}})(\boldsymbol{I}_{6N^2} - \boldsymbol{\Pi}_u^\star)(\boldsymbol{I}_N \otimes \boldsymbol{K})(\boldsymbol{P} \otimes \boldsymbol{I}_{12})^{-1}\tilde{\boldsymbol{\omega}}^\star.\end{aligned} \tag{7.33}$$

By exploiting the *mixed-product property* of the Kronecker product [30]

$$(\boldsymbol{D}_1 \otimes \boldsymbol{D}_2)(\boldsymbol{D}_3 \otimes \boldsymbol{D}_4) = (\boldsymbol{D}_1\boldsymbol{D}_3) \otimes (\boldsymbol{D}_2\boldsymbol{D}_4) \tag{7.34}$$

with matrices $\boldsymbol{D}_1, \boldsymbol{D}_2, \boldsymbol{D}_3$, and $\boldsymbol{D}_4$ of proper dimensions, the following chains of equalities hold

$$\begin{aligned}(\boldsymbol{P} \otimes \boldsymbol{I}_{12})(\boldsymbol{I}_N \otimes \bar{\boldsymbol{F}})(\boldsymbol{P} \otimes \boldsymbol{I}_{12})^{-1} &= (\boldsymbol{P} \otimes \boldsymbol{I}_{12})(\boldsymbol{I}_{N^2} \otimes \boldsymbol{F})(\boldsymbol{P}^{-1} \otimes \boldsymbol{I}_{12}) \\ &= (\boldsymbol{P} \otimes \boldsymbol{F})(\boldsymbol{P}^{-1} \otimes \boldsymbol{I}_{12}) = \boldsymbol{I}_{N^2} \otimes \boldsymbol{F} = \boldsymbol{I}_N \otimes \bar{\boldsymbol{F}},\end{aligned} \tag{7.35}$$

$$\begin{aligned}(\boldsymbol{P} \otimes \boldsymbol{I}_{12})(\boldsymbol{I}_N \otimes \bar{\boldsymbol{C}})(\boldsymbol{I}_{6N^2} - \boldsymbol{\Pi}_u^\star) &= (\boldsymbol{P} \otimes \boldsymbol{I}_{12})(\boldsymbol{I}_{N^2} \otimes \boldsymbol{C})(\boldsymbol{I}_{6N^2} - \boldsymbol{\Pi}_u^\star) \\ &= (\boldsymbol{P} \otimes \boldsymbol{C})(\boldsymbol{I}_{6N^2} - \boldsymbol{\Pi}_u^\star) = (\boldsymbol{I}_{N^2} \otimes \boldsymbol{C})(\boldsymbol{P} \otimes \boldsymbol{I}_6)(\boldsymbol{I}_{6N^2} - \boldsymbol{\Pi}_u^\star) \\ &= (\boldsymbol{I}_{N^2} \otimes \boldsymbol{C})\left[\frac{\boldsymbol{0}_{6N\times 6N^2}}{\boldsymbol{S}}\right] = \left[\frac{\boldsymbol{0}_{12N\times 6N^2}}{(\boldsymbol{I}_{N(N-1)} \otimes \boldsymbol{C})\boldsymbol{S}}\right].\end{aligned} \tag{7.36}$$

By considering (7.35), (7.36), and the structure of $\boldsymbol{I}_N \otimes \bar{\boldsymbol{F}}$, the system (7.33) can be partitioned as

$$\begin{aligned}\dot{\tilde{\boldsymbol{\omega}}}^\star &= \left[\begin{array}{c|c} \boldsymbol{I}_N \otimes \boldsymbol{F} & \boldsymbol{O}_{12N,12N(N-1)} \\ \hline \boldsymbol{O}_{12N(N-1),12N} & \boldsymbol{I}_{N(N-1)} \otimes \boldsymbol{F} \end{array}\right] \tilde{\boldsymbol{\omega}}^\star + \left[\begin{array}{c} \boldsymbol{O}_{12N\times 6N^2} \\ \hline (\boldsymbol{I}_{N(N-1)} \otimes \boldsymbol{C})\boldsymbol{S} \end{array}\right] (\boldsymbol{I}_N \otimes \boldsymbol{K})\tilde{\boldsymbol{z}}^\star \\ &= \left[\begin{array}{c|c} \boldsymbol{F}^u_\omega & \boldsymbol{O}_{12N,12N(N-1)} \\ \hline \boldsymbol{O}_{12N\times 6N^2} & \boldsymbol{F}^c_\omega \end{array}\right] \tilde{\boldsymbol{\omega}}^\star + \left[\begin{array}{c} \boldsymbol{O}_{12N\times 6N^2} \\ \hline \boldsymbol{C}^c_\omega \end{array}\right] (\boldsymbol{I}_N \otimes \boldsymbol{K})\tilde{\boldsymbol{z}}^\star. \qquad (7.37)\end{aligned}$$

System (7.37) can be viewed as a linear system with input $(\boldsymbol{I}_N \otimes \boldsymbol{K})\tilde{\boldsymbol{z}}^\star$; such a system is composed by the subsystem $\{\boldsymbol{F}^u_\omega, \boldsymbol{O}_{12N\times 6N^2}\}$ that is clearly uncontrollable, and the subsystem $\{\boldsymbol{F}^c_\omega, \boldsymbol{C}^c_\omega\}$, that is controllable if the system $\{\boldsymbol{F}, \boldsymbol{C}\}$ is controllable since the matrix $\boldsymbol{S}$ is a full row rank matrix. In order to make $\tilde{\boldsymbol{\omega}}^\star$ convergent to zero the following conditions must be met:

C1 The uncontrollable part $\{\boldsymbol{F}^u_\omega, \boldsymbol{O}_{6N\times 6N}\}$ must be asymptotically stable. It is straightforward to recognize that this implies that the dynamics of system (7.13) is asymptotically stable.

C2 The control input gain $(\boldsymbol{I}_N \otimes \boldsymbol{K})$ has to be chosen in such a way to stabilize the subsystem $\{\boldsymbol{F}^c_\omega, \boldsymbol{C}^c_\omega\}$.

In order to guarantee that the conditions C1 and C2 are satisfied, the following choices of the control matrices must be done

- The matrix gain $\boldsymbol{K}_s$ has to be chosen in such a way that $\boldsymbol{F} = \boldsymbol{A} + \boldsymbol{C}\boldsymbol{K}_s$ is a Hurwitz matrix. It is worth noticing that, since $(\boldsymbol{A},\boldsymbol{C})$ represents a controllable system, the existence of such a matrix is ensured.
- The matrix gain $\boldsymbol{K}$ has to be chosen in such a way to stabilize the subsystem $\{\boldsymbol{F}^c_\omega, \boldsymbol{C}^c_\omega\}$. Again, the controllability of $(\boldsymbol{A}, \boldsymbol{C})$ ensures that such a matrix exists.

It is worth noticing that the choice of both $\boldsymbol{K}_s$ and $\boldsymbol{K}$ does not depend on a particular topology; thus the stability condition can be checked off-line on the basis of the knowledge of the sole system (7.13). □

Finally, by resorting to the same arguments used for $\tilde{\boldsymbol{y}}^\star$, in the presence of bounded uncertainties $\boldsymbol{\xi}$, it can be proven that the system (7.26) is globally uniformly ultimately bounded [29].

7.4 Fault diagnosis and isolation scheme

In the presence of a fault-affecting one of the robots in the team, the first step is to devise a suitable strategy aimed at making each robot able to detect the occurrence of the fault and to isolate the faulty robot, even if there is no direct communication between them. The collective state observer devised in Section 7.3 can be exploited for defining a set of residual vectors; this allows one to avoid increasing the computational burden and the information exchange among the teammates.

To this aim, let us define the following residual vector for the ith robot as [19]

$$^{i}\boldsymbol{r} = \sum_{j \in \mathcal{N}_i} ({}^{j}\hat{\boldsymbol{y}} - {}^{i}\hat{\boldsymbol{y}}) + \boldsymbol{\Pi}_i(\boldsymbol{y} - \hat{\boldsymbol{y}}_i) \tag{7.38}$$

that can be rearranged as a stacked vector $^{i}\boldsymbol{r} = \begin{bmatrix} ^{i}\boldsymbol{r}_1^T & ^{i}\boldsymbol{r}_2^T & \cdots & ^{i}\boldsymbol{r}_N^T \end{bmatrix}^T \in \mathbb{R}^{12N}$, where each component $^{i}\boldsymbol{r}_k \in \mathbb{R}^{12}$ represents the residual computed by robot i relative to robot k, and that allows the robot i to monitor the healthy state of robot k.

7.4.1 Residuals in the absence of faults

Let us introduce the collective residual vector $\boldsymbol{r}^\star \in \mathbb{R}^{12N^2}$ defined as

$$\boldsymbol{r}^\star = \begin{bmatrix} ^{1}\boldsymbol{r} \\ ^{2}\boldsymbol{r} \\ \vdots \\ ^{N}\boldsymbol{r} \end{bmatrix} \in \mathbb{R}^{12N^2}, \tag{7.39}$$

that, from (7.19), is given by

$$\boldsymbol{r}^\star = \tilde{\boldsymbol{L}}^\star \tilde{\boldsymbol{y}}^\star. \tag{7.40}$$

Under the assumption of Lemma 7.1, it can be stated that such a vector converges exponentially to zero while, in the presence of bounded uncertainties, it is bounded as well. From (7.40) it is possible to select the vector collecting all the residuals relative to the kth robot, i.e., $\boldsymbol{r}_k^\star = \begin{bmatrix} ^{1}\boldsymbol{r}_k^{\star T}, \ldots, {}^{N}\boldsymbol{r}_k^{\star T} \end{bmatrix}^T \in \mathbb{R}^{12N}$ as

$$\boldsymbol{r}_k^\star = \mathrm{diag}\,\{\boldsymbol{\Gamma}_k, \boldsymbol{\Gamma}_k, \ldots, \boldsymbol{\Gamma}_k\}\, \boldsymbol{r}^\star = \boldsymbol{\Gamma}_k^\star \tilde{\boldsymbol{L}}^\star \tilde{\boldsymbol{y}}^\star = \tilde{\boldsymbol{L}}_k^\star \tilde{\boldsymbol{y}}_k^\star, \tag{7.41}$$

where $\tilde{\boldsymbol{L}}_k^\star = \boldsymbol{L} \otimes \boldsymbol{I}_{12} + \boldsymbol{\Pi}_k$ and the vector $\tilde{\boldsymbol{y}}_k^\star = \boldsymbol{\Gamma}_k^\star \tilde{\boldsymbol{y}}^\star = \begin{bmatrix} ^{1}\tilde{\boldsymbol{y}}_k^T, \ldots, {}^{N}\tilde{\boldsymbol{y}}_k^T \end{bmatrix}^T \in \mathbb{R}^{12N}$ collects the estimation errors $^{i}\tilde{\boldsymbol{y}}_k$ of the observers.
From (7.24), after some algebraic steps, the dynamics of $\tilde{\boldsymbol{y}}_k^\star$, in the absence of faults, is

$$\dot{\tilde{\boldsymbol{y}}}_k^\star = -k_o \boldsymbol{\Gamma}_k^\star \tilde{\boldsymbol{L}}^\star \tilde{\boldsymbol{y}}^\star - \boldsymbol{\Gamma}_k^\star \boldsymbol{\xi}^\star = -k_o \tilde{\boldsymbol{L}}_k^\star \tilde{\boldsymbol{y}}_k^\star - \boldsymbol{1}_N \otimes \boldsymbol{B}_k(\boldsymbol{z}_k) \boldsymbol{\xi}_k. \tag{7.42}$$

Since in the case of strongly connected communication graph $-\tilde{\boldsymbol{L}}^\star$ is Hurwitz [28], the following property can be easily proven:

Property 7.3. *The matrix* $-\tilde{\boldsymbol{L}}_k^\star$ *is Hurwitz; thus*

- *the dynamics in* (7.42) *is asymptotically stable;*
- *there exist two positive constants* $\kappa, \lambda > 0$ *such that [31]*

$$\left\| e^{-k_o \tilde{\boldsymbol{L}}_k^\star t} \right\| \leq \kappa e^{-\lambda t}. \tag{7.43}$$

Theorem 7.2. *In the absence of faults, the residual* $^{i}\boldsymbol{r}_k$*, i.e., the residual computed by robot i and relative to robot k, is norm-bounded by a time-varying threshold* $^{i}\mu_k(t)$ *depending on the initial state of the whole system and the communication graph.*

Proof. From (7.41) and by virtue of Property 7.3, the residual ${}^i\boldsymbol{r}_k$, in the absence of faults, can be written as

$$
{}^i\boldsymbol{r}_k = \boldsymbol{\Gamma}_i \tilde{\boldsymbol{L}}_k^\star \tilde{\boldsymbol{y}}_k^\star = \boldsymbol{\Gamma}_i \tilde{\boldsymbol{L}}_k^\star \left[e^{-k_o \tilde{\boldsymbol{L}}_k^\star t} \tilde{\boldsymbol{y}}_k^\star(0) - \int_0^t e^{-k_o \tilde{\boldsymbol{L}}_k^\star (t-\tau)} \left(\boldsymbol{1}_N \otimes \boldsymbol{B}_k(\boldsymbol{z}_k) \boldsymbol{\xi}_k(\tau) \right) d\tau \right]. \tag{7.44}
$$

Therefore, it can be upper bounded as

$$
\begin{aligned}
\left\| {}^i\boldsymbol{r}_k \right\| &\leq \left\| \boldsymbol{\Gamma}_i \tilde{\boldsymbol{L}}_k^\star \right\| \left[\left\| e^{-k_o \tilde{\boldsymbol{L}}_k^\star t} \tilde{\boldsymbol{y}}_k^\star(0) \right\| + \int_0^t \left\| e^{-k_o \tilde{\boldsymbol{L}}_k^\star (t-\tau)} \left(\boldsymbol{1}_N \otimes \boldsymbol{B}_k(\boldsymbol{z}_k) \boldsymbol{\xi}_k(\tau) \right) d\tau \right\| \right] \\
&\leq \left\| \boldsymbol{\Gamma}_i \tilde{\boldsymbol{L}}_k^\star \right\| \left[\left\| \tilde{\boldsymbol{y}}_k^\star(0) \right\| \kappa e^{-\lambda t} + \frac{\kappa \sqrt{N}\, \bar{\xi}}{\lambda\, \varepsilon_m(\boldsymbol{\Lambda}_k)} (1 - e^{-\lambda t}) \right],
\end{aligned} \tag{7.45}
$$

where $\varepsilon_m(\boldsymbol{\Lambda}_k)$ is the minimum eigenvalue of $\boldsymbol{\Lambda}_k$, while Property 7.3 and the following inequality

$$
\left\| \boldsymbol{1}_N \otimes \boldsymbol{B}_k(\boldsymbol{z}_k) \boldsymbol{\xi}_k \right\| \leq \frac{\sqrt{N}}{\varepsilon_m(\boldsymbol{\Lambda}_k)} \bar{\xi} \tag{7.46}
$$

have been taken into account. Moreover, it holds [19]

$$
\left\| \boldsymbol{\Gamma}_i \tilde{\boldsymbol{L}}_k^\star \right\| \leq \sqrt{12 d_i} + {}^i\delta_k, \tag{7.47}
$$

with d_i the in-degree of node i, and ${}^i\delta_k = 1$ if $i = k$ and ${}^i\delta_k = 0$ otherwise. Therefore, the right-hand member of (7.45) can be written as a time-varying threshold:

$$
{}^i\mu_k(t) = (\sqrt{12 d_i} + {}^i\delta_k) \left(\left\| \tilde{\boldsymbol{y}}_k^\star(0) \right\| \kappa e^{-\lambda t} + \frac{\kappa \sqrt{N}\, \bar{\xi}}{\lambda\, \varepsilon_m(\boldsymbol{\Lambda}_k)} (1 - e^{-\lambda t}) \right). \tag{7.48}
$$

Inequalities (7.45) and (7.48) prove the theorem. □

Remark 7.3. *The calculation of the thresholds ${}^i\mu_k$ requires a reliable estimate of $\|\tilde{\boldsymbol{y}}_k^\star(0)\|$, λ, and κ. The constant $\|\tilde{\boldsymbol{y}}_k^\star(0)\|$ can be estimated on the basis of approximate information about the initial state of the system by supposing that the robots are included on a known bounded area. Regarding λ and κ, if the Laplacian matrix of the system is known can be computed as in [31]; otherwise the Laplacian matrix can be estimated by considering the worst case scenario.*

7.4.2 Residuals in the presence of faults

Let us consider a fault $\boldsymbol{\phi}_k$ affecting the kth robot. The following theorem can be stated.

Theorem 7.3. *A fault occurring on the kth robot at time $t_f > t_0$ affects only the residuals components ${}^i\boldsymbol{r}_k$ ($\forall i = 1, 2, \ldots, N$) and not the residual components ${}^i\boldsymbol{r}_j$ ($\forall i, j = 1, 2, \ldots, N$ and $j \neq k$).*

Proof. In the presence of a fault affecting robot k, it holds $\|\boldsymbol{\phi}_k\| > 0$; therefore the dynamics of $\tilde{\boldsymbol{y}}_k^\star$ (7.42) becomes

$$\dot{\tilde{\boldsymbol{y}}}_k^\star = -k_o\tilde{\boldsymbol{L}}_k^\star\tilde{\boldsymbol{y}}_k^\star - \boldsymbol{1}_N \otimes \boldsymbol{B}_k(z_k)\left(\boldsymbol{\xi}_k + \boldsymbol{J}_{M,k}^T\boldsymbol{\phi}_k\right), \tag{7.49}$$

and, thus, by virtue of (7.44), the residual components ${}^i\boldsymbol{r}_k$ become

$$\begin{aligned} {}^i\boldsymbol{r}_k = \boldsymbol{\Gamma}_i\tilde{\boldsymbol{L}}_k^\star \Bigg[& e^{-k_o\tilde{\boldsymbol{L}}_k^\star t}\tilde{\boldsymbol{y}}_k^\star(0) - \int_0^t e^{-k_o\tilde{\boldsymbol{L}}_k^\star(t-\tau)}\left(\boldsymbol{1}_N \otimes \boldsymbol{B}_k(z_k)\boldsymbol{\xi}_k(\tau)\right)d\tau \\ & + \int_0^t e^{-k_o\tilde{\boldsymbol{L}}_k^\star(t-\tau)}(\boldsymbol{1}_N \otimes \boldsymbol{B}_k(z_k)\boldsymbol{J}_{M,k}^T\boldsymbol{\phi}_k(\tau))d\tau \Bigg]. \end{aligned} \tag{7.50}$$

Equation (7.50) proves that the residual ${}^i\boldsymbol{r}_k$ is affected by the fault. On the other side, by considering that the matrix $\boldsymbol{\Gamma}_j^\star$ selects only the components of the vector $\boldsymbol{\phi}_M$ associated with the jth robot, which are null, the dynamics of ${}^i\boldsymbol{y}_j^\star$ with $j \neq k$ is

$$\dot{\tilde{\boldsymbol{y}}}_j^\star = -k_o\tilde{\boldsymbol{L}}_j^\star\tilde{\boldsymbol{y}}_j^\star - \boldsymbol{1}_N \otimes \boldsymbol{B}_j(z_j)\boldsymbol{\xi}_j, \tag{7.51}$$

which is not affected by the faulty robot. This proves the theorem. □

7.4.3 Detection and isolation strategy

On the basis of Theorems 7.2 and 7.3, the following strategy can be implemented in order to detect the occurrence of a fault $\boldsymbol{\phi}_k$ affecting the robot k and to isolate the faulty robot. In detail, when the following holds, a fault on robot k is detected by robot i

$$\begin{cases} \exists\, t > t_f: & \|{}^i\boldsymbol{r}_k(t)\| > {}^i\mu_k(t) \\ \forall l \in (1,2,\ldots,N), l \neq k, \forall t > t_0, & \|{}^i\boldsymbol{r}_l(t)\| \leq {}^i\mu_l(t) \end{cases} \tag{7.52}$$

Note that, in the presence of nonzero initial observer estimation errors and model uncertainty, the residuals are different from zero even in the absence of faults. For this reason, in order to avoid the occurrence of false alarms, and by virtue of Theorem 7.2, the decision about the occurrence of a fault is made only when a residual exceeds the adaptive thresholds computed in (7.48). Furthermore, it is worth remarking that Theorem 7.3 ensures that only the residual ${}^i\boldsymbol{r}_k$ is affected by the fault $\boldsymbol{\phi}_k$ while the other residuals computed by robot i are insensitive to it. The condition (7.52) ensures that all the robots of the cell can detect a fault even if there is no direct communication with the faulty robot.

Moreover, a sufficient detectability condition can be stated for the fault $\boldsymbol{\phi}_k$, i.e.,

$$\exists\, t > t_f: \left\| \boldsymbol{\Gamma}_i\tilde{\boldsymbol{L}}_k^\star \int_0^t e^{-k_o\tilde{\boldsymbol{L}}_k^\star(t-\tau)}(\boldsymbol{1}_N \otimes \boldsymbol{B}_k(z_k)\boldsymbol{J}_{M,k}^T\boldsymbol{\phi}_k(\tau))d\tau \right\| \geq 2\,{}^i\mu_k. \tag{7.53}$$

Such a condition can be derived by the following chain of inequalities

$$
\begin{aligned}
\|{}^{i}\boldsymbol{r}_k\| &= \Bigg\| \boldsymbol{\Gamma}_i\tilde{\boldsymbol{L}}_k^{\star}\left(e^{-k_o\tilde{\boldsymbol{L}}_k^{\star}t}\tilde{\boldsymbol{y}}_k^{\star}(0) - \int_0^t e^{-k_o\tilde{\boldsymbol{L}}_k^{\star}(t-\tau)}\left(\mathbf{1}_N\otimes\boldsymbol{B}_k(\boldsymbol{z}_k)\boldsymbol{\xi}_k(\tau)\right)d\tau\right) \\
&\quad +\boldsymbol{\Gamma}_i\tilde{\boldsymbol{L}}_k^{\star}\left(\int_0^t e^{-k_o\tilde{\boldsymbol{L}}_k^{\star}(t-\tau)}(\mathbf{1}_N\otimes\boldsymbol{B}_k(\boldsymbol{z}_k)\boldsymbol{J}_{M,k}^{T}\boldsymbol{\phi}_k(\tau))d\tau\right)\Bigg\| \\
&\geq \left\|\boldsymbol{\Gamma}_i\tilde{\boldsymbol{L}}_k^{\star}\left(\int_0^t e^{-k_o\tilde{\boldsymbol{L}}_k^{\star}(t-\tau)}(\mathbf{1}_N\otimes\boldsymbol{B}_k(\boldsymbol{z}_k)\boldsymbol{J}_{M,k}^{T}\boldsymbol{\phi}_k(\tau))d\tau\right)\right\| \qquad (7.54)\\
&\quad -\left\|\boldsymbol{\Gamma}_i\tilde{\boldsymbol{L}}_k^{\star}\left(e^{-k_o\tilde{\boldsymbol{L}}_k^{\star}t}\tilde{\boldsymbol{y}}_k^{\star}(0) - \int_0^t e^{-k_o\tilde{\boldsymbol{L}}_k^{\star}(t-\tau)}\left(\mathbf{1}_N\otimes\boldsymbol{B}_k(\boldsymbol{z}_k)\boldsymbol{\xi}_k(\tau)\right)d\tau\right)\right\| \\
&\geq \left\|\boldsymbol{\Gamma}_i\tilde{\boldsymbol{L}}_k^{\star}\left(\int_0^t e^{-k_o\tilde{\boldsymbol{L}}_k^{\star}(t-\tau)}(\mathbf{1}_N\otimes\boldsymbol{B}_k(\boldsymbol{z}_k)\boldsymbol{J}_{M,k}^{T}\boldsymbol{\phi}_k(\tau))d\tau\right)\right\| - {}^{i}\mu_k \geq {}^{i}\mu_k.
\end{aligned}
$$

It is worth remarking that this condition is only sufficient but not necessary; thus a fault can be detected even if this condition is not met.

7.5 Experiments

In order to validate the devised framework, experiments on a real multi-manipulator setup have been carried out. A video of the experiment is available at the link*. As represented in Figure 7.2, a heterogeneous team of three robots ($N = 3$) is considered which is composed of the following:

1. A fixed-base 7 DOFs Kinova ultra lightweight Gen2 Jaco arm, for which it holds $n_1 = 7$.
2. A Kinova Movo mobile robot consisting of an omnidirectional base (3 DOFs), a variable height torso (1 DOF), and a Kinova Gen2 Jaco arm (7 DOFs), for which it holds $n_2 = 11$.
3. A Kinova Movo mobile robot consisting of an omnidirectional base (3 DOFs), a variable height torso (1 DOF), and a 2-link robot arm (2 DOFs), for which it holds $n_3 = 6$.

The Kinova Movo robots are also equipped with RGB-D sensors, in particular Microsoft Kinect v2, which allow one to possibly monitor the scene; this sensor is considered as end effector of the latter robot. In the following, we will refer to the aforementioned robots as fixed manipulator, mobile manipulator, and mobile camera, respectively.

*http://webuser.unicas.it/lai/robotica/video/SMC19.html

Figure 7.2 Experimental setup composed of three heterogeneous robots

Concerning the hardware specifications, Movo mobile robots are provided with two dedicated Intel NUC Kits NUC5i7RYH with Intel Core i7-5557U processor and 16 GB RAM, whereas the fixed manipulator is controlled by a standard PC with Intel Core i7-5500U processor and 8 GB RAM; moreover, Wi-Fi modules TP-Link TL-WN821N are used for each robot, which enable the intrarobot communication through a local network set on a TP-Link TD-W8960N router. Finally, the software architecture relies on ROS middleware and ArUco markers [32] have been introduced to initially localize mobile robots in the environment.

A cooperative service task is considered where the fixed-base manipulator is in charge of pouring the contents of a bottle into a glass, while the mobile manipulator is in charge of holding that glass; finally, the camera robot may be exploited to provide a different point of view for scene monitoring.

With regard to the cooperative task, the formulation proposed in [27] is adopted, based on which the team centroid and formation are defined. In detail, by introducing the stacked vector of the end effector configurations $\boldsymbol{x} = \left[\boldsymbol{x}_1^T \ \cdots \ \boldsymbol{x}_N^T\right] \in \mathbb{R}^{6N}$, the centroid $\boldsymbol{\sigma}_c \in \mathbb{R}^6$ is obtained as follows:

$$\boldsymbol{\sigma}_c(\boldsymbol{x}) = \frac{1}{N}\sum_{i=1}^{N}\boldsymbol{x}_i = \frac{1}{N}\left[\boldsymbol{I}_6 \ \cdots \ \boldsymbol{I}_6\right]\boldsymbol{x} = \boldsymbol{J}_c\boldsymbol{x}, \tag{7.55}$$

with $\boldsymbol{J}_c \in \mathbb{R}^{6\times 6N}$ the centroid task Jacobian matrix, while the formation $\boldsymbol{\sigma}_f \in \mathbb{R}^{6(N-1)}$ is defined as

$$\begin{aligned}\boldsymbol{\sigma}_f(\boldsymbol{x}) &= \left[(\boldsymbol{x}_2 - \boldsymbol{x}_1)^T \ (\boldsymbol{x}_3 - \boldsymbol{x}_2)^T \cdots (\boldsymbol{x}_N - \boldsymbol{x}_{N-1})^T\right]^T \\ &= \begin{bmatrix} -\boldsymbol{I}_6 & \boldsymbol{I}_6 & \boldsymbol{O}_6 & \cdots & \boldsymbol{O}_6 \\ \boldsymbol{O}_6 & -\boldsymbol{I}_6 & \boldsymbol{I}_6 & \cdots & \boldsymbol{O}_6 \\ \vdots & \vdots & \vdots & \vdots & \vdots \\ \boldsymbol{O}_6 & \boldsymbol{O}_6 \cdots & \boldsymbol{O}_6 & -\boldsymbol{I}_6 & \boldsymbol{I}_6 \end{bmatrix}\boldsymbol{x} = \boldsymbol{J}_f\boldsymbol{x},\end{aligned} \tag{7.56}$$

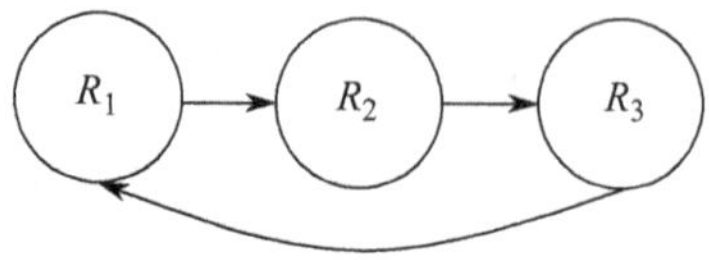

Figure 7.3 Communication graph

being $\boldsymbol{J}_f \in \mathbb{R}^{6(N-1)\times 6N}$ the formation task Jacobian matrix. The overall task function $\boldsymbol{\sigma} \in \mathbb{R}^{6N}$ is thus derived by combining (7.55) and (7.56)

$$\boldsymbol{\sigma}(\boldsymbol{x}) = \begin{bmatrix} \boldsymbol{\sigma}_c(\boldsymbol{x}) \\ \boldsymbol{\sigma}_f(\boldsymbol{x}) \end{bmatrix} = \begin{bmatrix} \boldsymbol{J}_c \\ \boldsymbol{J}_f \end{bmatrix} \boldsymbol{x} = \boldsymbol{J}_\sigma \boldsymbol{x},$$

with $\boldsymbol{J}_\sigma \in \mathbb{R}^{6N\times 6N}$ the overall task Jacobian matrix.

At this point, let us consider that a desired trajectory $\boldsymbol{\sigma}_d(t)$ ($\dot{\boldsymbol{\sigma}}_d(t)$,$\ddot{\boldsymbol{\sigma}}_d(t)$) is assigned. By resorting to a second-order closed loop inverse kinematics law, the central control law in (7.9) for the cooperative task space trajectory tracking is given by

$$\boldsymbol{u} = \boldsymbol{J}_\sigma^{-1}(\ddot{\boldsymbol{\sigma}}_d + \boldsymbol{K}_d(\dot{\boldsymbol{\sigma}}_d - \dot{\boldsymbol{\sigma}}(\dot{\boldsymbol{x}})) + \boldsymbol{K}_p(\boldsymbol{\sigma}_d - \boldsymbol{\sigma}(\boldsymbol{x}))), \tag{7.57}$$

with $\boldsymbol{K}_d, \boldsymbol{K}_p \in \mathbb{R}^{6N\times 6N}$ positive definite gain matrices, for which it is straightforward to show that the tracking error $\boldsymbol{\sigma}_d - \boldsymbol{\sigma}$ asymptotically converges to the origin. In accordance to the devised strategy, by replacing $\boldsymbol{x}$ and $\dot{\boldsymbol{x}}$ in (7.57) with the respective estimates, each robot i can compute the estimated global input ${}^i\hat{\boldsymbol{u}}_g$ as in (7.10) and, in turn, can derive the local auxiliary input $\hat{\boldsymbol{u}}_i$ as in (7.12).

The following set of gains is considered: $\boldsymbol{K}_p = \boldsymbol{I}_{18}$ and $\boldsymbol{K}_d = 2\boldsymbol{I}_{18}$ in (7.57) and $k_o = 5$ in (7.15). Concerning the communication graph, the directed strongly connected graph shown in Figure 7.3 has been adopted.

Starting from the initial configuration shown in Figure 7.4(a), the desired task consists of the following steps:

1. The mobile manipulator and the camera robot move closer to the fixed-base manipulator.
2. The mobile manipulator hangs out the glass as in Figure 7.4(b).
3. The fixed-base manipulator pours the contents of the bottle into the glass as in Figure 7.4(c).
4. The mobile base manipulator delivers the glass while the fixed-base manipulator returns in a configuration with the bottle vertical.

A fault occurs on the fixed-base manipulator (with index $i = 1$) during last phase. In particular, the following fault term is introduced at time $t \approx 35$ s

$$\boldsymbol{\phi}_1 = \boldsymbol{J}_{M,1}^\dagger \boldsymbol{\psi} \qquad \text{with} \quad \boldsymbol{\psi} = \begin{bmatrix} 0 & 0 & 1 & 0 & 0 & 0 \end{bmatrix}^T \text{ m/s}^2 \tag{7.58}$$

which induces a downwards motion of the manipulator's end effector as shown in Figure 7.4(d).

In this experiment, we consider that, when a robot detects the fault of any robot in the team, it performs a shutdown procedure and stops its motion.

Experimental results are reported in Figures 7.5–7.8. More specifically, the norm of the task tracking error $\tilde{\boldsymbol{\sigma}} = \boldsymbol{\sigma}_d - \boldsymbol{\sigma}$ and of its derivative $\dot{\tilde{\boldsymbol{\sigma}}} = \dot{\boldsymbol{\sigma}}_d - \dot{\boldsymbol{\sigma}}$ as well as

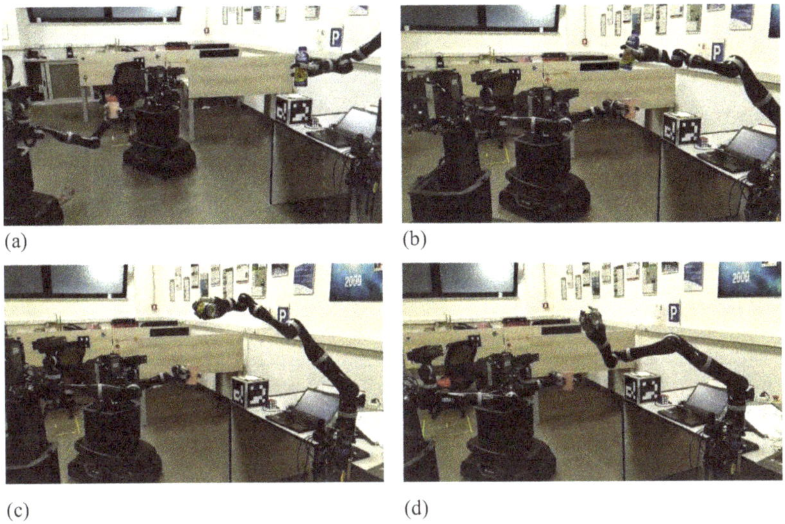

Figure 7.4 *Snapshots of the experiment. In detail, part (a) represents the starting configuration, part (b) shows the mobile manipulator hanging out the glass, part (c) is the pouring phase and part (d) represents the fault occurrence.*

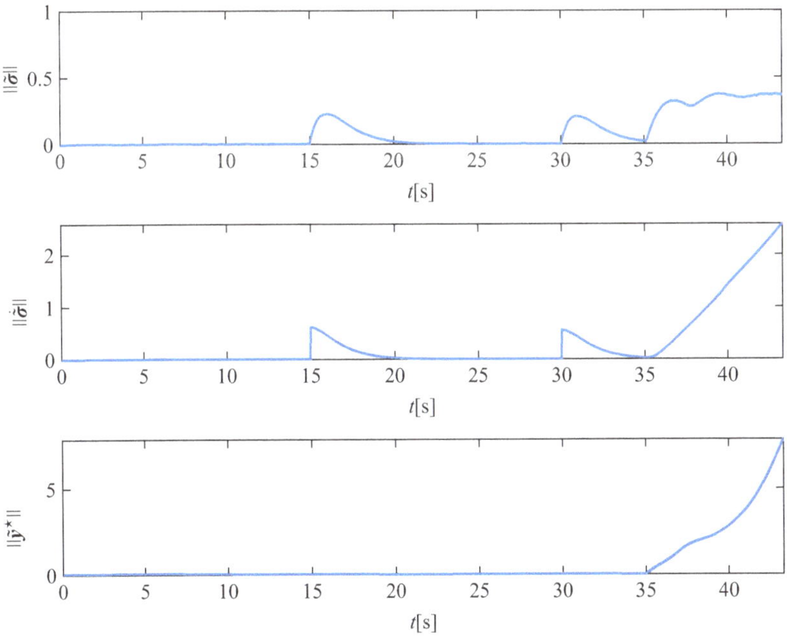

Figure 7.5 *Evolution of the norm of the task error* $||\tilde{\boldsymbol{\sigma}}||$ *(top), the norm of its derivative* $||\dot{\tilde{\boldsymbol{\sigma}}}||$ *(middle), and the norm of the auxiliary state estimation error* $||\tilde{\boldsymbol{y}}^{\star}||$ *(bottom)*

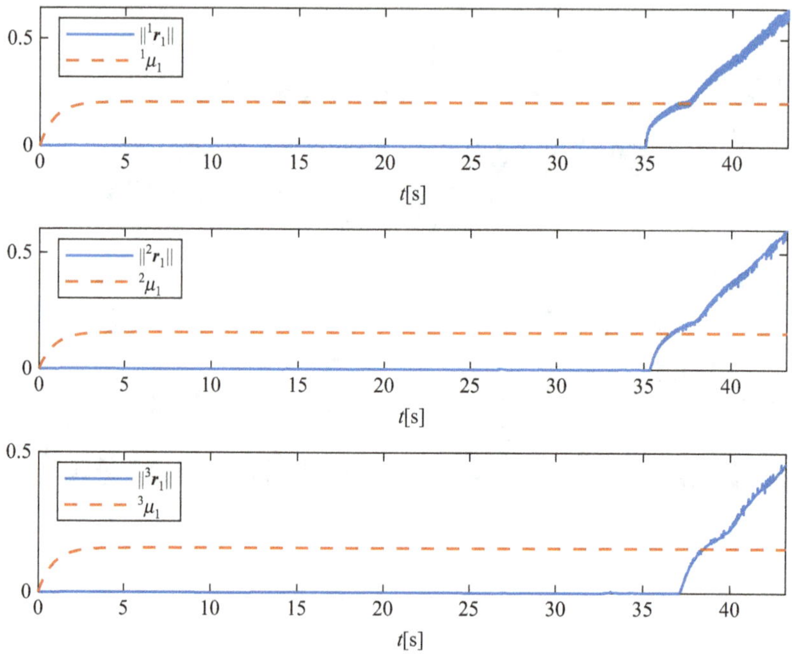

Figure 7.6 Evolution of the norm of residual components relative to robot 1 (faulty robot) computed by the three robots (solid line) compared with the respective adaptive thresholds (dashed line)

the norm of the auxiliary state error $\tilde{\boldsymbol{y}}^{\star}$ are reported in Figure 7.5 and are shown to increase in correspondence of the fault occurrence at $t \approx 35$ s. Figure 7.6 reports the norm of the residuals computed by all the robots and associated with the faulty robot 1 (solid lines), i.e., $\|{}^i\boldsymbol{r}_1\|$, $\forall\, i$, compared with the respective adaptive thresholds (dashed lines), i.e., ${}^i\mu_1\ \forall\, i$, which are computed according to (7.48) by considering $\bar{\xi} = 0.1$ and, without loss of generality, $\|\tilde{\boldsymbol{y}}_k^{\star}(0)\| = 0$. Hence, the figure shows that all the robots are able to detect the occurrence of the fault without the need for direct communication with the faulty robot, that is, all the robots verify the condition $\|{}^i\boldsymbol{r}_1\| \geq {}^i\mu_1\ \forall i$ after the fault occurrence. For the sake of completeness, the norms of the residuals associated with the healthy robots 2 and 3 (solid lines) are reported in Figures 7.7 and 7.8, respectively. The respective adaptive thresholds are also reported (dashed lines) which are shown to always be greater than the residual signals; this verifies that the fault of one robot does not affect residuals of the other teammates.

7.6 Conclusions

A distributed framework for detecting and isolating faults in multi-manipulator robotic systems has been presented. By considering that a desired cooperative mission is

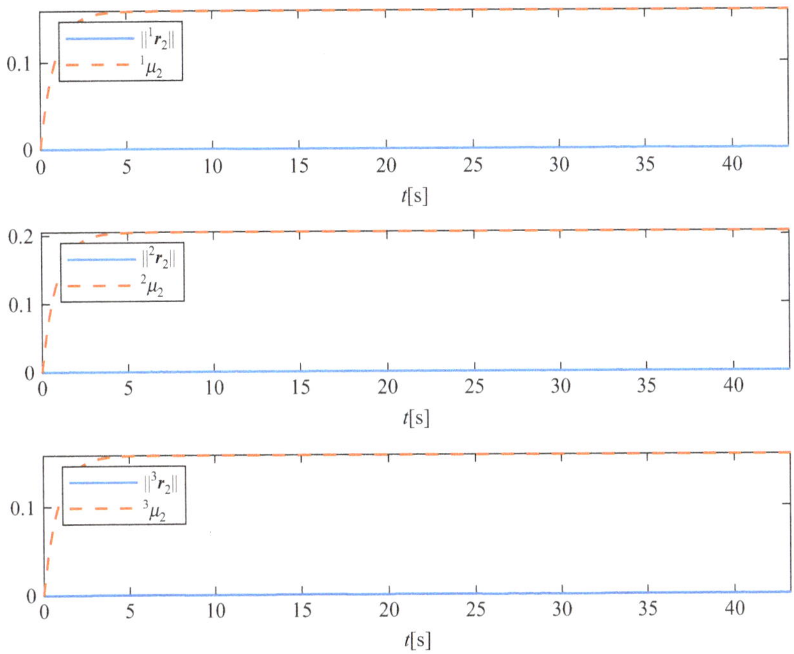

Figure 7.7 Evolution of the norm of residual components relative to robot 2 *(healthy robot) computed by the three robots (solid line) compared with the respective adaptive thresholds (dashed line)*

assigned to the team, an observer-controller architecture is adopted where the observer layer is in charge of estimating the state of all the robots of the team whereas the controller layer, given this estimate, is in charge of defining the local control law that allows one to achieve the desired global task. Furthermore, based on the information of the observer layer, residual signals and adaptive thresholds are defined which allow each robot to detect and identify faults of any teammate without the need of direct communication. Theoretical findings have been corroborated with experimental results on a setup composed of three heterogeneous robots. As future works, the approach will be extended to dual-arm mobile robots and will include accommodation strategies once the fault is detected and isolated.

Acknowledgment

This work is supported by the project Dipartimenti di Eccellenza granted to DIEI Department of the University of Cassino and Southern Lazio.

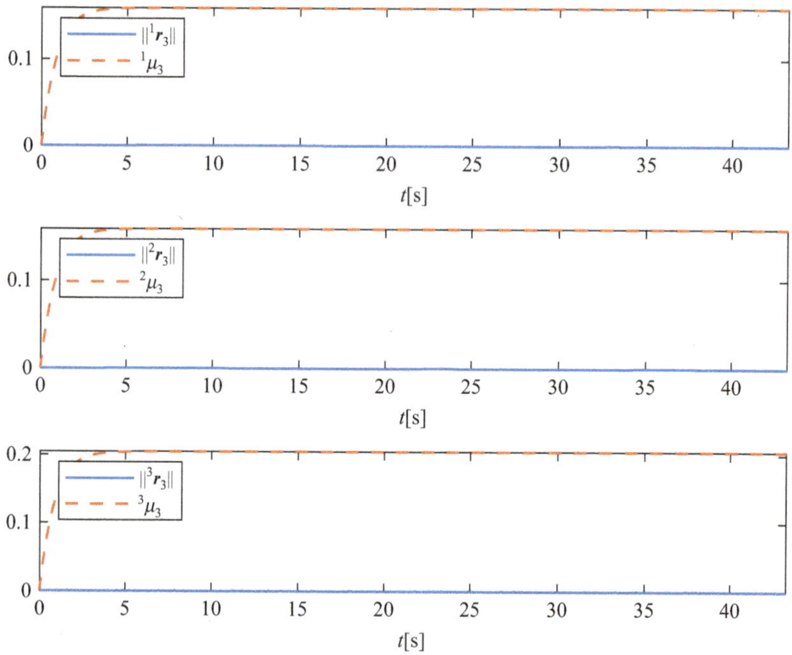

Figure 7.8 Evolution of the norm of residual components relative to robot 3 *(healthy robot) computed by the three robots (solid line) compared with the respective adaptive thresholds (dashed line)*

References

[1] D.K. Maczka, A.S. Gadre, and D.J. Stilwell. Implementation of a cooperative navigation algorithm on a platoon of autonomous underwater vehicles. In *Proceedings MTS/IEEE Conf. Oceans 2000*, pages 1–6, 2007.

[2] R.W. Beard, T.W. McLain, D.B. Nelson, D. Kingston, D. Johanson. Decentralized cooperative aerial surveillance using fixed-wing miniature UAVs. *Proceedings of the IEEE*, 94(7):1306–1324, 2006.

[3] A. Marino, F. Pierri. A two stage approach for distributed cooperative manipulation of an unknown object without explicit communication and unknown number of robots. *Robotics and Autonomous Systems*, 103:122–133, 2018.

[4] X. Zhang, M.M. Polycarpou, T. Parisini. A robust detection and isolation scheme for abrupt and incipient faults in nonlinear systems. *IEEE Transactions on Automatic Control*, 47(4):576–593, 2002.

[5] X. Zhang, T. Parisini, M.M. Polycarpou. Adaptive fault-tolerant control of nonlinear uncertain systems: An information-based diagnostic approach. *IEEE Transactions on Automatic Control*, 49(8):1259–1274, 2004.

[6] L. Liu, Z. Wang, H. Zhang. Adaptive fault-tolerant tracking control for MIMO discrete-time systems via reinforcement learning algorithm with less learning parameters. *IEEE Transactions on Automation Science and Engineering*, 14(1):299–313, 2017.

[7] X. Zhang, Q. Zhang. Distributed fault diagnosis in a class of interconnected nonlinear uncertain systems. *International Journal of Control*, 85(11):1644–1662, 2012.

[8] R.M.G. Ferrari, T. Parisini, M.M. Polycarpou. Distributed fault detection and isolation of large-scale discrete-time nonlinear systems: An adaptive approximation approach. *IEEE Transactions on Automatic Control*, 57(2):275–290, 2012.

[9] F. Boem, R. M. G. Ferrari, C. Keliris, T. Parisini, M.M. Polycarpou. A distributed networked approach for fault detection of large-scale systems. *IEEE Transactions on Automatic Control*, 62(1):18–33, 2017.

[10] I. Shames, A.M.H. Teixeira, H. Sandberg, K.H. Johansson. Distributed fault detection for interconnected second-order systems. *Automatica*, 47(12):2757–2764, 2011.

[11] A. Teixeira, I. Shames, H. Sandberg, K.H. Johansson. Distributed fault detection and isolation resilient to network model uncertainties. *IEEE Transactions on Cybernetics*, 44(11):2024–2037, 2014.

[12] M.R. Davoodi, K. Khorasani, H.A. Talebi, H.R. Momeni. Distributed fault detection and isolation filter design for a network of heterogeneous multiagent systems. *IEEE Transactions on Control Systems Technology*, 22(3):1061–1069, 2014.

[13] M. Davoodi, N. Meskin, K. Khorasani. Simultaneous fault detection and consensus control design for a network of multi-agent systems. *Automatica*, 66:185–194, 2016.

[14] M. Khalili, X. Zhang, M.M. Polycarpou, T. Parisini, Y. Cao. Distributed adaptive fault-tolerant control of uncertain multi-agent systems. *Automatica*, 87:142–151, 2018.

[15] C. Liu, B. Jiang, R.J. Patton, K. Zhang. Hierarchical-structure-based fault estimation and fault-tolerant control for multiagent systems. *IEEE Transactions on Control of Network Systems*, 6(2):586–597, 2019.

[16] A.R. Mehrabian, K. Khorasani. Distributed formation recovery control of heterogeneous multiagent Euler–Lagrange systems subject to network switching and diagnostic imperfections. *IEEE Transactions on Control Systems Technology*, 24(6):2158–2166, 2016.

[17] G. Chen, Y. Song, F.L. Lewis. Distributed fault-tolerant control of networked uncertain Euler–Lagrange systems under actuator faults. *IEEE Transactions on Cybernetics*, 47(7):1706–1718, 2017.

[18] L. Liu, J. Shan. Distributed formation control of networked Euler–Lagrange systems with fault diagnosis. *Journal of the Franklin Institute*, 352(3):952–973, 2015.

[19] F. Arrichiello, A. Marino, F. Pierri. Observer-based decentralized fault detection and isolation strategy for networked multirobot systems. *IEEE Transactions on Control Systems Technology*, 23(4):1465–1476, 2015.

[20] A. Marino, F. Pierri, F. Arrichiello. Distributed fault detection isolation and accommodation for homogeneous networked discrete-time linear systems. *IEEE Transactions on Automatic Control*, 62(9):4840–4847, 2017.

[21] O. Khatib. A unified approach for motion and force control of robot manipulators: The operational space formulation. *IEEE Journal on Robotics and Automation*, 3(1):43–53, 1987.

[22] J.A. Fax, R.M. Murray. Information flow and cooperative control of vehicle formations. *IEEE Transactions on Automatic Control*, 49(9):1465–1476, 2004.

[23] R. Olfati-Saber, R.M. Murray. Consensus problems in networks of agents with switching topology and time-delays. *IEEE Transactions on Automatic Control*, 49(9):1520–1533, 2004.

[24] M. Mesbahi, M. Egerstedt. *Graph theoretic methods in multiagent networks*. Princeton, NJ: Princeton University Press, 2010.

[25] W. Ren, R.W. Beard. *Distributed consensus in multi-vehicle cooperative control*. Communications and Control Engineering. Springer, Berlin, 2008.

[26] C. Lopez-Limon, J. Ruiz-Leon, A. Cervantes-Herrera, A. Ramirez-Trevino. Formation and trajectory tracking of discrete-time multi-agent systems using block control. In *IEEE 18th Conf. on Emerging Technologies Factory Automation (ETFA)*, Sep. 2013.

[27] G. Antonelli, F. Arrichiello, F. Caccavale, A. Marino. Decentralized time-varying formation control for multi-robot systems. *The International Journal of Robotics Research*, 33:1029–1043, 2014.

[28] G. Antonelli, F. Arrichiello, F. Caccavale, A. Marino. A decentralized controller-observer scheme for multi-agent weighted centroid tracking. *IEEE Transactions on Automatic Control*, 58(5):1310–1316, 2013.

[29] H.K. Khalil. *Nonlinear systems*. Upper Saddle River, NJ: Prentice-Hall, 2002.

[30] H. Roger, R.J. Charles. *Topics in matrix analysis*. Cambridge: Cambridge University Press, 1994.

[31] G. Hu, M. Liu. The weighted logarithmic matrix norm and bounds of the matrix exponential. *Linear Algebra and Its Applications*, 390:145–154, 2004.

[32] F. Romero Ramirez, R. Muñoz-Salinas, R. Medina-Carnicer. Speeded up detection of squared fiducial markers. *Image and Vision Computing*, 76:38–47, 2018.

Chapter 8

Nonlinear optimal control for aerial robotic manipulators

Gerasimos Rigatos[1], Masoud Abbaszadeh[2], and Patrice Wira[3]

The chapter proposes a nonlinear optimal control approach for aerial manipulators, that is, multi-DOF unmanned aerial vehicles (UAVs) that comprise a robotic arm with flexible joints. The method has been successfully tested so far on the control problem of several types of unmanned vehicles and the present chapter shows that it can also provide an optimal solution to the control problem of a 5-DOF aerial manipulator. To implement this control scheme, the state-space model of the aerial manipulator undergoes first approximate linearization around a temporary operating point, through the first-order Taylor series expansion and through the computation of the associated Jacobian matrices. To select the feedback gains of the H-infinity controller, an algebraic Riccati equation is repetitively solved at each time-step of the control method. The global stability and the robustness properties of the control loop are proven through Lyapunov analysis. Finally, to implement state estimation-based feedback control, the H-infinity Kalman filter is used as a robust state estimator.

8.1 Introduction

The use of aerial robotic manipulators, also known as rotor flying manipulators, is rapidly deploying in rescue, defense and ammunition or transportation tasks [1–4]. Aerial robotic manipulators comprise a multi-rotor UAV and a robotic manipulator mounted on it [5–8]. The dynamic and kinematic model of such aerial robots accumulates the complexity of the models of its constituents, that is, of the UAV and of the robotic manipulator. Actually, the dynamic model of such aerial robots is a multivariable and strongly nonlinear system and the solution of the related control problem is a nontrivial task [9–12]. Besides, this system is underactuated, which means that it has more degrees of freedom than its control inputs and consequently the solution of the

[1]Unit of Industrial Automation, Industrial Systems Institute, Rion Patras, Greece
[2]GE Global Research, General Electric, Niskayuna, NY, USA
[3]IRIMAS, Université d' Haute Alsace, Mulhouse, France

related control problem becomes far more complicated [13–16]. The control forces and torques that are exerted on the rotor flying manipulator come from the thrust of the rotors and from the motors that rotate the links of the robotic arm [17–20]. Usually, the manipulators in such robotic systems make use of rigid revolute joints. By substituting such joints with flexible ones, the aerial manipulators can perform grasping and compliance tasks with more safety and the related damage risk can be reduced. On the other side, the use of flexible joints introduces additional degrees of freedom in the dynamic model of the aerial manipulators and raises further difficulty for solving the associated control problem [21–23].

In this chapter, a nonlinear optimal (H-infinity) control approach is proposed for the problem of control of aerial robotic manipulators with flexible joints [24–26]. To this end, the chapter develops first the dynamic model of the robotic system that comprises the rotorcraft UAV and the manipulator with flexible joints. This is done by computing the Lagrangian of the robotic system and by applying the Euler–Lagrange method. Thus, a nonlinear multi-DOF and underactuated state-space model of the aerial robot is obtained which is finally written in an affine-in-the-input form. Next, the state-space description of the aerial robot undergoes approximate linearization around a time-varying operating point, which is updated at each time-step of the control method. The linearization point is defined by the present value of the state vector of the robotic system and by the last sampled value of the control inputs vector. The linearization relies on Taylor series expansion and on the computation of the associated Jacobian matrices [27–29]. For the approximately linearized model of the aerial robot an optimal (H-infinity) feedback controller is designed. The modeling error that is due to the truncation of higher order terms in the Taylor series expansion is considered to be a perturbation that is finally compensated by the robustness of the control algorithm.

The proposed H-infinity control method stands for the solution of the optimal control problem for the nonlinear dynamics of the aerial robotic manipulator, under model uncertainties and external perturbations. Actually, it represents the solution to a min–max differential game, in which the controller tries to minimize a quadratic cost function of the system's state-vector error, whereas the model uncertainty and external disturbances terms try to maximize this cost functional. To compute the stabilizing feedback gains of the H-infinity controller, an algebraic Riccati equation is repetitively solved at each iteration of the control algorithm [30,31]. The stability properties of the control scheme are proven through Lyapunov analysis. It is demonstrated that the control loop of the aerial robotic manipulator satisfies the H-infinity tracking performance criterion [1,32]. This signifies elevated robustness against model uncertainties and external perturbations. Besides, under moderate conditions it is proven that the control scheme is globally asymptotically stable. Finally, to implement feedback control without the need to measure the entire state vector of the aerial robotic system, the H-infinity Kalman filter is used as a robust state estimator [1,33].

The structure of the chapter is as follows: in Section 8.2 the Lagrangian of the aerial robotic system is computed and the associated state-space model is obtained by applying the Euler–Lagrange method. In Section 8.3 the dynamic model of the aerial robotic manipulator with flexible joints is subject to approximate linearization through

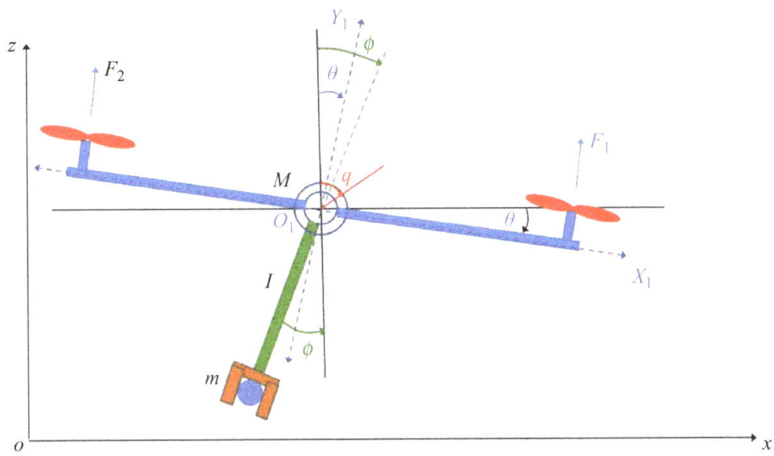

Figure 8.1 Reference frames for the aerial robotic manipulator with elastic joints

Taylor series expansion and the computation of the associated Jacobian matrices. In Section 8.4 the differential flatness properties of the dynamic model of the aerial robotic manipulator are proven. In Section 8.5 an H-infinity (optimal) controller is developed for the dynamic model of the aerial robotic manipulator. In Section 8.6 the stability properties of the control loop of the aerial robotic system are proven through Lyapunov analysis. In Section 8.7 the H-infinity Kalman filter is used as a robust state estimator that allows for implementing state estimation-based feedback control of the aerial robotic manipulator. In Section 8.8 the performance of the control scheme is tested through simulation experiments. Finally, in Section 8.9 concluding remarks are stated.

8.2 Dynamic model of the aerial robotic manipulator

The aerial robotic manipulator comprises a quadrotor UAV and a robotic arm with flexible joints, as shown in Figures 8.1 and 8.2. The dynamic model of the aerial robotic system will be obtained with the use of the Euler–Lagrange method.

The aerial robot is considered to move on the xz-plane as shown in Figures 8.1 and 8.2. The inertial coordinates system is denoted as XOZ. The body-fixed coordinates system is denoted as $X_1O_1Z_1$. The rotation (roll) angle of the UAV is denoted as θ. The rotation angle of the link of the robotic manipulator with respect to the vertical axis of the inertial system is denoted as ϕ. The joint at the basis of the link exhibits flexibility. As a result, the turn angle of the motor that provides rotation to the joint is not equal to ϕ and is denoted as q.

The forces and torques that are exerted on the aerial manipulator system are shown in the diagram of Figure 8.2: (i) there are lift forces generated by the rotors of the UAV, (ii) when the lift forces are uneven, a torque T_θ is generated, which makes

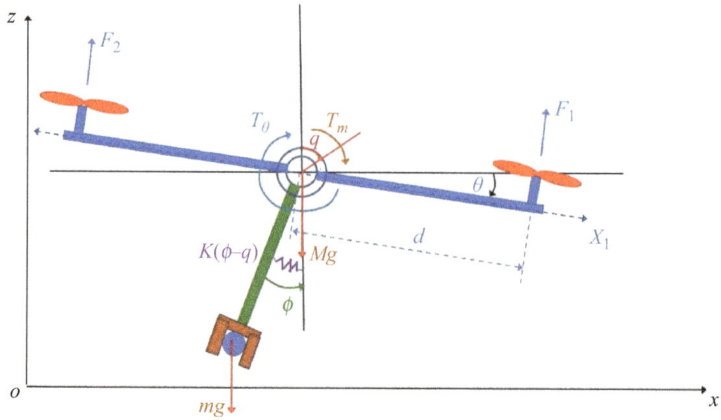

Figure 8.2 Forces and torques exerted on the aerial robotic manipulator with flexible joints

the UAV turn by the roll angle θ, (iii) the motor that is mounted on the basis of the robotic manipulator generates a torque T_m which makes the motor's rotor turn by an angle q. The transmission of this turn motion is performed through an elastic joint and the robotic link finally rotates by the aforementioned angle ϕ.

By denoting the mass of the UAV as M, its moment of inertia as I_U and the coordinates of its center of gravity in the inertial reference frame as (x, z), the kinetic energy of the UAV is given by

$$K_u = \tfrac{1}{2}I_u\dot{\theta}^2 + \tfrac{1}{2}M\dot{x}^2 + \tfrac{1}{2}M\dot{z}^2 \tag{8.1}$$

It is considered that the moment of inertia of the robotic link is I_m, and that its aggregate mass m is located at its end-effector, while the coordinates of the end-effector in the inertial frame are denoted as (x_m, y_m). Moreover, it is considered that the moment of inertia of the motor that rotates the flexible joint is J. Then, the kinetic energy of the robotic manipulator is given by

$$K_m = \tfrac{1}{2}I_m\dot{\theta}^2 + \tfrac{1}{2}m\dot{x}_m^2 + \tfrac{1}{2}m\dot{z}_m^2 + \tfrac{1}{2}J\dot{q}^2 \tag{8.2}$$

Considering that the length of the robotic link is l, the coordinates of the position and the velocity of the end-effector of the robotic manipulator in the inertial reference frame are given by

$$\begin{aligned} x_m &= x - l\sin(\phi) \Rightarrow \dot{x}_m = \dot{x} - l\cos(\phi)\dot{\phi} \\ z_m &= z - l\cos(\phi) \Rightarrow \dot{z}_m = \dot{z} + l\sin(\phi)\dot{\phi} \end{aligned} \tag{8.3}$$

Consequently, the aggregate kinetic energy of the aerial manipulator is

$$\begin{aligned} K = {} & \tfrac{1}{2}I_u\dot{\theta}^2 + \tfrac{1}{2}m\dot{x}^2 + \tfrac{1}{2}m\dot{z}^2 \\ & + \tfrac{1}{2}I_m\dot{\phi}^2 + \tfrac{1}{2}m(\dot{x} - l\cos(\phi)\dot{\phi})^2 + \tfrac{1}{2}m(\dot{z} + l\sin(\phi)\dot{\phi})^2 + \tfrac{1}{2}J\dot{q}^2 \end{aligned} \tag{8.4}$$

The potential energy of UAV is

$$P_u = Mgz \tag{8.5}$$

Considering that all mass of the manipulator is concentrated at its end-effector, its potential energy is

$$P_m = Mgz + mgz_m + \frac{1}{2}k(\phi - q)^2 \Rightarrow P_m = mg(z - l\cos(\phi)) + \frac{1}{2}k(\phi - q)^2 \tag{8.6}$$

The aggregate potential energy of the aerial robotic manipulator is

$$P = Mgz + mg(z - l\cos(\phi)) + \frac{1}{2}k(\phi - q)^2 \tag{8.7}$$

As a result of the above, the Lagrangian of the aerial robotic manipulator is

$$\begin{aligned} &L = K - P \Rightarrow \\ &L = \left[\tfrac{1}{2}I_u\dot{\theta}^2 + \tfrac{1}{2}m\dot{x}^2 + \tfrac{1}{2}m\dot{z}^2 \right.\\ &\quad \left. + \tfrac{1}{2}I_m\dot{\phi}^2 + \tfrac{1}{2}m(\dot{x} - l\cos(\phi)\dot{\phi})^2 + \tfrac{1}{2}m(\dot{z} + l\sin(\phi)\dot{\phi})^2 + \tfrac{1}{2}J\dot{q}^2\right] \\ &\quad - \left[Mgz + mg(z - l\cos(\phi)) + \tfrac{1}{2}k(\phi - q)^2\right] \end{aligned} \tag{8.8}$$

The state variables of the system are $[x, z, \theta, \phi, q]$ while their time derivatives are $[\dot{x}, \dot{z}, \dot{\theta}, \dot{\phi}, \dot{q}]$. By applying the Euler–Lagrange method one has

$$\begin{aligned} &\tfrac{\partial}{\partial t}\tfrac{\partial L}{\partial \dot{x}} - \tfrac{\partial L}{\partial x} = F_x \quad \tfrac{\partial}{\partial t}\tfrac{\partial L}{\partial \dot{z}} - \tfrac{\partial L}{\partial z} = F_z \\ &\tfrac{\partial}{\partial t}\tfrac{\partial L}{\partial \dot{\theta}} - \tfrac{\partial L}{\partial \theta} = T_\theta \quad \tfrac{\partial}{\partial t}\tfrac{\partial L}{\partial \dot{\phi}} - \tfrac{\partial L}{\partial \phi} = 0 \\ &\tfrac{\partial}{\partial t}\tfrac{\partial L}{\partial \dot{q}} - \tfrac{\partial L}{\partial q} = T_m \end{aligned} \tag{8.9}$$

where $F_x = (F_1 + F_2)\cos(\theta)$, $F_z = (F_1 + F_2)\sin(\theta)$, $T_\theta = (F_1 - F_2)d$, with d standing for the distance of the rotor from the UAV's center of gravity.

From the Euler–Lagrange equation with respect to x, and after intermediate operations one obtains

$$\begin{aligned} &\tfrac{\partial}{\partial t}\tfrac{\partial L}{\partial \dot{x}} - \tfrac{\partial L}{\partial x} = F_x \Rightarrow \\ &(M + m)\ddot{x} + ml\sin(\phi)\dot{\phi}^2 - ml\cos(\phi)\ddot{\phi} = F_x \end{aligned} \tag{8.10}$$

From the Euler–Lagrange equation with respect to z, and after intermediate operations one obtains

$$\begin{aligned} &\tfrac{\partial}{\partial t}\tfrac{\partial L}{\partial \dot{z}} - \tfrac{\partial L}{\partial z} = F_z \Rightarrow \\ &(M + m)\ddot{z} + ml\cos(\phi)\dot{\phi}^2 + ml\sin(\phi)\ddot{\phi} - Mg - mg = F_z \end{aligned} \tag{8.11}$$

From the Euler–Lagrange equation with respect to θ, and after intermediate operations one obtains

$$\begin{aligned} &\tfrac{\partial}{\partial t}\tfrac{\partial L}{\partial \dot{\theta}} - \tfrac{\partial L}{\partial \theta} = T_\theta \Rightarrow \\ &I_u\ddot{\theta} = T_\theta \end{aligned} \tag{8.12}$$

From the Euler–Lagrange equation with respect to ϕ, and after intermediate operations, one obtains

$$\begin{gathered}\frac{\partial}{\partial t}\frac{\partial L}{\partial \dot{\phi}}-\frac{\partial L}{\partial \phi}=0\Rightarrow\\ (I_m+ml^2)\ddot{\phi}-m\ddot{x}l\cos(\phi)+m\ddot{z}l\sin(\phi)-mgl\sin(\phi)-k(\phi-q)=0\end{gathered} \tag{8.13}$$

From the Euler–Lagrange equation with respect to q, and after intermediate operations, one obtains

$$\begin{aligned}&\frac{\partial}{\partial t}\frac{\partial L}{\partial \dot{q}}-\frac{\partial L}{\partial q}=0\Rightarrow\\ &J\ddot{q}+k(\phi-q)=T_m\end{aligned} \tag{8.14}$$

Equation (8.12) is independent from the rest of the state-space equations of the aerial robotic manipulator with elastic joints. This signifies that the roll angle θ of the UAV can be directly controlled by a control input of the form $T_\theta=(1/I_u)[\ddot{\theta}_d-k_d(\dot{\theta}-\dot{\theta}_d)-k_p(\theta-\theta_d)]$, where the control gains $k_d>0$, $k_p>0$ are chosen so as the characteristic polynomial $p(s)=s^2+k_ds+k_p$ to be a Hurwitz one.

Next, the state-space model of the aerial robotic system becomes

$$\begin{aligned}(M+m)\ddot{x}+ml\sin(\phi)\dot{\phi}^2-ml\cos(\phi)\ddot{\phi}&=F_x\\ (M+m)\ddot{z}+ml\cos(\phi)\dot{\phi}^2+ml\sin(\phi)\ddot{\phi}-Mg-mg&=F_z\\ (I_m+ml^2)\ddot{\phi}-m\ddot{x}l\cos(\phi)+m\ddot{z}l\sin(\phi)-mgl\sin(\phi)-k(\phi-q)&=0\\ J\ddot{q}+k(\phi-q)&=T_m\end{aligned} \tag{8.15}$$

By denoting the control inputs $u_1=F_x\Rightarrow u_1=(F_1+F_2)\cos(\phi)$, $u_2=F_Z+(M+m)g\Rightarrow u_2=(F_1+F_2)\sin(\phi)+(M+m)g$ and $u_3=T_m$ it can be seen that by solving the control problem for the aerial robotic manipulator, one also arrives at the following relation for the lift forces F_1, F_2:

$$(F_1+F_2)=\{u_1^2+[u_2-(M+m)g]^2\}^{1/2} \tag{8.16}$$

Besides, using that the distance between the center of gravity of the UAV and the position of its rotors is equal to d, (8.12) gives

$$(F_1-F_2)=\frac{I_u}{d}\ddot{\theta} \tag{8.17}$$

From (8.16) and (8.17) one can find the values of the lift forces of the rotors. Moreover, from $u_3=T_m$ one can find the value of the torque provided by the motor of the robotic link.

Next, using (8.15) the state-space model of the aerial robotic manipulator is formulated in matrix form. Actually, one has

$$\begin{pmatrix} (M+m) & 0 & -ml\cos(\phi) & 0 \\ 0 & (M+m) & ml\sin(\phi) & 0 \\ -ml\cos(\phi) & ml\sin(\phi) & (I_m+ml^2) & 0 \\ 0 & 0 & 0 & J \end{pmatrix} \begin{pmatrix} \ddot{x} \\ \ddot{z} \\ \ddot{\phi} \\ \ddot{q} \end{pmatrix} + \begin{pmatrix} ml\sin(\phi)\dot{\phi}^2 \\ -ml\cos(\phi)\dot{\phi}^2 \\ -mgl\sin(\phi)-k(\phi-q) \\ k(\phi-q) \end{pmatrix} = \begin{pmatrix} 1 & 0 & 0 \\ 0 & 1 & 0 \\ 0 & 0 & 0 \\ 0 & 0 & 1 \end{pmatrix} \begin{pmatrix} u_1 \\ u_2 \\ u_3 \end{pmatrix} \tag{8.18}$$

By denoting the system's state vector as $\tilde{q} = [x, z, \theta, \phi, q]^T$, the state-space description of the aerial robotic manipulator is written in the concise form

$$M(\tilde{q})\ddot{\tilde{q}} + C(\tilde{q}, \dot{\tilde{q}}) = G_m u \tag{8.19}$$

where $M(\tilde{q}) \in R^{4\times 4}$ is the inertia matrix, $C(\tilde{q}, \dot{\tilde{q}}) \in R^{4\times 1}$ is the Coriolis and centrifugal forces matrix, and $G_m \in R^{4\times 3}$ is the control inputs gains matrix. Next, the inverse of the inertia matrix $M(\tilde{q})^{-1}$ is computed as follows:

$$M(\tilde{q})^{-1} = \frac{1}{detM} \begin{pmatrix} M_{11} & -M_{21} & M_{31} & -M_{41} \\ -M_{12} & M_{22} & -M_{32} & M_{42} \\ M_{13} & -M23 & M_{33} & -M_{43} \\ -M_{14} & M_{24} & -M_{34} & M_{44} \end{pmatrix} \tag{8.20}$$

where $detM = J[(M+m)^2(I_m+ml^2) - (M+m)m^2l^2]$ and the sub-determinants of this matrix are defined as

$$\begin{aligned} &M_{11} = J[(M+m)(I_m) + ml^2] - m^2l^2\sin^2(\phi), \;\; M_{12} = J[m^2l^2\sin(\phi)\cos(\phi)], \\ &M_{13} = J[(M+m)ml\cos(\phi)], \;\; M_{14} = 0, \;\; M_{21} = J[m^2l^2\sin(\phi)\cos(\phi)], \\ &M_{22} = J[(M+m)(I_m+ml^2) - m^2l^2\cos^2(\phi)], \;\; M_{23} = J[(M+m)ml\sin(\phi)] \\ &M_{24} = 0, M_{31} = J[(M+m)ml\cos(\phi)] \;\; M_{32} = J[(M+m)ml\sin(\phi)], \\ &M_{33} = J(M+m)^2, \;\; M_{34} = 0, \;\; M_{41} = 0, \;\; M_{42} = 0, \;\; M_{43} = 0 \text{ and} \\ &M_{44} = (M+m)[(M+m)(I_m+ml^2) - m^2l^2]. \end{aligned}$$

Next, the state-space model of the aerial robotic manipulator is written as

$$\begin{aligned} &M(\tilde{q})\ddot{\tilde{q}} + C(\tilde{q}, \dot{\tilde{q}}) = G_m u \Rightarrow \\ &\qquad \ddot{\tilde{q}} = -M(\tilde{q})^{-1}C(\tilde{q}, \dot{\tilde{q}}) + M(\tilde{q})^{-1}G_m u \end{aligned} \tag{8.21}$$

Using that $C(\tilde{q}, \dot{\tilde{q}}) = [c_{11}, c_{21}, c_{31}, c_{41}]^T$, the product term $-M(\tilde{q})^{-1}C(\tilde{q}, \dot{\tilde{q}})$ becomes

$$-M(\tilde{q})^{-1}C(\tilde{q}, \dot{\tilde{q}}) = -\frac{1}{detM} \begin{pmatrix} M_{11}c_{11} - M_{21}c_{21} + M_{31}c_{31} - M_{41}c_{41} \\ -M_{12}c_{11} + M_{22}c_{21} - M_{32}c_{31} + M_{42}c_{41} \\ M_{13}c_{11} - M_{23}c_{21} + M_{33}c_{31} - M_{43}c_{41} \\ -M_{14}c_{11} + M_{24}c_{21} - M_{34}c_{31} + M_{44}c_{41} \end{pmatrix} \tag{8.22}$$

Moreover, about the product term $M(\tilde{q})^{-1}G_m$ one has

$$M(\tilde{q})^{-1}G_m = \frac{1}{detM}\begin{pmatrix} M_{11} & -M_{21} & -M_{41} \\ -M_{12} & M_{22} & M_{42} \\ M_{13} & -M_{23} & -M_{13} \\ -M_{14} & M_{24} & M_{44} \end{pmatrix} \tag{8.23}$$

Next, by defining the state variables of the aerial robotic system as $x_1 = x$, $x_2 = \dot{x}$, $x_3 = z$, $x_4 = \dot{z}$, $x_5 = \phi$, $x_6 = \dot{\phi}$, $x_7 = q$ and $x_8 = \dot{q}$, one obtains the following state-space description:

$$\dot{x}_1 = x_2 \tag{8.24}$$

$$\begin{aligned} \dot{x}_2 = & -\tfrac{1}{detM}(M_{11}c_{11} - M_{21}c_{21} + M_{31}c_{31} - M_{41}c_{41}) \\ & + \tfrac{M_{11}}{detM}u_1 - \tfrac{M_{21}}{detM}u_2 - \tfrac{M_{41}}{detM}u_3 \end{aligned} \tag{8.25}$$

$$\dot{x}_3 = x_4 \tag{8.26}$$

$$\begin{aligned} \dot{x}_4 = & -\tfrac{1}{detM}(-M_{12}c_{11} + M_{22}c_{21} - M_{32}c_{31} + M_{42}c_{41}) \\ & - \tfrac{M_{12}}{detM}u_1 + \tfrac{M_{22}}{detM}u_2 + \tfrac{M_{42}}{detM}u_3 \end{aligned} \tag{8.27}$$

$$\dot{x}_5 = x_6 \tag{8.28}$$

$$\begin{aligned} \dot{x}_6 = & -\tfrac{1}{detM}(M_{13}c_{11} - M_{23}c_{21} + M_{33}c_{31} - M_{43}c_{41}) \\ & + \tfrac{M_{13}}{detM}u_1 - \tfrac{M_{23}}{detM}u_2 + \tfrac{M_{43}}{detM}u_3 \end{aligned} \tag{8.29}$$

$$\dot{x}_7 = x_8 \tag{8.30}$$

$$\begin{aligned} \dot{x}_8 = & -\tfrac{1}{detM}(-M_{14}c_{11} + M_{24}c_{21} - M_{34}c_{31} + M_{44}c_{41}) \\ & - \tfrac{M_{14}}{detM}u_1 + \tfrac{M_{24}}{detM}u_2 + \tfrac{M_{44}}{detM}u_3 \end{aligned} \tag{8.31}$$

Equivalently, the state-space model of the aerial robotic manipulator can be written in the following affine-in-the-input form:

$$\begin{pmatrix} \dot{x}_1 \\ \dot{x}_2 \\ \dot{x}_3 \\ \dot{x}_4 \\ \dot{x}_5 \\ \dot{x}_6 \\ \dot{x}_7 \\ \dot{x}_8 \end{pmatrix} = \begin{pmatrix} f_1 \\ f_2 \\ f_3 \\ f_4 \\ f_5 \\ f_6 \\ f_7 \\ f_8 \end{pmatrix} + \begin{pmatrix} g_{11} \\ g_{21} \\ g_{31} \\ g_{41} \\ g_{51} \\ g_{61} \\ g_{71} \\ g_{81} \end{pmatrix} u_1 + \begin{pmatrix} g_{12} \\ g_{22} \\ g_{32} \\ g_{42} \\ g_{52} \\ g_{62} \\ g_{72} \\ g_{82} \end{pmatrix} u_2 + \begin{pmatrix} g_{13} \\ g_{23} \\ g_{33} \\ g_{43} \\ g_{53} \\ g_{63} \\ g_{73} \\ g_{83} \end{pmatrix} u_3 \tag{8.32}$$

which is also written concisely as

$$\dot{x} = f(x) + g_1(x)u_1 + g_2(x)u_2 + g_3(x)u_3 \tag{8.33}$$

with $x\in R^{8\times 1}$, $u\in R^{3\times 1}$, $f(x)\in R^{8\times 1}$, $g_1(x)\in R^{8\times 1}$, $g_2(x)\in R^{8\times 1}$ and $g_3(x)\in R^{8\times 1}$, while the elements of the aforementioned vector fields are defined as

$$f_1(x) = x_2, \quad g_{11}(x) = 0, \quad g_{12}(x) = 0, \quad g_{13}(x) = 0$$

$$f_2(x) = -\frac{1}{detM}(M_{11}c_{11} - M_{21}c_{21} + M_{31}c_{31} - M_{41}c_{41}), \quad g_{21}(x) = \frac{M_{11}}{detM},$$
$$g_{22}(x) = -\frac{M_{21}}{detM}, \quad g_{23}(x) = -\frac{M_{41}}{detM}$$

$$f_3(x) = x_4, \quad g_{31}(x) = 0, \quad g_{32}(x) = 0, \quad g_{33}(x) = 0$$

$$f_4(x) = -\frac{1}{detM}(-M_{12}c_{11} + M_{22}c_{21} - M_{32}c_{31} + M_{42}c_{41}), \quad g_{41}(x) = -\frac{M_{12}}{detM},$$
$$g_{42}(x) = \frac{M_{22}}{detM}, \quad g_{43}(x) = \frac{M_{42}}{detM}$$

$$f_5(x) = x_6, \quad g_{51}(x) = 0, \quad g_{52}(x) = 0, \quad g_{53}(x) = 0$$

$$f_6(x) = -\frac{1}{detM}(M_{13}c_{11} - M_{23}c_{21} + M_{33}c_{31} - M_{43}c_{41}), \quad g_{61}(x) = \frac{M_{13}}{detM},$$
$$g_{62}(x) = \frac{M_{23}}{detM}, \quad g_{63}(x) = \frac{M_{43}}{detM}$$

$$f_7(x) = x_8, \quad g_{71}(x) = 0, \quad g_{72}(x) = 0, \quad g_{73}(x) = 0$$

$$f_8(x) = -\frac{1}{detM}(-M_{14}c_{11} + M_{24}c_{21} - M_{34}c_{31} + M_{44}c_{41}), \quad g_{81}(x) = -\frac{M_{14}}{detM},$$
$$g_{82}(x) = \frac{M_{24}}{detM}, \quad g_{83}(x) = \frac{M_{44}}{detM}$$

8.3 Approximate linearization of the model of the aerial robotic manipulator

The dynamic model of the aerial robotic manipulator, previously written in the state-space form

$$\dot{x} = f(x) + g_1(x)u_1 + g_2(x)u_2 + g_3(x)u_3 \tag{8.34}$$

undergoes approximate linearization around a time-varying operating point (x^*, u^*), where x^* is the present value of the system's state vector and u^* is the last sampled value of the control inputs vector. The linearization relies on Taylor series expansion and on the computation of the associated Jacobian matrices. The linearization procedure gives the following model:

$$\dot{x} = Ax + Bu + \tilde{d} \tag{8.35}$$

where $\tilde{d}$ is the cumulative disturbance term which is due to the modeling error. This is induced by the truncation of higher order terms in the Taylor series expansion. It may also comprise the effects of the external perturbations. About matrices A and B one has

$$\begin{aligned} A = {} & [\nabla_x f(x)] \mid_{(x^*,u^*)} \\ & + [g_1(x)]u_1 \mid_{(x^*,u^*)} + [g_2(x)]u_1 \mid_{(x^*,u^*)} + [g_3(x)]u_3 \mid_{(x^*,u^*)} \end{aligned} \tag{8.36}$$

$$B = [g_1(x)\, g_2(x)\, g_3(x)] \mid_{(x^*,u^*)} \tag{8.37}$$

To compute the aforementioned Jacobian matrices, the following partial derivatives are given first:

$\frac{\partial M_{11}}{\partial x_i}=0$ for $i=1,2,3,4,6,7,8$ and $\frac{\partial M_{11}}{\partial x_5}=-2x_i^2l^2\sin(\phi)\cos(\phi)$, $\frac{\partial M_{12}}{\partial x_i}=0$ for $i=1,2,3,4,6,7,8$ and $\frac{\partial M_{12}}{\partial x_5}=Jm^2l^2[\cos^2(\phi)-\sin(\phi)]$, $\frac{\partial M_{13}}{\partial x_i}=0$ for $i=1,2,3,4,6,7,8$ and $\frac{\partial M_{13}}{\partial x_5}=J(M+m)ml(-\sin(\phi))$, $\frac{\partial M_{14}}{\partial x_i}=0$ for $i=1,2,3,4,5,6,7,8$

$\frac{\partial M_{21}}{\partial x_i}=0$ for $i=1,2,3,4,6,7,8$ and $\frac{\partial M_{21}}{\partial x_5}=Jm^2l^2[\cos(\phi)-\sin^2(\phi)]$, $\frac{\partial M_{22}}{\partial x_i}=0$ for $i=1,2,3,4,6,7,8$ and $\frac{\partial M_{22}}{\partial x_5}=Jm^2l^2 2\cos(\phi)\sin(\phi)$, $\frac{\partial M_{23}}{\partial x_i}=0$ for $i=1,2,3,4,6,7,8$ and $\frac{\partial M_{23}}{\partial x_5}=J(M+m)ml\cos(\phi)$, $\frac{\partial M_{24}}{\partial x_i}=0$ for $i=1,2,3,4,5,6,7,8$

$\frac{\partial M_{31}}{\partial x_i}=0$ for $i=1,2,3,4,6,7,8$ and $\frac{\partial M_{31}}{\partial x_5}=J(M+m)(ml)(-\sin(\phi))$, $\frac{\partial M_{32}}{\partial x_i}=0$ for $i=1,2,3,4,6,7,8$ and $\frac{\partial M_{32}}{\partial x_5}=J(M+m)(ml)\cos(\phi)$, $\frac{\partial M_{33}}{\partial x_i}=0$ for $i=1,2,3,4,5,6,7,8$, $\frac{\partial M_{34}}{\partial x_i}=0$ for $i=1,2,3,4,5,6,7,8$

$\frac{\partial M_{41}}{\partial x_i}=0$ for $i=1,2,3,4,5,6,7,8$, $\frac{\partial M_{42}}{\partial x_i}=0$ for $i=1,2,3,4,5,6,7,8$, $\frac{\partial M_{43}}{\partial x_i}=0$ for $i=1,2,3,4,5,6,7,8$, and $\frac{\partial M_{44}}{\partial x_i}=0$ for $i=1,2,3,4,5,6,7,8$

In a similar manner one computes

$$\frac{\partial c_{11}}{\partial x_i}=0 \text{ for } i=1,2,3,4,7,8,\ \frac{\partial c_{11}}{\partial x_5}=ml\cos(\phi)\dot{\phi}^2 \text{ and } \frac{\partial c_{11}}{\partial x_6}=2ml\sin(\phi)\dot{\phi}.$$

$$\frac{\partial c_{21}}{\partial x_i}=0 \text{ for } i=1,2,3,4,7,8,\ \frac{\partial c_{21}}{\partial x_5}=ml\sin(\phi)\dot{\phi}^2 \text{ and } \frac{\partial c_{21}}{\partial x_6}=-2ml\cos(\phi)\dot{\phi}.$$

$$\frac{\partial c_{31}}{\partial x_i}=0 \text{ for } i=1,2,3,4,6,8,\ \frac{\partial c_{31}}{\partial x_5}=-mgl\cos(\phi)-k \text{ and } \frac{\partial c_{31}}{\partial x_7}=k.$$

$$\frac{\partial c_{41}}{\partial x_i}=0 \text{ for } i=1,2,3,4,6,8,\ \frac{\partial c_{41}}{\partial x_5}=k \text{ and } \frac{\partial c_{41}}{\partial x_7}=-k.$$

The computation of the Jacobian matrices $\nabla_x f(x)\mid_{(x^*,u^*)}$, $\nabla_x g_1(x)\mid_{(x^*,u^*)}$, $\nabla_x g_2(x)\mid_{(x^*,u^*)}$, and $\nabla_x g_3(x)\mid_{(x^*,u^*)}$ is as follows:

First row of the Jacobian matrix $\nabla_x\mid_{(x^*,u^*)}$: $\frac{\partial f_1}{\partial x_1}=0$, $\frac{\partial f_1}{\partial x_2}=1$, $\frac{\partial f_1}{\partial x_3}=0$, $\frac{\partial f_1}{\partial x_4}=0$, $\frac{\partial f_1}{\partial x_5}=0$, $\frac{\partial f_1}{\partial x_6}=0$, $\frac{\partial f_1}{\partial x_7}=0$, $\frac{\partial f_1}{\partial x_8}=0$.

Second row of the Jacobian matrix $\nabla_x\mid_{(x^*,u^*)}$: for $i=1,2,\ldots,8$

$$\begin{aligned}\frac{\partial f_2}{\partial x_i}=-\frac{1}{detM}\cdot\Big[&\frac{\partial M_{11}}{\partial x_i}c_{11}+M_{11}\frac{\partial c_{11}}{\partial x_i}-\frac{\partial M_{21}}{\partial x_i}c_{21}-M_{21}\frac{\partial c_{21}}{\partial x_i}\\&+\frac{\partial M_{31}}{\partial x_i}c_{31}+M_{31}\frac{\partial c_{31}}{\partial x_i}-\frac{\partial M_{41}}{\partial x_i}c_{41}-M_{41}\frac{\partial c_{41}}{\partial x_i}\Big]\end{aligned} \tag{8.38}$$

Third row of the Jacobian matrix $\nabla_x\mid_{(x^*,u^*)}$: $\frac{\partial f_3}{\partial x_1}=0$, $\frac{\partial f_3}{\partial x_2}=0$, $\frac{\partial f_3}{\partial x_3}=0$, $\frac{\partial f_3}{\partial x_4}=1$, $\frac{\partial f_3}{\partial x_5}=0$, $\frac{\partial f_3}{\partial x_6}=0$, $\frac{\partial f_3}{\partial x_7}=0$, $\frac{\partial f_3}{\partial x_8}=0$.

Fourth row of the Jacobian matrix $\nabla_x \mid_{(x^*,u^*)}$: for $i = 1, 2, \ldots, 8$

$$\begin{aligned}\frac{\partial f_4}{\partial x_i} &= -\frac{1}{detM}\cdot\Big[-\frac{\partial M_{12}}{\partial x_i}c_{11} - M_{12}\frac{\partial c_{11}}{\partial x_i} + \frac{\partial M_{22}}{\partial x_i}c_{21} + M_{22}\frac{\partial c_{21}}{\partial x_i} \\ &\quad - \frac{\partial M_{32}}{\partial x_i}c_{31} - M_{32}\frac{\partial c_{31}}{\partial x_i} + \frac{\partial M_{42}}{\partial x_i}c_{41} + M_{42}\frac{\partial c_{41}}{\partial x_i}\Big]\end{aligned} \tag{8.39}$$

Fifth row of the Jacobian matrix $\nabla_x \mid_{(x^*,u^*)}$: $\frac{\partial f_5}{\partial x_1} = 0$, $\frac{\partial f_5}{\partial x_2} = 0$, $\frac{\partial f_5}{\partial x_3} = 0$, $\frac{\partial f_5}{\partial x_4} = 0$, $\frac{\partial f_5}{\partial x_5} = 0$, $\frac{\partial f_5}{\partial x_6} = 1$, $\frac{\partial f_5}{\partial x_7} = 0$, $\frac{\partial f_5}{\partial x_8} = 0$.

Sixth row of the Jacobian matrix $\nabla_x \mid_{(x^*,u^*)}$: for $i = 1, 2, \ldots, 8$

$$\begin{aligned}\frac{\partial f_6}{\partial x_i} &= -\frac{1}{detM}\cdot\Big[\frac{\partial M_{13}}{\partial x_i}c_{11} + M_{13}\frac{\partial c_{11}}{\partial x_i} - \frac{\partial M_{23}}{\partial x_i}c_{21} - M_{23}\frac{\partial c_{21}}{\partial x_i} \\ &\quad + \frac{\partial M_{33}}{\partial x_i}c_{31} + M_{33}\frac{\partial c_{31}}{\partial x_i} - \frac{\partial M_{44}}{\partial x_i}c_{41} + M_{43}\frac{\partial c_{41}}{\partial x_i}\Big]\end{aligned} \tag{8.40}$$

Seventh row of the Jacobian matrix $\nabla_x \mid_{(x^*,u^*)}$: $\frac{\partial f_7}{\partial x_1} = 0$, $\frac{\partial f_7}{\partial x_2} = 0$, $\frac{\partial f_7}{\partial x_3} = 0$, $\frac{\partial f_7}{\partial x_4} = 0$, $\frac{\partial f_7}{\partial x_5} = 0$, $\frac{\partial f_7}{\partial x_6} = 0$, $\frac{\partial f_7}{\partial x_7} = 0$, $\frac{\partial f_7}{\partial x_8} = 1$.

Eighth row of the Jacobian matrix $\nabla_x \mid_{(x^*,u^*)}$: for $i = 1, 2, \ldots, 8$

$$\begin{aligned}\frac{\partial f_8}{\partial x_i} &= -\frac{1}{detM}\cdot\Big[-\frac{\partial M_{14}}{\partial x_i}c_{11} - M_{14}\frac{\partial c_{11}}{\partial x_i} + \frac{\partial M_{24}}{\partial x_i}c_{21} + M_{24}\frac{\partial c_{21}}{\partial x_i} \\ &\quad - \frac{\partial M_{34}}{\partial x_i}c_{31} - M_{34}\frac{\partial c_{31}}{\partial x_i} + \frac{\partial M_{44}}{\partial x_i}c_{41} + M_{44}\frac{\partial c_{41}}{\partial x_i}\Big]\end{aligned} \tag{8.41}$$

In a concise form, the Jacobian matrix $\nabla_x f(x) \mid_{(x^*,u^*)}$

$$\nabla_x f(x) = \begin{pmatrix} \frac{\partial f_1}{\partial x_1} & \frac{\partial f_1}{\partial x_2} & \cdots & \frac{\partial f_1}{\partial x_8} \\ \frac{\partial f_2}{\partial x_1} & \frac{\partial f_2}{\partial x_2} & \cdots & \frac{\partial f_2}{\partial x_8} \\ \cdots & \cdots & \cdots & \cdots \\ \frac{\partial f_8}{\partial x_1} & \frac{\partial f_8}{\partial x_2} & \cdots & \frac{\partial f_8}{\partial x_8} \end{pmatrix} \tag{8.42}$$

Computation of the Jacobian matrix $\nabla_x g_1(x) \mid_{(x^*,u^*}$):

First row of the Jacobian matrix $\nabla_x g_1(x) \mid_{(x^*,u^*)}$: $\frac{\partial g_{11}(x)}{\partial x_i} = 0$ for $i = 1, 2, \ldots, 8$.

Second row of the Jacobian matrix $\nabla_x g_1(x) \mid_{(x^*,u^*)}$: $\frac{\partial g_{21}(x)}{\partial x_i} = \frac{\partial M_{11}}{\partial x_i}$ for $i = 1, 2, \ldots, 8$

Third row of the Jacobian matrix $\nabla_x g_1(x) \mid_{(x^*,u^*)}$: $\frac{\partial g_{31}(x)}{\partial x_i} = 0$ for $i = 1, 2, \ldots, 8$.

Fourth row of the Jacobian matrix $\nabla_x g_1(x) \mid_{(x^*,u^*)}$: $\frac{\partial g_{41}(x)}{\partial x_i} = -\frac{\partial M_{12}}{\partial x_i}$ for $i = 1, 2, \ldots, 8$.

Fifth row of the Jacobian matrix $\nabla_x g_1(x) \mid_{(x^*,u^*)}$: $\frac{\partial g_{51}(x)}{\partial x_i} = 0$ for $i = 1, 2, \ldots, 8$.

Sixth row of the Jacobian matrix $\nabla_x g_1(x) \mid_{(x^*,u^*)}$: $\frac{\partial g_{61}(x)}{\partial x_i} = \frac{\partial M_{13}}{\partial x_i}$ for $i = 1, 2, \ldots, 8$.

Seventh row of the Jacobian matrix $\nabla_x g_1(x)\mid_{(x^*,u^*)}$: $\frac{\partial g_{71}(x)}{\partial x_i} = 0$ for $i = 1, 2, \ldots, 8$.

Eighth row of the Jacobian matrix $\nabla_x g_1(x)\mid_{(x^*,u^*)}$: $\frac{\partial g_{81}(x)}{\partial x_i} = -\frac{\partial M_{14}}{\partial x_i}$ for $i = 1, 2, \ldots, 8$.

In a concise form, the Jacobian matrix $\nabla_x g_1(x)\mid_{(x^*,u^*)}$

$$\nabla_x g_1(x) = \begin{pmatrix} \frac{\partial g_{11}}{\partial x_1} & \frac{\partial g_{11}}{\partial x_2} & \cdots & \frac{\partial g_{11}}{\partial x_8} \\ \frac{\partial g_{21}}{\partial x_1} & \frac{\partial g_{21}}{\partial x_2} & \cdots & \frac{\partial g_{21}}{\partial x_8} \\ \cdots & \cdots & \cdots & \cdots \\ \frac{\partial g_{81}}{\partial x_1} & \frac{\partial g_{81}}{\partial x_2} & \cdots & \frac{\partial g_{81}}{\partial x_8} \end{pmatrix} \tag{8.43}$$

Computation of the Jacobian matrix $\nabla_x g_2(x)\mid_{(x^*,u^*}$):

First row of the Jacobian matrix $\nabla_x g_2(x)\mid_{(x^*,u^*)}$: $\frac{\partial g_{12}(x)}{\partial x_i} = 0$ for $i = 1, 2, \ldots, 8$.

Second row of the Jacobian matrix $\nabla_x g_2(x)\mid_{(x^*,u^*)}$: $\frac{\partial g_{22}(x)}{\partial x_i} = -\frac{\partial M_{21}}{\partial x_i}$ for $i = 1, 2, \ldots, 8$.

Third row of the Jacobian matrix $\nabla_x g_2(x)\mid_{(x^*,u^*)}$: $\frac{\partial g_{32}(x)}{\partial x_i} = 0$ for $i = 1, 2, \ldots, 8$.

Fourth row of the Jacobian matrix $\nabla_x g_2(x)\mid_{(x^*,u^*)}$: $\frac{\partial g_{42}(x)}{\partial x_i} = \frac{\partial M_{22}}{\partial x_i}$ for $i = 1, 2, \ldots, 8$

Fifth row of the Jacobian matrix $\nabla_x g_2(x)\mid_{(x^*,u^*)}$: $\frac{\partial g_{52}(x)}{\partial x_i} = 0$ for $i = 1, 2, \ldots, 8$.

Sixth row of the Jacobian matrix $\nabla_x g_2(x)\mid_{(x^*,u^*)}$: $\frac{\partial g_{62}(x)}{\partial x_i} = -\frac{\partial M_{23}}{\partial x_i}$ for $i = 1, 2, \ldots, 8$.

Seventh row of the Jacobian matrix $\nabla_x g_2(x)\mid_{(x^*,u^*)}$: $\frac{\partial g_{72}(x)}{\partial x_i} = 0$ for $i = 1, 2, \ldots, 8$.

Eighth row of the Jacobian matrix $\nabla_x g_2(x)\mid_{(x^*,u^*)}$: $\frac{\partial g_{82}(x)}{\partial x_i} = \frac{\partial M_{24}}{\partial x_i}$ for $i = 1, 2, \ldots, 8$.

In a concise form, the Jacobian matrix $\nabla_x g_2(x)\mid_{(x^*,u^*)}$

$$\nabla_x g_2(x) = \begin{pmatrix} \frac{\partial g_{12}}{\partial x_1} & \frac{\partial g_{12}}{\partial x_2} & \cdots & \frac{\partial g_{12}}{\partial x_8} \\ \frac{\partial g_{22}}{\partial x_1} & \frac{\partial g_{22}}{\partial x_2} & \cdots & \frac{\partial g_{22}}{\partial x_8} \\ \cdots & \cdots & \cdots & \cdots \\ \frac{\partial g_{82}}{\partial x_1} & \frac{\partial g_{82}}{\partial x_2} & \cdots & \frac{\partial g_{82}}{\partial x_8} \end{pmatrix} \tag{8.44}$$

Computation of the Jacobian matrix $\nabla_x g_3(x)\mid_{(x^*,u^*}$):

First row of the Jacobian matrix $\nabla_x g_3(x)\mid_{(x^*,u^*)}$: $\frac{\partial g_{13}(x)}{\partial x_i} = 0$ for $i = 1, 2, \ldots, 8$.

Second row of the Jacobian matrix $\nabla_x g_3(x)\mid_{(x^*,u^*)}$: $\frac{\partial g_{23}(x)}{\partial x_i} = -\frac{\partial M_{41}}{\partial x_i}$ for $i = 1, 2, \ldots, 8$.

Third row of the Jacobian matrix $\nabla_x g_3(x)\mid_{(x^*,u^*)}$: $\frac{\partial g_{33}(x)}{\partial x_i} = 0$ for $i = 1, 2, \ldots, 8$.

Fourth row of the Jacobian matrix $\nabla_x g_3(x)\mid_{(x^*,u^*)}$: $\frac{\partial g_{43}(x)}{\partial x_i} = \frac{\partial M_{42}}{\partial x_i}$ for $i = 1, 2, \ldots, 8$.

Fifth row of the Jacobian matrix $\nabla_x g_3(x) \mid_{(x^*,u^*)}$: $\frac{\partial g_{53}(x)}{\partial x_i} = 0$ for $i = 1, 2, \dots, 8$.

Sixth row of the Jacobian matrix $\nabla_x g_3(x) \mid_{(x^*,u^*)}$: $\frac{\partial g_{63}(x)}{\partial x_i} = -\frac{\partial M_{43}}{\partial x_i}$ for $i = 1, 2, \dots, 8$.

Seventh row of the Jacobian matrix $\nabla_x g_3(x) \mid_{(x^*,u^*)}$: $\frac{\partial g_{73}(x)}{\partial x_i} = 0$ for $i = 1, 2, \dots, 8$.

Eighth row of the Jacobian matrix $\nabla_x g_3(x) \mid_{(x^*,u^*)}$: $\frac{\partial g_{83}(x)}{\partial x_i} = \frac{\partial M_{44}}{\partial x_i}$ for $i = 1, 2, \dots, 8$.

In a concise form, the Jacobian matrix $\nabla_x g_3(x) \mid_{(x^*,u^*)}$

$$\nabla_x g_3(x) = \begin{pmatrix} \frac{\partial g_{13}}{\partial x_1} & \frac{\partial g_{13}}{\partial x_2} & \cdots & \frac{\partial g_{13}}{\partial x_8} \\ \frac{\partial g_{23}}{\partial x_1} & \frac{\partial g_{23}}{\partial x_2} & \cdots & \frac{\partial g_{23}}{\partial x_8} \\ \cdots & \cdots & \cdots & \cdots \\ \frac{\partial g_{83}}{\partial x_1} & \frac{\partial g_{83}}{\partial x_2} & \cdots & \frac{\partial g_{83}}{\partial x_8} \end{pmatrix} \tag{8.45}$$

It should be noted that without loss of generality, in the preceding dynamic model of the aerial robotic system, the robotic manipulator consisted of one link and one flexible joint. The entire robotic system exhibits already 5 degrees of freedom, denoted by its state variables, that is, $[x, y, \theta, \phi, q]$. It is straightforward to extend the modeling approach and the related design of the nonlinear optimal controller to the case of a robotic manipulator with more links and flexible joints.

8.4 Differential flatness properties of the aerial robotic manipulator

The flat outputs vector of the aerial robotic manipulator is taken to be $y = [y_1, y_2, y_3, y_4]^T$ or $y = [x, y, \phi, q]^T$. The control inputs vector, initially consisting of u_i, $i = 1, 2, 3$, is now extended with the inclusion of the joint elasticity parameter k as an additional control input. Thus, the extended control inputs vector is u_1, u_2, u_3, k. This allows one to satisfy the condition that in a multivariable differentially flat system the number of flat outputs should be equal to the number of the control inputs [31].

From (8.24) one has

$$x_2 = \dot{x}_1 \Rightarrow x_2 = \dot{y}_1 \tag{8.46}$$

Consequently, state variable x_2 is a differential function of the system's flat outputs. From (8.26) one has

$$x_4 = \dot{x}_3 \Rightarrow x_2 = \dot{y}_2 \tag{8.47}$$

Consequently, state variable x_4 is a differential function of the system's flat outputs. From (8.28) one has

$$x_6 = \dot{x}_5 \Rightarrow x_6 = \dot{y}_3 \tag{8.48}$$

Consequently, state variable x_6 is a differential function of the system's flat outputs. From (8.30) one has

$$x_8 = \dot{x}_7 \Rightarrow x_8 = \dot{y}_4 \tag{8.49}$$

Consequently, state variable x_8 is a differential function of the system's flat outputs. Moreover, from (8.25), (8.27), (8.29) and (8.31) one has a system of four equations with unknown elements of the control inputs vector. As a result, the control inputs of the aerial robotic manipulator can also be written as differential functions of the system's flat outputs:

$$u_i = f_u(y, \dot{y}) \tag{8.50}$$

This signifies that the aerial robotic manipulator with flexible joints is a differentially flat system.

8.5 The nonlinear H-infinity control

8.5.1 Tracking error dynamics

Next, a nonlinear optimal (H-infinity) controller will be developed for the aerial robotic manipulator with flexible joints. The initial nonlinear model of the aerial robotic system is in the form

$$\dot{x} = f(x, u) \quad x \in R^n, \ u \in R^m \tag{8.51}$$

Linearization is performed at each iteration of the control algorithm around its present operating point $(x^*, u^*) = (x(t), u(t - T_s))$. The linearized equivalent of the aerial robotic system is described by

$$\dot{x} = Ax + Bu + L\tilde{d} \quad x \in R^n, \ u \in R^m, \ \tilde{d} \in R^q \tag{8.52}$$

Thus, after linearization around its current operating point, the model of the aerial robotic manipulator with flexible joints is written as

$$\dot{x} = Ax + Bu + d_1 \tag{8.53}$$

Parameter d_1 stands for the linearization error in the model of the aerial robot appearing in (8.53). The reference set points for the system are denoted by $\mathbf{x_d} = [x_1^d, \ldots, x_4^d]$. Tracking of this trajectory is achieved after applying the control input u^*. At every time instant the control input u^* is assumed to differ from the control input u appearing in (8.53) by an amount equal to Δu, that is, $u^* = u + \Delta u$. One can write

$$\dot{x}_d = Ax_d + Bu^* + d_2 \tag{8.54}$$

The dynamics of the controlled system described in (8.54) can be also written as

$$\dot{x} = Ax + Bu + Bu^* - Bu^* + d_1 \tag{8.55}$$

and by denoting $d_3 = -Bu^* + d_1$ as an aggregate disturbance term, one obtains

$$\dot{x} = Ax + Bu + Bu^* + d_3 \tag{8.56}$$

By subtracting (8.54) from (8.56) one has

$$\dot{x} - \dot{x}_d = A(x - x_d) + Bu + d_3 - d_2 \tag{8.57}$$

By denoting the tracking error as $e = x - x_d$ and the aggregate disturbance term as $\tilde{d} = d_3 - d_2$, the tracking error dynamics becomes

$$\dot{e} = Ae + Bu + \tilde{d} \tag{8.58}$$

The above-linearized form of the model of the aerial robotic manipulator with flexible joints can be efficiently controlled after applying an H-infinity feedback control scheme.

8.5.2 Min–max control and disturbance rejection

The initial nonlinear model of the aerial robotic manipulator with flexible joints is in the form

$$\dot{x} = f(x, u) \;\; x \in R^n, \;\; u \in R^m \tag{8.59}$$

Linearization of the model of the aerial robotic manipulator is performed at each iteration of the control algorithm round its present operating point $(x^*, u^*) = (x(t), u(t - T_s))$. The linearized equivalent of the system is described by

$$\dot{x} = Ax + Bu + L\tilde{d} \;\; x \in R^n, \;\; u \in R^m, \;\; \tilde{d} \in R^q \tag{8.60}$$

where matrices A and B are obtained from the computation of the previously defined Jacobians and vector $\tilde{d}$ denotes disturbance terms due to linearization errors. The problem of disturbance rejection for the linearized model that is described by

$$\begin{aligned} \dot{x} &= Ax + Bu + L\tilde{d} \\ y &= Cx \end{aligned} \tag{8.61}$$

where $x \in R^n$, $u \in R^m$, $\tilde{d} \in R^q$ and $y \in R^p$, cannot be handled efficiently if the classical LQR (linear quadratic regulator) control scheme is applied. This is because of the existence of the perturbation term $\tilde{d}$. The disturbance term $\tilde{d}$ can represent (i) modeling (parametric) uncertainty and external perturbation terms that affect the aerial robotic system, (ii) noise terms of any distribution.

In the H-infinity control approach, a feedback control scheme is designed for trajectory tracking by the system's state vector and simultaneous disturbance rejection, considering that the disturbance affects the system in the worst possible manner. The effects that disturbances have on the aerial robotic manipulator are incorporated in the following quadratic cost function:

$$J(t) = \frac{1}{2}\int_0^T [y^T(t)y(t) + ru^T(t)u(t) - \rho^2\tilde{d}^T(t)\tilde{d}(t)]dt, \;\; r, \rho > 0 \tag{8.62}$$

Equation (8.62) denotes a min–max differential game taking place between disturbance and control inputs. Actually, the control inputs try to minimize this cost function while the disturbance inputs try to maximize it. Then, the optimal feedback control law is given by (Figure 8.3)

$$u(t) = -Kx(t) \tag{8.63}$$

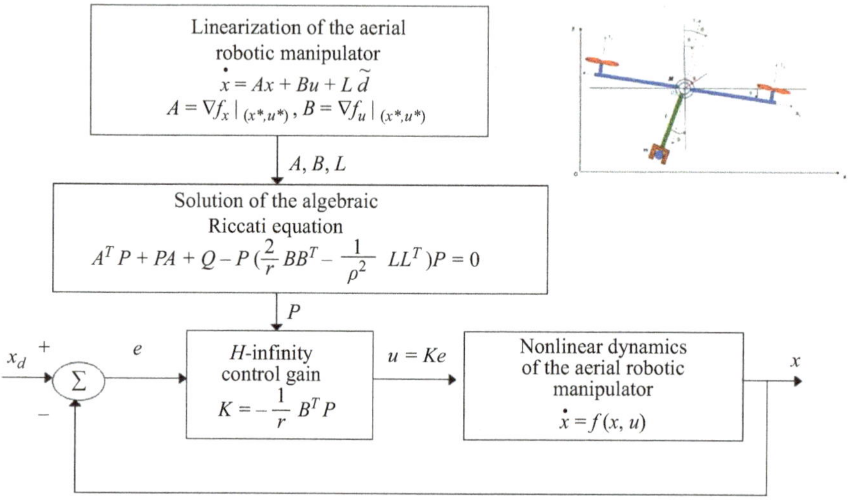

Figure 8.3 Diagram of the control scheme for the aerial robotic manipulator with elastic joints

with $K = (1/r)B^T P$, while P is a positive semi-definite symmetric matrix which is obtained from the solution of the Riccati equation of the form

$$A^T P + PA = -Q + P\Big(\frac{2}{r}BB^T - \frac{1}{\rho^2}LL^T\Big)P \tag{8.64}$$

The transients of the control algorithm are determined by matrix Q and also by gains r and ρ. The latter gain is the H-infinity attenuation coefficient, and its minimum value that allows the solution of (8.64) is the one that provides maximum robustness to the control algorithm for the aerial robotic manipulator.

8.6 Lyapunov stability analysis

Through Lyapunov stability analysis it will be shown that the proposed nonlinear control scheme assures H-infinity tracking performance for the aerial robotic manipulator, and that under moderate conditions about the disturbance terms, asymptotic convergence to the reference set points is achieved. The tracking error dynamics for the aerial robotic manipulator is written in the form

$$\dot{e} = Ae + Bu + L\tilde{d} \tag{8.65}$$

where in the aerial robotic manipulator's case $L = I \in R^{8\times 8}$ with I being the identity matrix. Variable $\tilde{d}$ denotes model uncertainties and external disturbances of the aerial manipulator's model. The following Lyapunov equation is considered:

$$V = \tfrac{1}{2}e^T Pe \tag{8.66}$$

where $e = x - x_d$ is the tracking error. By differentiating with respect to time, one obtains

$$\dot{V} = \tfrac{1}{2}\dot{e}^T Pe + \tfrac{1}{2}eP\dot{e} \Rightarrow$$
$$\dot{V} = \tfrac{1}{2}[Ae + Bu + L\tilde{d}]^T P + \tfrac{1}{2}e^T P[Ae + Bu + L\tilde{d}] \Rightarrow \tag{8.67}$$

$$\dot{V} = \tfrac{1}{2}[e^T A^T + u^T B^T + \tilde{d}^T L^T]Pe$$
$$+ \tfrac{1}{2}e^T P[Ae + Bu + L\tilde{d}] \Rightarrow \tag{8.68}$$

$$\dot{V} = \tfrac{1}{2}e^T A^T Pe + \tfrac{1}{2}u^T B^T Pe + \tfrac{1}{2}\tilde{d}^T L^T Pe$$
$$+ \tfrac{1}{2}e^T PAe + \tfrac{1}{2}e^T PBu + \tfrac{1}{2}e^T PL\tilde{d} \tag{8.69}$$

The previous equation is rewritten as

$$\dot{V} = \tfrac{1}{2}e^T(A^T P + PA)e + \left(\tfrac{1}{2}u^T B^T Pe + \tfrac{1}{2}e^T PBu\right)$$
$$+ \left(\tfrac{1}{2}\tilde{d}^T L^T Pe + \tfrac{1}{2}e^T PL\tilde{d}\right) \tag{8.70}$$

Assumption 8.1. For given positive definite matrix Q and coefficients r and ρ, there exists a positive definite matrix P, which is the solution of the following matrix equation:

$$A^T P + PA = -Q + P\left(\tfrac{2}{r}BB^T - \tfrac{1}{\rho^2}LL^T\right)P \tag{8.71}$$

Moreover, the following feedback control law is applied to the system

$$u = -\tfrac{1}{r}B^T Pe \tag{8.72}$$

By substituting (8.71) and (8.72) one obtains

$$\dot{V} = \tfrac{1}{2}e^T\left[-Q + P\left(\tfrac{2}{r}BB^T - \tfrac{1}{2\rho^2}LL^T\right)P\right]e$$
$$+ e^T PB\left(-\tfrac{1}{r}B^T Pe\right) + e^T PL\tilde{d} \Rightarrow \tag{8.73}$$

$$\dot{V} = -\tfrac{1}{2}e^T Qe + \left(\tfrac{1}{r}e^T PBB^T Pe - \tfrac{1}{2\rho^2}e^T PLL^T Pe\right.$$
$$\left. - \tfrac{1}{r}e^T PBB^T Pe\right) + e^T PL\tilde{d} \tag{8.74}$$

which after intermediate operations gives

$$\dot{V} = -\frac{1}{2}e^T Qe - \frac{1}{2\rho^2}e^T PLL^T Pe + e^T PL\tilde{d} \tag{8.75}$$

or, equivalently

$$\dot{V} = -\tfrac{1}{2}e^T Qe - \tfrac{1}{2\rho^2}e^T PLL^T Pe +$$
$$+ \tfrac{1}{2}e^T PL\tilde{d} + \tfrac{1}{2}\tilde{d}^T L^T Pe \tag{8.76}$$

Lemma 1. *The following inequality holds*

$$\tfrac{1}{2}e^T PL\tilde{d} + \tfrac{1}{2}\tilde{d}L^T Pe - \tfrac{1}{2\rho^2}e^T PLL^T Pe \leq \tfrac{1}{2}\rho^2\tilde{d}^T\tilde{d} \tag{8.77}$$

Proof. The binomial $\left(\rho\alpha - \frac{1}{\rho}b\right)^2$ is considered. Expanding the left part of the above inequality one gets

$$\begin{array}{c} \rho^2 a^2 + \frac{1}{\rho^2}b^2 - 2ab \geq 0 \Rightarrow \frac{1}{2}\rho^2 a^2 + \frac{1}{2\rho^2}b^2 - ab \geq 0 \Rightarrow \\ ab - \frac{1}{2\rho^2}b^2 \leq \frac{1}{2}\rho^2 a^2 \Rightarrow \frac{1}{2}ab + \frac{1}{2}ab - \frac{1}{2\rho^2}b^2 \leq \frac{1}{2}\rho^2 a^2 \end{array} \tag{8.78}$$

The following substitutions are carried out: $a = \tilde{d}$ and $b = e^T PL$ and the previous relation becomes

$$\frac{1}{2}\tilde{d}^T L^T Pe + \frac{1}{2}e^T PL\tilde{d} - \frac{1}{2\rho^2}e^T PLL^T Pe \leq \frac{1}{2}\rho^2 \tilde{d}^T \tilde{d} \tag{8.79}$$

Equation (8.79) is substituted in (8.76) and the inequality is enforced, thus giving

$$\dot{V} \leq -\frac{1}{2}e^T Qe + \frac{1}{2}\rho^2 \tilde{d}^T \tilde{d} \tag{8.80}$$

Equation (8.80) shows that the H-infinity tracking performance criterion is satisfied. The integration of $\dot{V}$ from 0 to T gives

$$\begin{array}{c} \int_0^T \dot{V}(t)dt \leq -\frac{1}{2}\int_0^T ||e||_Q^2 dt + \frac{1}{2}\rho^2 \int_0^T ||\tilde{d}||^2 dt \Rightarrow \\ 2V(T) + \int_0^T ||e||_Q^2 dt \leq 2V(0) + \rho^2 \int_0^T ||\tilde{d}||^2 dt \end{array} \tag{8.81}$$

Moreover, if there exists a positive constant $M_d > 0$ such that

$$\int_0^\infty ||\tilde{d}||^2 dt \leq M_d \tag{8.82}$$

then one gets

$$\int_0^\infty ||e||_Q^2 dt \leq 2V(0) + \rho^2 M_d \tag{8.83}$$

Thus, the integral $\int_0^\infty ||e||_Q^2 dt$ is bounded. Moreover, $V(T)$ is bounded and from the definition of the Lyapunov function V in (8.66) it becomes clear that $e(t)$ will be also bounded since $e(t) \in \Omega_e = \{e | e^T Pe \leq 2V(0) + \rho^2 M_d\}$. □

According to the above and with the use of Barbalat's Lemma one obtains $\lim_{t\to\infty} e(t) = 0$.

The outline of the global stability proof is that at each iteration of the control algorithm the state vector of the aerial robotic manipulator converges toward the temporary equilibrium and the temporary equilibrium in turn converges toward the reference trajectory [1]. Thus, the control scheme exhibits global asymptotic stability properties and not local stability. Assume the ith iteration of the control algorithm and the ith time interval about which a positive definite symmetric matrix P is obtained from the solution of the Riccati equation appearing in (8.71). By following the stages of the stability proof, one arrives at (8.80) which shows that the H-infinity tracking performance criterion holds. By selecting the attenuation coefficient ρ to be sufficiently small and in particular to satisfy $\rho^2 < ||e||_Q^2 / ||\tilde{d}||^2$ one has that the first derivative of the Lyapunov function is upper bounded by 0. Therefore, for the ith time interval it is proven that the Lyapunov function defined in (8.66) is a decreasing one. This signifies that between the beginning and the end of the ith time interval, there will be a drop of the value of the Lyapunov function and since matrix P is a positive definite one, the only way for this to happen is the Euclidean norm of the state vector error e to be decreasing. This means that comparing to the beginning of each time interval, the distance of the state vector error from 0 at the end of the time interval has diminished. Consequently as the iterations of the control algorithm advance the tracking error will approach zero, and this is a global asymptotic stability condition.

8.7 Robust state estimation with the use of the H-infinity Kalman filter

The control loop can be implemented with the use of information provided by a small number of sensors and by processing only a small number of state variables. To reconstruct the missing information about the state vector of the aerial robotic manipulator, it is proposed to use a filtering scheme and based on it to apply state estimation-based control [31,33]. The recursion of the H-infinity Kalman filter, for the model of the aerial robotic system, can be formulated in terms of a *measurement update* and a *time update* part.

Measurement update:

$$\begin{aligned} D(k) &= [I - \theta W(k)P^{-}(k) + C^{T}(k)R(k)^{-1}C(k)P^{-}(k)]^{-1} \\ K(k) &= P^{-}(k)D(k)C^{T}(k)R(k)^{-1} \\ \hat{x}(k) &= \hat{x}^{-}(k) + K(k)[y(k) - C\hat{x}^{-}(k)] \end{aligned} \tag{8.84}$$

Time update:

$$\begin{aligned} \hat{x}^{-}(k+1) &= A(k)x(k) + B(k)u(k) \\ P^{-}(k+1) &= A(k)P^{-}(k)D(k)A^{T}(k) + Q(k) \end{aligned} \tag{8.85}$$

where it is assumed that parameter θ is sufficiently small to assure that the covariance matrix $P^{-}(k)^{-1} - \theta W(k) + C^{T}(k)R(k)^{-1}C(k)$ will be positive definite. When $\theta = 0$ the H-infinity Kalman filter becomes equivalent to the standard Kalman filter. One can measure that only a part of the state vector of the UAV and suspended payload system, such as state variables $x_1 = y$, $x_3 = z$, $x_5 = \theta$ and $x_7 = \phi$, and can estimate through filtering the rest of the state vector elements. Moreover, the proposed Kalman filtering method can be used for sensor fusion purposes.

8.8 Simulation tests

The performance of the proposed nonlinear optimal (H-infinity) control method for the model of the 5-DOF aerial robotic manipulator was tested through simulation experiments. For the computation of the controller's feedback gains, the algebraic Riccati equation given in (8.71) had to be solved at each time-step of the control algorithm. The obtained results are depicted in Figures 8.4–8.15 and confirm fast and accurate tracking of the reference set points while also assuring that the control inputs will remain within the permitted ranges and that actuators saturation is avoided. In the obtained results, the real value of the system's state vector variables is depicted

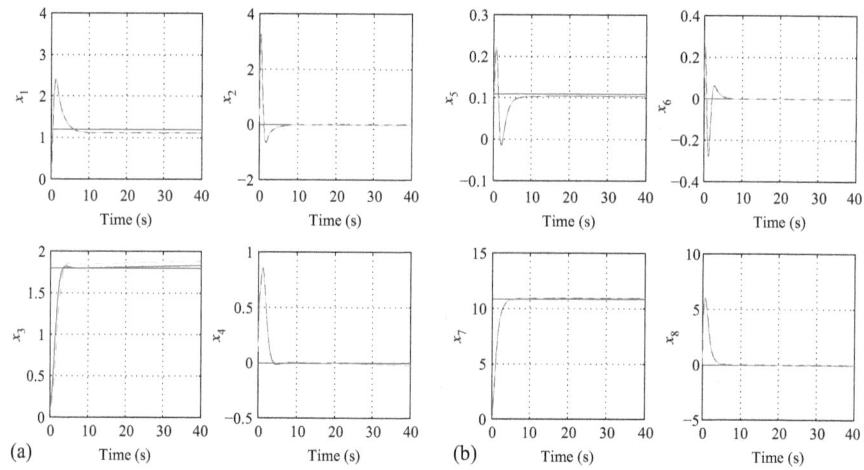

Figure 8.4 Tracking of set point 1 for the aerial robotic manipulator: (a) convergence of state variables x_1 (x-axis position of the UAV), x_2 (x-axis velocity of the UAV), x_3 (z-axis position of the UAV) and x_4 (z-axis velocity of the UAV) to their reference set points (red line: set point, blue line: real value, green line: estimated value); (b) convergence of state variables x_5 (rotation angle of the manipulator's link), x_6 (rotational speed of the manipulator's link), x_7 (rotation angle of the motor at the basis of the manipulator) and x_8 (rotation speed of motor at the basis of the manipulator) to their reference set points (red line: set point, blue line: real value, green line: estimated value)

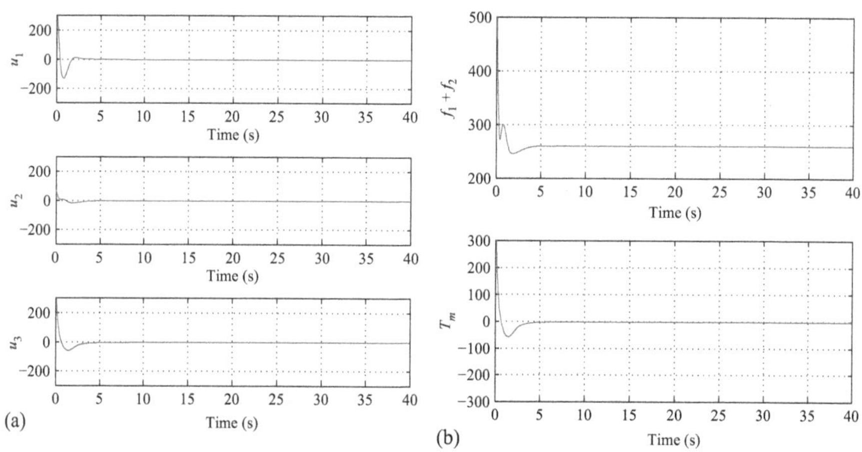

Figure 8.5 Tracking of set point 1 for the aerial robotic manipulator: (a) control inputs u_i $i = 1, 2, 3$ computed through the solution of the nonlinear optimal control problem; (b) control inputs $f_1 + f_2$ (aggregate lift force of the UAV) and T_m (torque of the motor mounted at the basis of the manipulator)

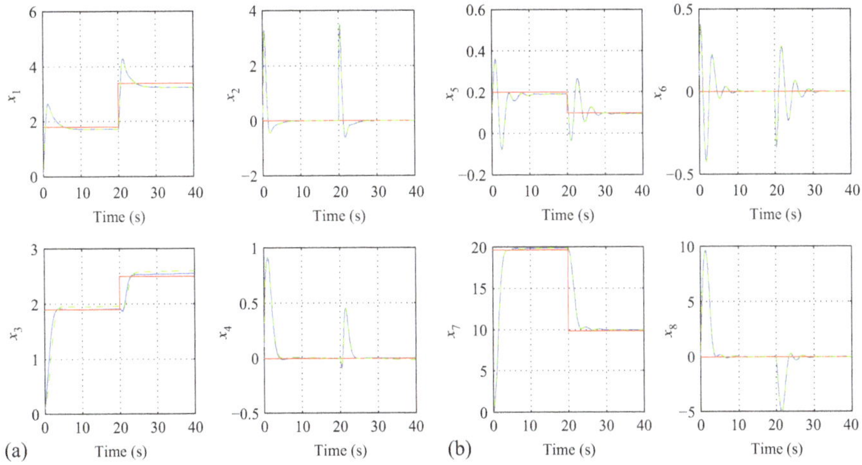

Figure 8.6 *Tracking of set point 2 for the aerial robotic manipulator: (a) convergence of state variables* x_1 *(x-axis position of the UAV),* x_2 *(x-axis velocity of the UAV),* x_3 *(z-axis position of the UAV) and* x_4 *(z-axis velocity of the UAV) to their reference set points (red line: set point, blue line: real value, green line: estimated value); (b) convergence of state variables* x_5 *(rotation angle of the manipulator's link),* x_6 *(rotational speed of the manipulator's link),* x_7 *(rotation angle of the motor at the basis of the manipulator) and* x_8 *(rotation speed of motor at the basis of the manipulator) to their reference set points (red line: set point, blue line: real value, green line: estimated value)*

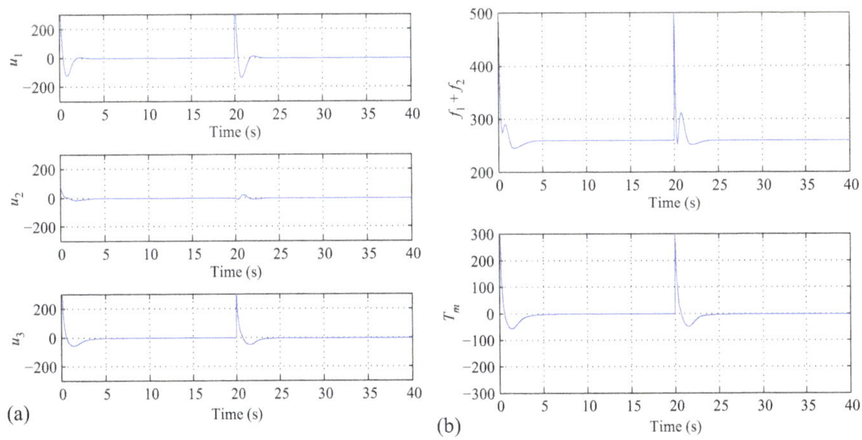

Figure 8.7 *Tracking of set point 2 for the aerial robotic manipulator: (a) control inputs* u_i $i = 1, 2, 3$ *computed through the solution of the nonlinear optimal control problem; (b) control inputs* $f_1 + f_2$ *(aggregate lift force of the UAV's motors) and* T_m *(torque of the motor mounted at the basis of the manipulator)*

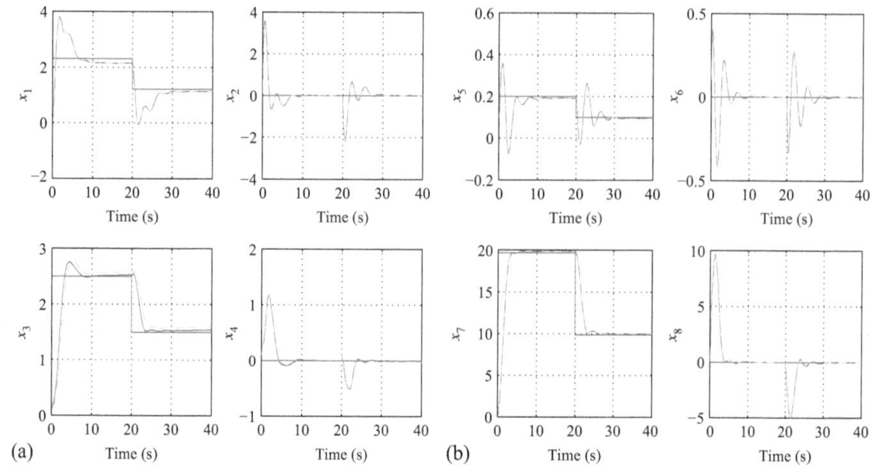

Figure 8.8 Tracking of set point 3 for the aerial robotic manipulator: (a) convergence of state variables x_1 (x-axis position of the UAV), x_2 (x-axis velocity of the UAV), x_3 (z-axis position of the UAV) and x_4 (z-axis velocity of the UAV) to their reference set points (red line: set point, blue line: real value, green line: estimated value); (b) convergence of state variables x_5 (rotation angle of the manipulator's link), x_6 (rotational speed of the manipulator's link), x_7 (rotation angle of the motor at the basis of the manipulator) and x_8 (rotation speed of motor at the basis of the manipulator) to their reference set points (red line: set point, blue line: real value, green line: estimated value)

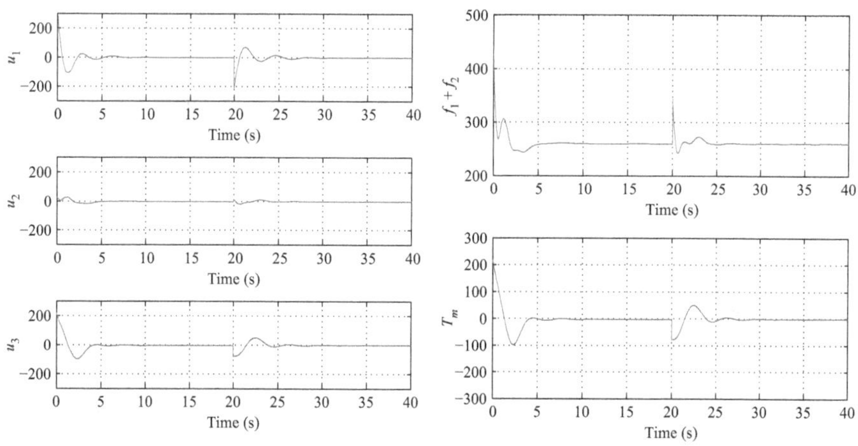

Figure 8.9 Tracking of set point 3 for the aerial robotic manipulator: (a) control inputs u_i $i = 1, 2, 3$ computed through the solution of the nonlinear optimal control problem; (b) control inputs $f_1 + f_2$ (aggregate lift force of the UAV's motors) and T_m (torque of the motor mounted at the basis of the manipulator)

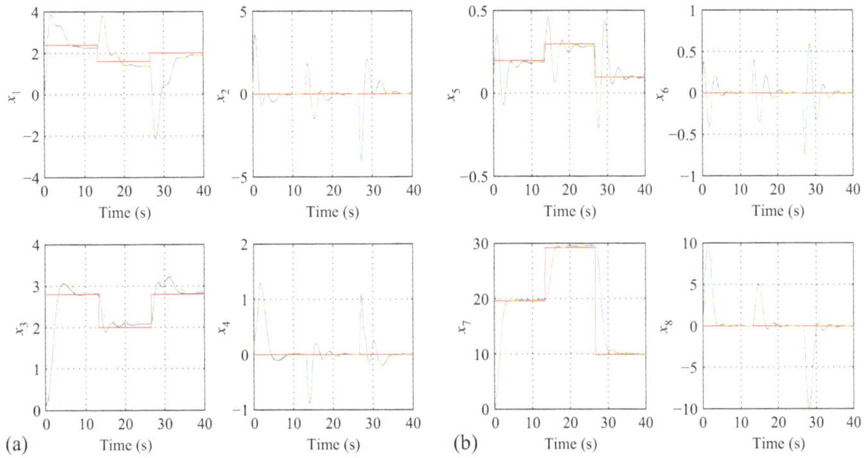

Figure 8.10 Tracking of set point 4 for the aerial robotic manipulator: (a) convergence of state variables x_1 (x-axis position of the UAV), x_2 (x-axis velocity of the UAV), x_3 (z-axis position of the UAV) and x_4 (z-axis velocity of the UAV) to their reference set points (red line: set point, blue line: real value, green line: estimated value); (b) convergence of state variables x_5 (rotation angle of the manipulator's link), x_6 (rotational speed of the manipulator's link), x_7 (rotation angle of the motor at the basis of the manipulator) and x_8 (rotation speed of motor at the basis of the manipulator) to their reference set points (red line: set point, blue line: real value, green line: estimated value)

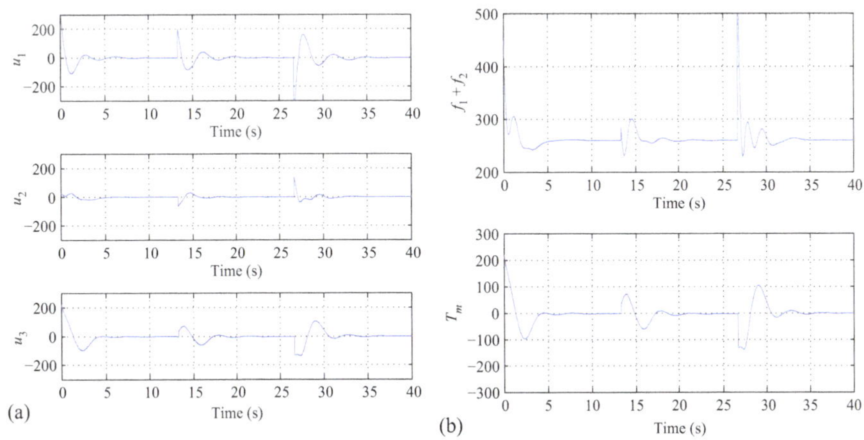

Figure 8.11 Tracking of set point 4 for the aerial robotic manipulator: (a) control inputs u_i $i = 1, 2, 3$ computed through the solution of the nonlinear optimal control problem; (b) control inputs $f_1 + f_2$ (aggregate lift force of the UAV's motors) and T_m (torque of the motor mounted at the basis of the manipulator)

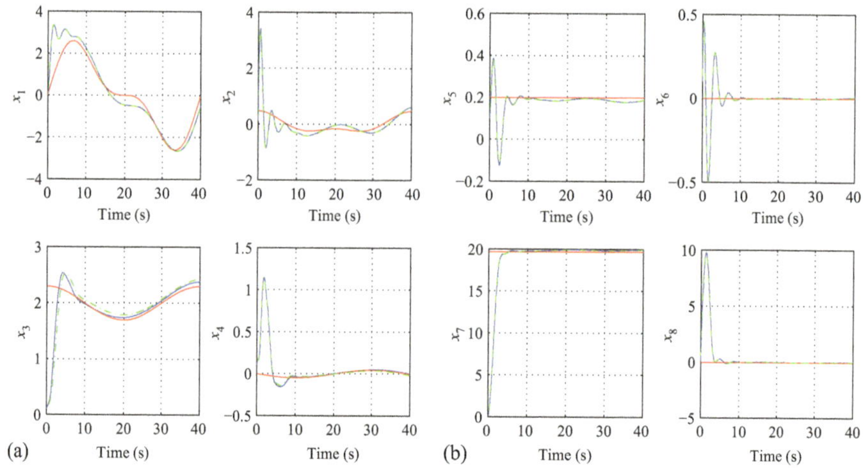

Figure 8.12 Tracking of set point 5 for the aerial robotic manipulator: (a) convergence of state variables x_1 (x-axis position of the UAV), x_2 (x-axis velocity of the UAV), x_3 (z-axis position of the UAV) and x_4 (z-axis velocity of the UAV) to their reference set points (red line: set point, blue line: real value, green line: estimated value); (b) convergence of state variables x_5 (rotation angle of the manipulator's link), x_6 (rotational speed of the manipulator's link), x_7 (rotation angle of the motor at the basis of the manipulator) and x_8 (rotation speed of motor at the basis of the manipulator) to their reference set points (red line: set point, blue line: real value, green line: estimated value)

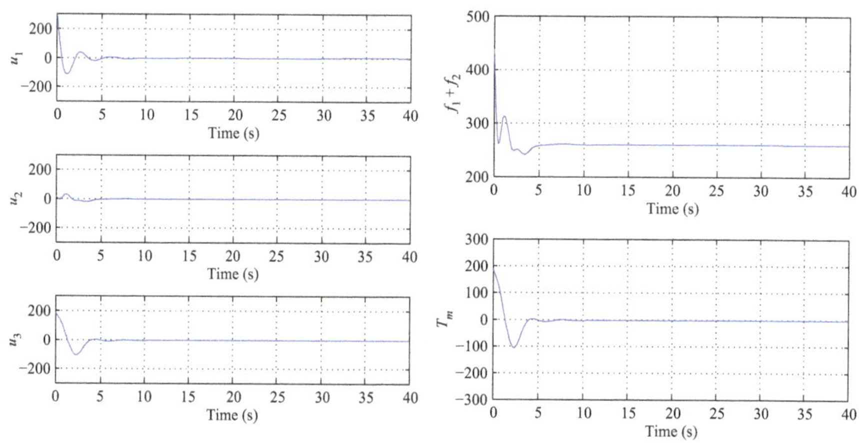

Figure 8.13 Tracking of set point 5 for the aerial robotic manipulator: (a) control inputs u_i $i = 1, 2, 3$ computed through the solution of the nonlinear optimal control problem; (b) control inputs $f_1 + f_2$ (aggregate lift force of the UAV's motors) and T_m (torque of the motor mounted at the basis of the manipulator)

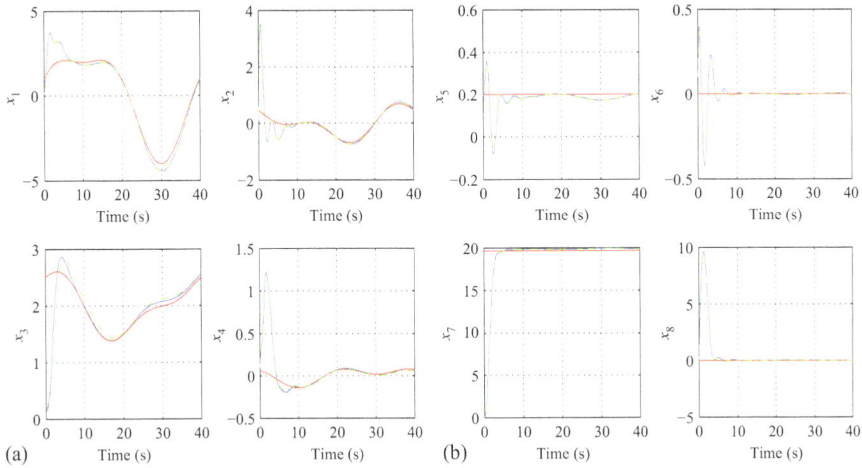

Figure 8.14 Tracking of set point 6 for the aerial robotic manipulator: (a) convergence of state variables x_1 (x-axis position of the UAV), x_2 (x-axis velocity of the UAV), x_3 (z-axis position of the UAV) and x_4 (z-axis velocity of the UAV) to their reference set points (red line: set point, blue line: real value, green line: estimated value): (b) convergence of state variables x_5 (rotation angle of the manipulator's link), x_6 (rotational speed of the manipulator's link), x_7 (rotation angle of the motor at the basis of the manipulator) and x_8 (rotation speed of motor at the basis of the manipulator) to their reference set points (red line: set point, blue line: real value, green line: estimated value)

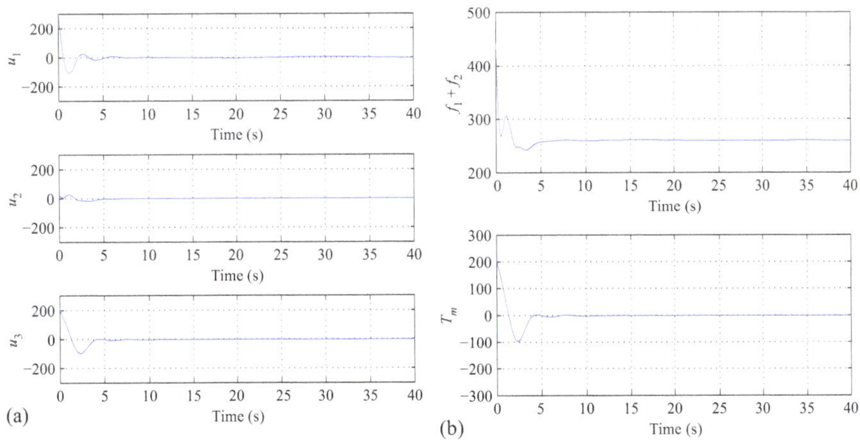

Figure 8.15 Tracking of set point 6 for the aerial robotic manipulator: (a) control inputs u_i $i = 1, 2, 3$ computed through the solution of the nonlinear optimal control problem; (b) control inputs $f_1 + f_2$ (aggregate lift force of the UAV's motors) and T_m (torque of the motor mounted at the basis of the manipulator)

in blue, the estimated value provided by the H-infinity Kalman filter is depicted in green, while the associated reference setpoints are plotted in red.

The transient performance of the control algorithm relies on parameters r, ρ and Q appearing in the previously noted algebraic Riccati equation. The values of r and Q determine the speed of convergence to the reference set points, while the value of ρ determines the robustness of the control algorithm. Actually, the smallest value of ρ for which the aforementioned Riccati equation admits as solution a positive definite matrix P is the one that provides maximum robustness to the control loop. Besides, the use of the H-infinity Kalman filter as a robust state estimator enables one to implement feedback control without the need to measure the entire state vector of the aerial robotic manipulator. Actually, one can measure only state variable $x_1 = x$, $x_3 = z$ and $x_5 = \phi$ and can estimate the rest five state variables of the system through the filter's recursion.

Comparing to other control schemes for the model of the aerial robotic manipulator, the proposed nonlinear optimal control method is assessed as follows: (i) it avoids complicated state variables transformations which are met in global linearization-based control approaches, (ii) it is applied directly on the nonlinear state-space model of the aerial robotic manipulator, thus avoiding inverse transformations and the related singularity problems which are met in global linearization-based control methods, (iii) it retains the advantages of typical optimal control, that is, fast and accurate tracking of the reference set points under moderate variations of the control inputs, (iv) unlike PID control it exhibits global stability properties and assures the reliable functioning of the control loop even under changes of the operating points (e.g., Cartesian coordinates and orientation angle set points of the UAV or turn angles of the links of the robotic manipulator), (v) unlike popular approaches for implementing optimal control such as MPC (model predictive control) the application of the control method is not hindered by the system's nonlinearities, (vi) unlike popular approaches for implementing optimal control such as NMPC (nonlinear model predictive control) there is proven convergence to an optimum without dependence on any initialization, (vii) unlike sliding-mode control or backstepping control, the use of the proposed control approach is feasible even if state-space transformations into convenient forms are not previously applied.

8.9 Conclusions

The chapter has presented a nonlinear optimal (H-infinity) control method for aerial robotic manipulators with flexible joints, that is, for aerial robotic systems that comprise a multi-rotor UAV and a robotic manipulator with flexible joints. First, the Lagrangian of the robotic system was computed and the associated 5-DOF dynamic model was obtained by applying the Euler–Lagrange principle. Next, the nonlinear dynamic model of the aerial robotic system was subject to approximate linearization around a time-varying operating point that was recomputed at each iteration of the control algorithm. The linearization point was defined by the present value of the system's state vector and by the last sampled value of the control inputs vector.

The linearization procedure relied on Taylor series expansion and on the computation of the system's Jacobian matrices. The modeling error which was due to truncation of higher order terms in the Taylor series expansion was considered to be a perturbation that was finally compensated by the robustness of the control algorithm.

For the approximately linearized model of the aerial robotic manipulator an H-infinity optimal feedback controller was designed. This controller achieves the solution of the optimal control problem for the aerial robotic system under model uncertainties and external perturbations. To find the stabilizing feedback gains of the controller, an algebraic Riccati equation had to be repetitively solved at each time-step of the control method. The stability properties of the control scheme were proven through Lyapunov analysis. First, it was demonstrated that the control loop satisfied the H-infinity tracking performance criterion, which signified elevated robustness against model uncertainty and external perturbations. Moreover, under moderate conditions it was proven that the control scheme is globally asymptotically stable. Finally, to implement state estimation-based control without the need to measure the entire state vector of the system, the H-infinity Kalman filter has been used as a robust state estimator.

References

[1] Rigatos G., Busawon K. Robotic manipulators and vehicles: Control, Estimation and Filtering. Cham: Springer. 2018.

[2] Acosta J.A., de Cos C.R., Ollero A. A robust decentralized strategy for multi-task control of unmanned aerial systems: Application on underactuated aerial manipulator. 2016 Intl. Conf. on Unmanned Aircraft Systems. IEEE ICUAS 2016. Jun. 2016, Arlington, VA.

[3] de Cos C.R., Acosta J.A., Ollero A. Relation-pose optimization for robust and nonlinear control of unmanned aerial manipulators. IEEE ICUAS 2017, IEEE 2017 Intl. Conf. on Unmanned Aerial Systems. Jun. 2017, Miami, FL.

[4] Lumni D., Santamariti-Navarro A., Rossi R., Rocco P., Bascetta L., Andrade-Cetto T. Nonlinear model predictive control for aerial manipulation. IEEE ICUAS 2017, IEEE 2017 Intl. Conf. on Unmanned Aerial Systems. Jun. 2017, Miami, FL.

[5] Beikzadeh H., Liu G. Trajectory tracking of quadrotor flying manipulators using L_1 adaptive control. Journal of the Franklin Institute. 2018, 355(4):6239–6261.

[6] Jimenez-Cano A.E., Martin J., Heredia G., Ollero A., Cano R. Control of an aerial robot with multi-link arm for assembly tasks. IEEE ICRA 2013, IEEE 2013 Intl. Conf. on Robotics and Automation. Oct. 2013, Karlsruhe, Germany.

[7] Forte F., Naldi R., Macchelli A., Marconi L. Impedance control of an aerial manipulator. IEEE ACC 2012, IEEE 2012 American Control Conference. Jun. 2012, Montreal, Canada.

[8] Lippiello V., Ruggiero F. Cartesian impedance control of a UAV with a robotic arm. IFAC SYROCO 2012, 10th IFAC Symposium on Robot Control. Sep. 2012, Dubrovnik, Croatia.

[9] Suarez A., Jimenez-Cano A.E., Vega V.M., Heredia G., Rodriguez-Castano A., Ollero A. Design of a lightweight dual-arm system for aerial manipulation. Mechatronics. 2018, 50:30–44.

[10] Acosta J.A., Sanchez M.I., Ollero A. Robust control of underactuated aerial manipulators via IDA-PBC. IEEE CDC 2014, 53rd IEEE Conf. on Decision and Control. Dec. 2014, Los Angeles, CA.

[11] Rugierro F., Trajillo M.A., Cano R., *et al.* A multi-layer control for multi-rotor UAVs equipped with a servo robot arm. IEEE ICRA 2015, IEEE 2015 Intl. Conf. on Robotics and Automation. May 2015, Seattle, WA.

[12] Meng X., He Y., Gu F., *et al.* Design and implementation of rotor aerial manipulator system. IEEE 2016 Intl. Conf. on Robotics and Biomimetics. Dec. 2016, Qingdao, China.

[13] Orsag M., Corpela C., Bogdan S., Oh P. Dexterous aerial robots in mobile manipulation using unmanned aerial systems. IEEE Transactions on Robotics. 2017, 33(6):1453–1476.

[14] Yang, B., He, Y., Han J., Liu, G. Modelling and control of rotor-flying multi-joint manipulator. Proc. 19th IFAC World Congress. Aug. 2014, Cape Town, South Africa.

[15] Caccavale F., Giglio G., Muscio G., Pierri F. Adaptive control for UAVs equipped with a robotic arm. Proc. IFAC World Congress. Aug. 2014, Cape Town, South Africa.

[16] Escareno J., Castillo J., Abassi W., Flores G., Camarillo K. Navigation strategy in flight retrieving and transportation for a rotorcraft MAV. 4th Workshop on Research, Education and Development of Unmanned Aerial Systems, REF-UAS 2017. Oct. 2017, Linköping, Sweden.

[17] Suarez A., Heredia G., Ollero A. Physical virtual impedance control in ultra-lightweight and compliant dual-arm aerial manipulators. IEEE Robotics and Automation Letters. 2018, 3:2553–2560.

[18] Khalifa A., Fanni M., Namerikawa T. MPC and DOb-based robust optimal control of a new quadrotor manipulation system. ECC 2016, European Control Conference 2016. Jun. 2016, Aalborg, Denmark.

[19] Garimella G., Kobilarov M. Towards model-predictive control for aerial pick and place. IEEE ICRA 2015, IEEE 2015 Intl. Conf. on Robotics and Automation. Jul. 2013, Seattle, WA.

[20] Lippiello V., Ruggiero F. Cartesian impedance control of a UAV with a robotic arm. 10th IFAC Symposium on Robot Control. Sep. 2012, Dubrovnik, Croatia.

[21] Yuksel B., Mahboubi S., Secchi C., Buhthoff H.H., Franchi A. Design, identification and experimental testing of a light-weight flexible-joint arm for aerial physical interaction. IEEE ICRA 2015, IEEE 2015 Intl. Conf. on Robotics and Automation. May 2015, Seattle, WA.

[22] Yuksel B., Buondonno G., Franchi A. Differential flatness and control of protocentric aerial manipulators with any number of arms and mixed rigid/flexible joints. IEEE IROS 2016, IEEE 2016 Intl. Conf. on Intelligent Robots and Systems. Oct. 2016, Daejeon, South Korea.
[23] Yuksel B., Staub N. Franchi A. Aerial robots with rigid/elastic joint arms: Single joint controllability study and preliminary experiments. IEEE IROS 2016, IEEE 2016 Intl. Conf. on Intelligent Robots and Systems. Oct. 2016, Daejeon, South Korea.
[24] Rigatos G. Intelligent renewable energy systems: Modelling and control. Cham: Springer. 2016.
[25] Rigatos G., Siano P., Cecati C. A new non-linear H-infinity feedback control approach for three-phase voltage source converters. Electric Power Components and Systems. 2015, 44(3):302–312.
[26] Rigatos G., Siano P., Wira P., Profumo F. Nonlinear H-infinity feedback control for asynchronous motors of electric trains. Journal of Intelligent Industrial Systems. 2015, 1(2):85–98.
[27] Rigatos G.G., Tzafestas S.G. Extended Kalman filtering for fuzzy modelling and multi-sensor fusion. Mathematical and Computer Modelling of Dynamical Systems. 2007, 13:251–266.
[28] Basseville M., Nikiforov I. Detection of abrupt changes: Theory and Applications. Upper Saddle River, NJ: Prentice-Hall. 1993.
[29] Rigatos G., Zhang Q. Fuzzy model validation using the local statistical approach. Fuzzy Sets and Systems. 2009, 60(7):882–904.
[30] Rigatos G. Modelling and control for intelligent industrial systems: Adaptive algorithms in robotics and industrial engineering. Berlin, Heidelberg: Springer. 2011.
[31] Rigatos G. Nonlinear control and filtering using differential flatness theory approaches: Applications to electromechanical systems. Cham: Springer. 2015.
[32] Toussaint G.J., Basar T., Bullo F. H_∞ optimal tracking control techniques for nonlinear underactuated systems. Proc. IEEE CDC 2000, 39th IEEE Conference on Decision and Control. 2000, Sydney, Australia.
[33] Gibbs B.P. Advanced Kalman filtering, least squares and modelling: A practical handbook. Hoboken, NJ: John Wiley & Sons, Inc. 2011.

Chapter 9
Fault diagnosis and fault-tolerant control techniques for aircraft systems

Paolo Castaldi[1], Nicola Mimmo[1], and Silvio Simani[2]

This chapter analyses and discusses an active fault-tolerant control (FTC) system for avionic applications. The approach applies to an aircraft longitudinal autopilot in the presence of faults affecting the system actuators. The key point of the developed FTC scheme consists of its active feature, since the fault diagnosis module provides a robust and reliable estimation of the fault signals, which are thus compensated. The design technique, relying on a nonlinear geometric approach (NLGA), i.e. a differential geometry tool, allows one to achieve adaptive filters (AFs), which provides both disturbance-decoupled fault estimates and fault isolation features. The chapter also shows how these fault estimates are thus exploited for control accommodation. In particular, by means of this NLGA, it is possible to obtain fault reconstructions decoupled from the wind components of the considered aircraft application. It is shown how this solution provides very good robustness characteristics and performances to the overall system. Finally, the effectiveness of the considered scheme is analysed by means of a high fidelity flight simulator, in different conditions and in the presence of actuator faults, turbulence, measurement noise and modelling errors.

9.1 Introduction

A conventional feedback control design for a complex system may lead to unsatisfactory performance, or even instability, in the event of malfunctions affecting actuators, sensors or other system components. This is particularly important for safety-critical systems, such as aircraft applications. In these cases, the effect of a minor fault in a system component, in particular, the actuators, can lead to catastrophic consequences.

To overcome these drawbacks, FTC systems have been developed in order to tolerate component malfunctions, while maintaining desirable stability, and performance properties. In general, FTC methods are classified into two types, i.e. passive fault-tolerant control scheme (PFTCS) and active fault-tolerant control scheme (AFTCS)

[1] Department of Electrical, Electronic and Information Engineering (DEI), University of Bologna, Bologna, Italy
[2] Department of Engineering, University of Ferrara, Ferrara, Italy

[1–3]. In PFTCS, controllers are fixed and designed to be robust against a class of presumed faults. This approach, which offers only limited fault-tolerant capabilities, does not need any fault estimate (or detection) or controller reconfiguration. In contrast to PFTCS, AFTCS reacts to the faults actively by reconfiguring control actions, so that stability and acceptable performance of the entire system can be maintained. AFTCS relies heavily on real-time fault detection and diagnosis (FDD) schemes, which are exploited for providing the most up-to-date information about the true status of the system. Usually, this information can be used from a logic-based switching controller or a feedback of the fault estimate. The approach proposed in this chapter relies on the latter strategy.

Over the last decades, many FDD techniques have been developed, see, e.g. the most important survey works [4–7]. Regarding the AFTCS design, it was argued that effective FDD is needed [1,3]. Moreover, it was claimed that, for the system to react properly to a fault, timely and accurate detection and location of the fault itself is needed. Fault detection and isolation (FDI) is the area where research studies have mostly been explored. On the other hand, FDD schemes represent a challenging topic because they must also provide the fault estimate. Unfortunately, disturbance affecting the system can cause false alarms, missed faults and wrong isolations; thus, robustnesses of FDI and FDD schemes represent a crucial factor [2,4,5,8].

This chapter summarises NLGA results in the field of AFTCS for aerospace systems for which the standard NLGA procedure presented in [9] has been extended to the input fault scenario in the presence of fault estimation feedback.

The filter structure is derived using the coordinate change of the NLGA theory developed in [9], which is only the starting point for the filter design. The application of the NLGA to the aircraft longitudinal model is investigated, in order to obtain fault estimates decoupled from disturbance and/or other faults. In this chapter, the actuators fault estimation is accomplished by AFs which, designed by using the NLGA, are analytically decoupled from relevant wind components. For the case of the design of the FDI (not FDD) module via the NLGA, see the paper [10].

The fault estimates provided by the proposed NLGA-AFs are unbiased via the above-mentioned disturbance decoupling. The AFs not using the proposed NLGA procedure are not decoupled from disturbances and/or other fault. The proposed FDD module thus increases the reliability of the overall AFTCS.

The AFTCS is obtained by adding a further feedback loop to the already built-in controller, which has been previously designed on the nominal fault-free plant. This approach, once properly tuned to be faster than the plant dynamics, does not degrade the stability properties guaranteed by the design of the nominal controller.

The differential geometry tool and NLGA-based aircraft AFTCS have been tested on a high-fidelity simulator. It implements realistic disturbance, such as sensor measurement noise and wind, thus showing the effectiveness and the good performance of the proposed AFTCS.

The chapter is organised as follows. Section 9.3 describes the structure of the proposed AFTCS: the theoretical and practical designs of the NLGA-AF, the estimation properties and convergence proof are provided in Section 9.3.1. The stability properties of the complete AFTCS are also considered in Section 9.3.2. Section 9.2 provides

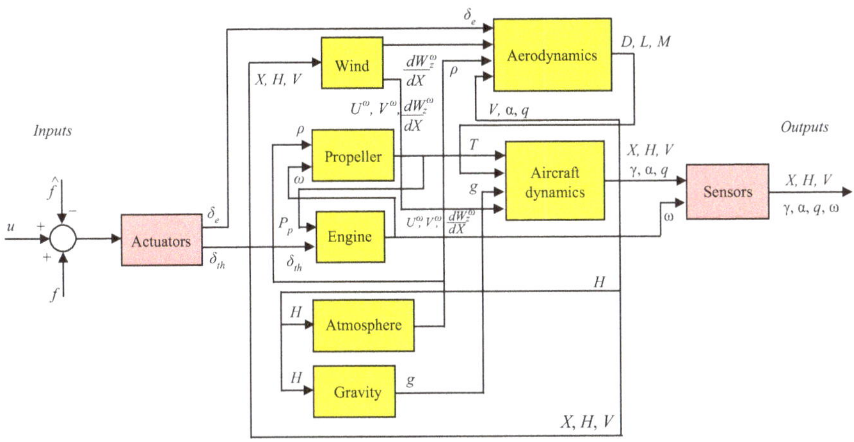

Figure 9.1 Schematic diagram of the aircraft simulator with its functional subsystems

more details regarding the flight simulator, whilst Section 9.4 shows the effectiveness and robustness features of the AFTCS strategy by extensive simulations. Concluding remarks are finally drawn in Section 9.5.

9.2 Aircraft model simulator

The simulated aircraft represents a Piper PA-30 for which detailed NASA and Lycoming technical data are available. The NASA Technical Notes [11–13] describing the aircraft and propeller aerodynamics and the engine manual [14] for engine modelling have been implemented for simulation purposes. Moreover, the model of the sensors is taken in agreement with [15].

The diagram of the aircraft simulator considered in this chapter is depicted in Figure 9.1 and the mathematical expressions representative of its blocks have been listed in order to help the interested reader to reproduce the proposed simulations.

The aircraft model is described by the relations of (9.1) [16] in which the time dependency has been omitted:

$$\begin{aligned}
\dot{X} &= V \cos\gamma + U^w \\
\dot{H} &= V \sin\gamma - W^w \\
\dot{V} &= \tfrac{1}{m}\left[T\cos\alpha - D - mg\sin\gamma\right] + V\sin\gamma\cos\gamma\,\tfrac{\partial W^w}{\partial X} \\
\dot{\gamma} &= \tfrac{1}{mV}\left[T\sin\alpha + L - mg\cos\gamma\right] + \cos^2\gamma\,\tfrac{\partial W^w}{\partial X} \\
\dot{\alpha} &= q - \dot{\gamma} \\
\dot{q} &= \tfrac{d_T}{I_y}T + \tfrac{M}{I_y}
\end{aligned} \tag{9.1}$$

The variables of the aircraft model are summarised in Table 9.1.

Table 9.1 Parameters of the aircraft model

Variable	Description	Unit
X	X inertial coordinate	m
H	Altitude	m
V	Airspeed	m/s
γ	Air ramp angle	rad
U^w	Horizontal wind	m/s
W^w	Vertical wind	m/s
m	Aircraft mass	kg
α	Attack angle	rad
g	Gravity constant	m/s^2
q	Body pitch rate	rad/s
d_T	Thrust arm	m
I_y	y-inertia momentum	kg m^2

Table 9.2 Aerodynamic coefficients of the aircraft model

Variable	Description	Unit
D	Drag force	N
L	Lift force	N
M	Pitch momentum	N m
ρ	Air density	kg/m^3
$\bar{c}$	Mean aerodynamic chord	m
$C_{D\#}$	Drag coefficients	–
$C_{L\#}$	Lift coefficients	–
$C_{m\#}$	Momentum coefficients	–
δ_e	Elevator	rad

The aircraft aerodynamics and its coefficients are summarised in Table 9.2 and described by the relations of (9.2) [16]:

$$\begin{aligned}
D &= \tfrac{1}{2}\rho V^2 S C_D \\
L &= \tfrac{1}{2}\rho V^2 S C_L \\
M &= \tfrac{1}{2}\rho V^2 S \bar{c} C_m \\
C_D &= C_{D_0} + C_{D_\alpha}\alpha \\
C_L &= C_{L_0} + C_{L_\alpha}\alpha + C_{L_q}\frac{\bar{c}}{2V}\left(q + \tfrac{\partial W^w}{\partial X}\right) + C_{L_{\delta_e}}\delta_e \\
C_m &= C_{m_0} + C_{m_\alpha}\alpha + C_{m_q}\frac{\bar{c}}{2V}\left(q + \tfrac{\partial W^w}{\partial X}\right) + C_{m_{\delta_e}}\delta_e
\end{aligned} \tag{9.2}$$

Table 9.3 *Engine model of the aircraft system*

Variable	Description	Unit
ω	Propeller angular rate	rad/s
Q_f	Shaft friction torque	N m
P_E	Engine power	W
I	Engine–propeller inertia	kg m^2
$C_\#$	Engine coefficients	*various*
δ_{th}	Throttle	–
J_v	Shaft friction coefficient	kg m^2 s

Table 9.4 *Propeller model of the aircraft system*

Variable	Description	Unit
P_P	Propeller power	W
T	Thrust	N
C_P	Propeller power coefficient	rad^{-3}
D_{PR}	Propeller diameter	m
η	Propeller efficiency	–

Regarding the aircraft engine, it is described by the relations of (9.3) [14], whose parameters are summarised in Table 9.3:

$$\begin{aligned} \dot{\omega} &= -\frac{Q_f}{I} + \frac{P_E}{I\omega} - \frac{P_P}{I\omega} \\ P_E &= C_1 H + C_2\omega\left[\delta_{th}\left(C_3 - C_4 H\right) - C_5\right] \\ Q_f &= J_v\omega^3 \end{aligned} \tag{9.3}$$

Finally, the model of the aircraft propeller is described by the relations of (9.4) [13] and its parameters are listed in Table 9.4:

$$\begin{aligned} P_P &= C_P\left(\frac{V\cos\alpha}{\omega D}\right)\rho\omega^3 D_{PR}^5 \\ T &= \frac{2}{V\cos\alpha}\eta\left(\frac{V\cos\alpha}{\omega D}\right)P_P \end{aligned} \tag{9.4}$$

The wind gusts U^w, W^w and $\partial W^w/\partial X$ are generated by means of a discrete wind gust block available in the MATLAB® and Simulink® environments. In particular, the simulated vertical wind gusts W^w are described into the inertial frame. Furthermore, a variable vertical wind always implies a rotational wind field, i.e. $\partial W^w/\partial X$ which causes a pitching moment. Also this effect has been taken into account during simulations. Finally, also Dryden turbulence model (translational and rotational) has been implemented by using the Aerospace Blockset of the MATLAB environment.

As highlighted in Figure 9.1, the aircraft simulator also implements the model of the measurements system consisting of the following modules:

- The command deflection angles, δ_e and δ_{th}, are acquired by potentiometers whose errors are modelled by white noises.
- The angular rate measurement, q, is given by a gyroscope. The corresponding modelled errors are the following: non-unitary scale factor, alignment error (random), g-sensitivity, additive white noise and gyro drift.
- The climb angle, γ, measurement is provided by digital filtering system based on a digital signal processor (DSP) processing both angular rate and accelerations provided by the IMU. The corresponding errors are the following: a systematic error generated by the apparent vertical. A coloured noise due to system structure and the environment influences.
- An air data computer, providing the speed V, the altitude H and the angle of attack α, is influenced by
  - errors affecting the determination of the airspeed, V: calibration error of differential pressure sensor, additive coloured noise induced by wind gusts and atmospheric turbulence, additive white noise;
  - errors affecting altitude, H: calibration error of the static pressure sensor, additive white noise;
  - errors affecting attack, α: calibration error affecting the wing boom sensors, additive white noise.
- The ground position, X, is provided by a GPS receiver: the error is modelled by means of a white noise.
- The engine shaft speed, ω, is measured by means of a tone-wheel which is typically affected by an error consisting in a white noise.

More details regarding the considered measurement system can be found in [15].

Finally, as shown in Figure 9.2, the aircraft simulator considered in this chapter includes an altitude and airspeed autopilot which has been given in [17,18] for the fault-free plant. This controller assures local asymptotic stability, in the sense of Lyapunov, of any cruise equilibrium point.

9.3 Active fault-tolerant control system design

Figure 9.2 describes the adopted structure of the AFTCS. In particular, u_r represents the reference input, u is the actuated input, y is the measured output, f is the actuator fault, whilst $\hat{f}$ is the estimated actuator fault.

Therefore, Figure 9.2 shows that the AFTCS is obtained by integrating the FDD module with the original control system. This estimated signal is injected into the control loop, in order to compensate the effect of the actuator fault.

The FDD module, which consists of a bank of NLGA-AF, asymptotically provides the correct fault estimation with a guaranteed and uniform rate of convergence [15,19].

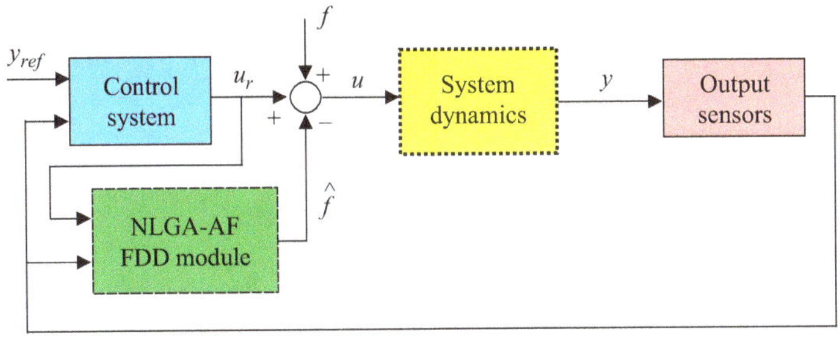

Figure 9.2 Logic diagram of the proposed AFTCS

In conclusion, thanks to the validity of the separation principle, the controller can be easily designed considering the fault-free plant, which represents an important benefit of the adopted fault-tolerance technique.

9.3.1 Fault diagnosis module

The model of the aircraft built on (9.1)–(9.4) can be rewritten in the compact form of (9.5):

$$\begin{aligned} \dot{x} &= n(x) + g(x)\left(u_r - \hat{f} + f\right) + p_d(x)\,d \\ y &= h(x) \end{aligned} \tag{9.5}$$

where the state vector $x \in \mathcal{X} \subseteq \mathbb{R}^7$, $u_r \in \mathcal{U} \subseteq \mathbb{R}^2$ is the control input vector, $f \in \mathcal{F} \subseteq \mathbb{R}^2$ is the fault vector, $d \in \mathbb{R}^3$ is the disturbance vector, $y \in \mathbb{R}^7$ is the output vector, whilst $n(x)$ is the columns of $g(x)$, and $p_d(x)$ are smooth vector fields, with $h(x)$ is a smooth map. In detail,

$$\begin{aligned} x &= \mathrm{col}\,(X,\ H,\ V,\ \gamma,\ \alpha,\ q,\omega) \\ u &= \mathrm{col}\,(\delta_e,\ \delta_{th}) \\ f &= \mathrm{col}\,(f_{\delta_e}, f_{\delta_{th}}) \\ d &= \mathrm{col}\left(U^w,\ W^w,\ \frac{\partial W^w}{\partial X}\right) \\ y &= \mathrm{col}\,(X,\ H,\ V,\ \gamma,\ \alpha,\ q,\omega) \end{aligned} \tag{9.6}$$

and

$$n(x) = \begin{bmatrix} V\cos\gamma \\ V\sin\gamma \\ \frac{1}{m}\left[T\cos\alpha - \frac{\rho V^2 S}{2}\left(C_{D_0} + C_{D_\alpha}\alpha\right) - mg\sin\gamma\right] \\ \frac{1}{mV}\left[T\sin\alpha + \frac{\rho V^2 S}{2}\left(C_{L_0} + C_{L_\alpha}\alpha + C_{L_q}\frac{\bar{c}}{2V}q\right) - mg\cos\gamma\right] \\ q - \frac{1}{mV}\left[T\sin\alpha + \frac{\rho V^2 S}{2}\left(C_{L_0} + C_{L_\alpha}\alpha + C_{L_q}\frac{\bar{c}}{2V}q\right) - mg\cos\gamma\right] \\ \frac{d_T}{I_y}T + \frac{\rho V^2 S\bar{c}}{2I_y}\left(C_{m_0} + C_{m_\alpha}\alpha + C_{m_q}\frac{\bar{c}}{2V}q\right) \\ -\frac{J_v\omega^3}{I} + \frac{C_1 H - C_5 C_2\omega}{I\omega} - \frac{P_P}{I\omega} \end{bmatrix} \tag{9.7}$$

$$g(x) = \begin{bmatrix} 0 & 0 \\ 0 & 0 \\ 0 & 0 \\ \frac{\rho SVC_{L_{\delta_e}}}{2m} & 0 \\ -\frac{\rho SVC_{L_{\delta_e}}}{2m} & 0 \\ \frac{\rho SV^2\bar{c}C_{m_{\delta_e}}}{2I_y} & 0 \\ 0 & \frac{C_2(C_3-C_4H)}{I} \end{bmatrix}, \quad p(x) = \begin{bmatrix} 1 & 0 & 0 \\ 0 & -1 & 0 \\ 0 & 0 & V\sin\gamma\cos\gamma \\ 0 & 0 & \frac{\rho SC_{L_q}\bar{c}}{4m} + \cos^2\gamma \\ 0 & 0 & -\frac{\rho SC_{L_q}\bar{c}}{4m} - \cos^2\gamma \\ 0 & 0 & \frac{\rho VS\bar{c}^2 C_{m_q}}{4I_y} \\ 0 & 0 & 0 \end{bmatrix} \tag{9.8}$$

The FDI scheme presented in this chapter is based on the evaluation of a bank of residuals, in this case two. The design of these residuals starts with the identification of the fault which should be isolated, i.e. the fault which should influence the residual. Given the fault vector f, select the jth component f_j and rewrite the system of (9.5) as follows:

$$\begin{aligned} \dot{x} &= n(x) + g(x)\left(u_r - \hat{f}\right) + g_j(x)f_j + p_j(x)\,d_j \\ y &= h(x) \end{aligned} \tag{9.9}$$

where $g_j(x)$ indicates the jth column of $g(x)$, $p_j(x) = [p(x),\ g_k(x)]$ and $d_k(x) = \mathrm{col}\,(d, f_k)$, for $k \neq j$ with $j, k \in 1, 2$. The design strategy for the diagnosis of the fault f_s via the NLGA method for disturbance decoupling is shown in [9] and summarised in the following:

- computation of Σ_*^P, i.e. the minimal conditioned invariant distribution containing P (where P is the distribution spanned by the columns of $p_j(x)$);
- computation of Ω^*, i.e. the maximal observability codistribution contained in $(\Sigma_*^P)^\perp$;
- if $g_s(x) \notin (\Omega^*)^\perp$ the so-called fault detectability condition is fulfilled, the fault f_j is detectable despite the presence of d_j.

If (and only if) the fault detectability condition is verified, a coordinate change in the state space and in the output space, $\Phi(x)$ and $\Psi(y)$, respectively, can be determined. They consist of a surjection Ψ_1 and a function Φ_1 such that $\Omega^* \cap \mathrm{span}\,\{d\,h\} = \mathrm{span}\,\{d\,(\Psi_1 \circ h)\}$ and $\Omega^* = \mathrm{span}\,\{d\Phi_1\}$, where

$$\begin{aligned} \Phi(x) &= \mathrm{col}\,(\bar{x}_1, \bar{x}_2, \bar{x}_3) &&= \mathrm{col}\,(\Phi_1(x),\ H_2\,h(x),\ \Phi_3(x)) \\ \Psi(y) &= \mathrm{col}\,(\bar{y}_1, \bar{y}_2) &&= \mathrm{col}\,(\Psi_1(y), H_2\,y) \end{aligned} \tag{9.10}$$

are (local) diffeomorphisms, whilst H_2 is a selection matrix, i.e. its rows are a subset of the rows of the identity matrix. It is worth observing that the first subsystem, whose dynamics is caught by $\bar{x}_1$, is always decoupled from the disturbance vector d_s and affected by the fault f_j as shown in the following equation:

$$\begin{aligned} \dot{\bar{x}}_1 &= n_1\,(\bar{x}_1, \bar{x}_2) + g_1\,(\bar{x}_1, \bar{x}_2)\left(u_r - \hat{f}\right) + l_1\,(\bar{x}_1, \bar{x}_2, \bar{x}_3)\,f_j \\ \bar{y}_1 &= h_1(\bar{x}_1) \end{aligned} \tag{9.11}$$

A second important advantage of the NLGA is represented by the direct availability of $\bar{x}_2$ at output through $\bar{y}_2$, i.e. $\bar{y}_2 = \bar{x}_2$. The third remarkable aspect stays in the absence of the component $\bar{x}_3$ in the case of state x fully available at measurement y. Indeed, if the output map of (9.9) is $h(x) = x$, i.e. if $y = x + v$, the NLGA algorithm leads to

- a minimal conditioned invariant distribution containing P, $\Sigma_*^P = \bar{P}$;
- maximal observability codistribution contained in$(\Sigma_*^P)^\perp$, $\Omega^* = (\bar{P})^\perp$;
- a *fault detectability condition* fulfilled if $g_s(x) \notin \bar{P}$;
- input and output transformations $\Phi(x) = \Psi(y)$, thus implying the absence of the subsystem $\bar{x}_3$;
- a change of coordinate span $\{d\Phi_1\} = \text{span}\{d\Psi_1\} = \bar{P}^\perp$ implying that $\bar{y}_1 = \bar{x}_1$;
- a new $\bar{x}_1$ subsystem whose dynamics is represented by

$$\begin{aligned} \dot{\bar{x}}_1 &= n_1(\bar{x}_1, \bar{y}_2) + g_1(\bar{x}_1, \bar{y}_2)\left(u_r - \hat{f}\right) + l_1(\bar{x}_1, \bar{y}_2) f_j \\ \bar{y}_1 &= \bar{x}_1 \end{aligned} \tag{9.12}$$

Given the state $\bar{x}_1 \in \mathbb{R}^{n_1}$ and the output $\bar{y}_2 \in \mathbb{R}^{n_2}$, choose the sth component of $\bar{x}_1$ such that

$$\max_{i=1,\ldots,n_1} |l_{1i}(\bar{x}_1, \bar{y}_2)| = |l_{1s}(\bar{x}_1, \bar{y}_2)| > 0 \quad \forall \bar{x}_1, \bar{y}_2 \in \mathbb{R}^{n_1} \times \mathbb{R}^{n_2}$$

where l_{1i} indicates the ith row of l_1. The dynamics of the state $\bar{x}_{1s}$ is described by the following equation [19]:

$$\begin{aligned} \dot{\bar{x}}_{1s} &= M_1\left(f_j - \hat{f}_j\right) + M_2 \\ \bar{y}_{1s} &= \bar{x}_{1s} \end{aligned} \tag{9.13}$$

where

$$\begin{aligned} M_1 &= l_{1s}(\bar{y}_1, \bar{y}_2) \\ M_2 &= n_{1s}(\bar{y}_1, \bar{y}_2) + g_{1s}(\bar{y}_1, \bar{y}_2)\left(u_r - \hat{\bar{f}}_j\right) \end{aligned}$$

where $\bar{f}_j$ is defined as a column vector $\bar{f}_j = \text{col}(f_1, f_2)$ in which the jth component is null. It is worth noting that the terms M_1 and M_2 are known, for each time instant, due to the direct availability at outputs of the states $\bar{x}_1$ and $\bar{x}_2$ through the measurements $\bar{y}_1$ and $\bar{y}_2$, respectively, which can be considered as a direct function of time. Also, the control input, u_r, as well as the estimation of faults, $\hat{\bar{f}}_j$, are considered measured (or known) and described as a function of time. Furthermore, thanks to the choice of l_{1s} the term $M_1 \neq 0, \forall t \geq 0$. The design of the NLGA-AF is achieved, with reference to the system model of (9.13), in order to provide a fault estimation $\hat{f}_j(t)$, which asymptotically converges to the magnitude of the fault f_j. Therefore, the signal f_j is estimated by means of an AF based on the least-squares algorithm with forgetting factor [19]. In the following, the function dependence on time t will be omitted.

The AF is given by the expression of the following equation:

$$\begin{cases} \dot{P} = \beta P - \dfrac{\breve{M}_1^2}{1+\breve{M}_1^2} P^2 & P(0) = P_0 > 0 \\ \dot{\hat{f}}_j = P \dfrac{\breve{M}_1}{1+\breve{M}_1^2} \left(\bar{y}_{1s} - \breve{M}_1 \hat{f}_j - \breve{M}_3 - \lambda \breve{\bar{y}}_{1s} \right) & \hat{f}_j(0) = 0 \\ \dot{\breve{M}}_1 = -\lambda \breve{M}_1 + M_1 & \breve{M}_1(0) = M_1(0) \\ \dot{\breve{M}}_3 = -\lambda \breve{M}_3 + M_2 - M_1 \hat{f}_j & \breve{M}_3(0) = 0 \\ \dot{\breve{\bar{y}}}_{1s} = -\lambda \breve{\bar{y}}_{1s} + \bar{y}_{1s} & \breve{\bar{y}}_{1s}(0) = \bar{y}_{1s}(0) \end{cases} \tag{9.14}$$

It has been proven in [19] that the estimation error, defined as $e_f = f_j - \hat{f}_j$, asymptotically converges to zero with a guaranteed rate of convergence. To prove this result, it is necessary to show that the term

$$\lim_{t\to\infty} \left(\bar{y}_{1s} - \breve{M}_1 \hat{f}_j - \breve{M}_3 - \lambda \breve{\bar{y}}_{1s} \right) = \lim_{t\to\infty} \left(\breve{M}_1 e_f \right) \tag{9.15}$$

Indeed, given an auxiliary system defined in the form of the following equation:

$$\begin{cases} \dot{y}_1 = -\lambda y_1 + \dot{\bar{y}}_{1s} & y_1(0) = 0 \\ \dot{y}_2 = -\lambda y_2 + \lambda \bar{y}_{1s} & y_2(0) = 0 \\ y = y_1 + y_2 \end{cases} \tag{9.16}$$

the relation of the following equation holds

$$\begin{aligned} y_1 &= \breve{M}_1 f_s + \breve{M}_3 \\ y_2 &= \lambda \breve{\bar{y}}_{1s} \end{aligned} \tag{9.17}$$

Consider now the following function:

$$V = \frac{1}{2} \left(y - \bar{y}_{1s} \right)^2 \tag{9.18}$$

which is trivially positive definite, and radially unbounded. Moreover, its first time derivative has the following form:

$$\dot{V} = -\lambda \left(y - \bar{y}_{1s} \right)^2 \tag{9.19}$$

Since $\dot{V}$ is trivially negative definite $\forall y \neq \bar{y}_{1s}$, V represents a Lyapunov function that globally asymptotically tends to zero. Moreover, from the relation of (9.17), the following relation holds

$$\lim_{t\to\infty} \bar{y}_{1s} = \lim_{t\to\infty} y = \lim_{t\to\infty} \left(\breve{M}_1 f_s + \breve{M}_3 + \lambda \breve{\bar{y}}_{1s} \right) \tag{9.20}$$

By means of the expression of (9.20), the relation of the following equation holds

$$\lim_{t\to\infty} \left(\bar{y}_{1s} - \breve{M}_1 \hat{f}_j - \breve{M}_3 - \lambda \breve{\bar{y}}_{1s} \right) = \lim_{t\to\infty} \left(\breve{M}_1 (f_s - \hat{f}_s) \right) = \lim_{t\to\infty} \left(\breve{M}_1 e_f \right) \tag{9.21}$$

This result implies that the asymptotic behaviour of the dynamics of the fault estimation error e_f is represented by

$$\lim_{t\to\infty} \dot{e}_f = -\lim_{t\to\infty} P \frac{\breve{M}_1^2}{1+\breve{M}_1^2} e_f \tag{9.22}$$

The asymptotic dynamics of the fault estimation error is represented by a stable linear time-varying autonomous system and, thanks positiveness of P and $\breve{M}_1^2/(1+\breve{M}_1^2)$, assures the asymptotic convergence of the fault estimation error to the origin. Indeed, $P(t) > 0\ \forall t > 0$ and, moreover, the following property holds

$$\lim_{t\to\infty} P^{-1}(t) = \lim_{t\to\infty} e^{-\beta t} \int_0^t e^{\beta\tau} \frac{\breve{M}_1^2(\tau)}{1+\breve{M}_1^2(\tau)} d\tau \tag{9.23}$$

The dynamics of the filter of (9.14) can be tuned to be sufficiently faster than the dynamics of the plant. In detail, the dynamics of $\breve{M}_1$ and P can be sped up, by means of the parameters λ and β, to consider a quasi-static plant and so a quasi-static M_1. In these conditions, the dynamics of the estimation error becomes independent of the plant:

$$\begin{aligned}\lim_{t\to\infty} \dot{e}_f \approx -\lim_{t\to\infty} \frac{e^{\beta t}\left({M_1}^2/(\lambda^2+{M_1}^2)\right)}{\int_0^t e^{\beta\tau}\left({M_1}^2/(\lambda^2+{M_1}^2)\right) d\tau} e_f &= -\lim_{t\to\infty} \frac{e^{\beta t}}{\int_0^t e^{\beta\tau}\, d\tau} e_f \\ &= -\beta e_f\end{aligned} \tag{9.24}$$

9.3.2 Fault accommodation strategy

Finally, the local asymptotic stability properties of the plant, guaranteed by the already in place controller, are kept valid by choosing λ and β such that the limit of (9.24) is valid. The autonomous dynamics of the fault estimation error $\dot{e}_f = -\beta e_f$ is interpreted as an exogenous and vanishing signal driving the plant

$$\dot{x} = n(x) + g(x)\left(u + e_f\right) + p(x)d$$

which is 0-input locally asymptotically stable (by hypothesis) at certain x_0. Thanks to Theorem 10.3.1 in [20] the overall system has got a local asymptotically stable point at $(x, e_f) = (x_0, 0)$.

9.4 Simulation results

This section is divided into three parts. Section 9.4.1 illustrates the analytical computation of the NLGA-AF for FDD purpose. Section 9.4.2 shows the achieved performance of the NLGA-AF, whilst Section 9.4.3 analyses the performance of the overall AFTCS exploiting, including the control system recalled in Section 9.2 and the NLGA-AF scheme reported in Section 9.3.

9.4.1 Fault diagnosis filter design

The NLGA methodology showed in Section 9.3.1 is directly applicable only to systems that are affine with respect to both inputs and disturbances. Moreover, a single fault can be isolated from disturbances and other faults, if an observable subsystem can be determined. Therefore, this section shows how the NLGA-AF methodology can be applied to the aircraft case. Moreover, the hypothesis assumed for the aircraft model approximation is highlighted, and the AFs thus obtained are shown.

To determine the estimation filter for the fault affecting the elevator, $\hat{f}_{\delta_e}$, the following variables are defined:

- the fault $f_j = f_{\delta_e}$;
- the fault $f_k = f_{\delta_{th}}$;
- the generalised disturbance $d_j = \text{col}\left(d, f_{\delta_{th}}\right)$;
- the input vector field associated with f_k, $g_k(x) = \text{col}(0, 0, 0, 0, 0, 0, (C_2(C_3 - C_4H)/I))$;
- the input vector field associated with the generalised disturbance $p_j(x) = [p(x), g_k(x)]$;
- the involutive closure of $p_j(x)$, i.e. $\bar{P}(x) = p_j(x)$.

Since the output map is $h(x) = x$, the changes of coordinates are found by determining the orthogonal to $\bar{P}(x)$ as an exact differential of $\Phi_1(x)$. In particular, considering that standard cruises are made at small climb angles such that $\cos\gamma \approx 1 - \gamma^2/2$ and $\sin\gamma \approx \gamma$, the following diffeomorphism is found:

$$\Phi_1(x) = \frac{\rho S\bar{c}^2 C_{m_q}}{4I_y} V\frac{\gamma}{2} - q\frac{1 + (\rho S C_{L_q}\bar{c}/4m)}{2}$$

The filter, designed adopting $\Phi_1(x)$ as change of coordinate, is decoupled from both aerodynamic disturbances and the other fault (i.e. the throttle). It is possible to specify the particular expression of the fault dynamics of (9.13) for $f_s = f_{\delta_e}$:

$$\begin{cases} \dot{\bar{y}}_{1s,\delta_e} = M_{1,\delta_e} f_{\delta_e} + M_{2,\delta_e} \\[2ex] \bar{y}_{1s,\delta_e} = \dfrac{\rho S\bar{c}^2 C_{m_q}}{8I_y} V\gamma - q\left(\dfrac{1}{2} + \dfrac{\rho S C_{L_q}\bar{c}}{8m}\right) \\[2ex] M_{1,\delta_e} = \dfrac{V^2\rho S\bar{c}}{4I_y}\left(\dfrac{\rho S\bar{c}^2\left(C_{L_{\delta_e}} C_{m_q} - C_{L_q} C_{m_{\delta_e}}\right)}{4m} - C_{m_{\delta_e}}\right) \\[2ex] M_{2,\delta_e} = \dfrac{\rho S\bar{c}^2 C_{m_q}}{8I_y}\gamma n_3(x) + \dfrac{\rho S\bar{c}^2 C_{m_q}}{8I_y} V n_4(x) - \left(\dfrac{1}{2} + \dfrac{\rho S C_{L_q}\bar{c}}{8m}\right) n_6(x) \end{cases} \tag{9.25}$$

where $n_i(x)$ indicates the ith row of $n(x)$ in (9.7). It is worth observing that $M_{1,\delta_e} \neq 0$ $\forall t \geq 0$, since V is always strictly positive in all the flight conditions.

On the other hand, in order to determine the filter for the diagnosis and the isolation of the throttle fault, $f_{\delta_{th}}$, from other faults (i.e. the elevator) and disturbance (i.e.

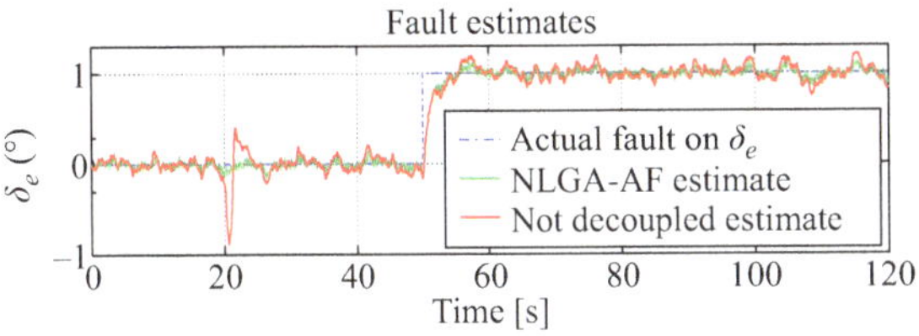

Figure 9.3 Estimates $\hat{f}_{\delta_e}$ of the fault f_{δ_e}

the wind gusts), the following model has to be considered: to determine the estimation filter for the fault affecting the elevator, $\hat{f}_{\delta_e}$, the following variables are defined:

- the fault $f_j = f_{\delta_{th}}$;
- the fault $f_k = f_{\delta_e}$;
- the generalised disturbance $d_j = \text{col}\left(d, f_{\delta_e}\right)$;
- the input vector field associated with f_k, $g_k(x) = (\rho SV/2)\text{col}(0, 0, 0, C_{L_{\delta_e}}/m, -C_{L_{\delta_e}}/m, V\bar{c}C_{m_{\delta_e}}/2I_y, 0)$;
- the input vector field associated with the generalised disturbance $p_j(x) = [p(x), g_k(x)]$;
- the involutive closure of $p_j(x)$, i.e. $\bar{P}(x) = p_j(x)$.

The filter for the diagnosis of $f_{\delta_{th}}$ has been obtained by following the same procedure exploited for determining the filter of the elevator fault f_{δ_e}. However, this case does not require any model approximation. This filter is described by the relations of (9.26).

$$\begin{cases} \dot{\bar{y}}_{1s,\delta_{th}} &= M_{1,\delta_{th}} f_{\delta_{th}} + M_{2,\delta_{th}} \\ \bar{y}_{1s,\delta_{th}} &= \omega \\ M_{1,\delta_{th}} &= \frac{C_2(C_3 - C_4 H)}{I} \\ M_{2,\delta_{th}} &= -\frac{Q_f}{I} + \frac{P_E}{I\omega} - \frac{P_P}{I\omega} \end{cases} \tag{9.26}$$

Note finally that the term $M_{1,\delta_{th}}$ results always greater than zero for the altitude simulated in the aircraft model of Section 9.2.

9.4.2 NLGA-AF simulation results

In this case study, the NLGA-AF simulation results described in Section 9.3.1 are decoupled from both aerodynamic disturbances, i.e. wind, and the other fault. In particular, Figure 9.3 shows a fault on the elevator (blue dotted line) of size $f_{\delta_e} = 1°$ and its estimation (green bold line) during an altitude holds flight phase.

Figure 9.3 shows that the fault is detected and estimated with a time delay smaller than the characteristic flight dynamics period, and the convergence of the estimate to the actual fault size can be observed. The fault commence at time $t = 50$ s. Moreover, Figure 9.3 also highlights the features of the NLGA-AF design, which provides a fault estimation decoupled from wind disturbances. In fact, without any disturbance

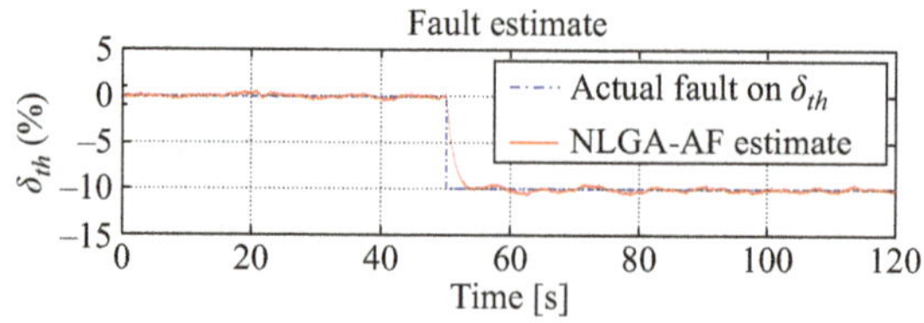

Figure 9.4 Estimate $\hat{f}_{\delta_{th}}$ of the fault $f_{\delta_{th}}$

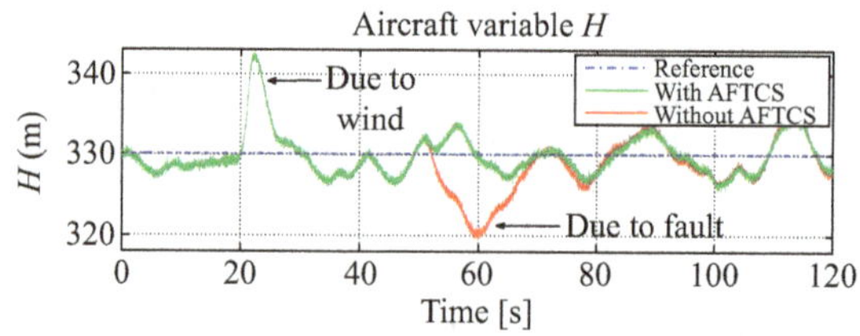

Figure 9.5 Aircraft altitude H with and without AFTCS

decoupling, the wind gust commencing at $t = 20$ s would affect the fault estimation $\hat{f}_{\delta_e}$ (red bold line).

The same considerations hold for Figure 9.4, which shows the case of a fault affecting the throttle δ_{th} of size $f_{\delta_{th}} = -10\%$.

9.4.3 AFTCS performance

This section summarises the simulation results concerning the performance of the overall AFTCS when applied to the proposed aircraft simulator. The performances of the controlled aircraft with or without the estimated fault feedback are compared. In this way, the advantages given by the proposed AFTCS strategy are highlighted.

The following simulations performed in the presence of wind and noise on both the input and output sensors mostly serve to highlight the advantages of the fault recovery procedure obtained using the fault estimate feedback. In fact, as the fault is detected and estimated with a time delay smaller than the characteristic flight dynamics period, the fault-free flight conditions can be maintained without any performance degradation.

In particular, Figure 9.5 compares the effects of both the wind gust and the elevator fault f_{δ_e} on the altitude variable H, thus highlighting the effectiveness of the proposed AFTCS strategy.

On the other hand, Figure 9.6 depicts the aircraft variable V during the transient after the occurrence of the fault $f_{\delta_{th}}$. It is worth observing that the results refer to the same conditions of Figure 9.5 (noises, disturbances, etc.).

Finally, due to the presence of disturbance and uncertainty, the proposed simulations highlight the robustness properties of the developed FDD and AFTCS solutions, when applied to the high-fidelity aircraft simulator.

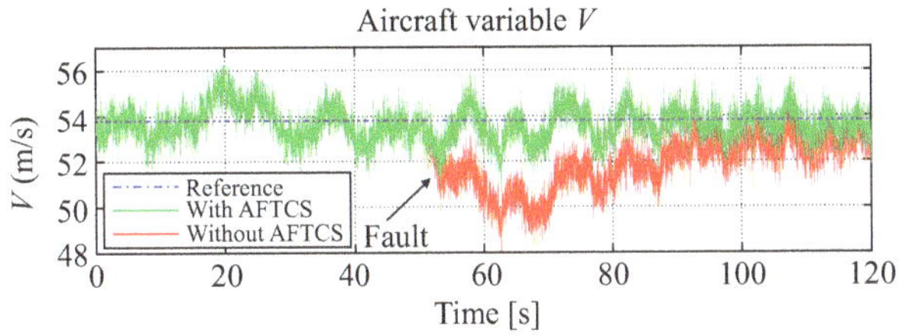

Figure 9.6 Aircraft speed V with and without AFTCS

9.5 Conclusion

In this chapter, an active fault-tolerant control system was proposed and applied to an aircraft model. In particular, the developed solution was based on an FDD module providing the fault estimate and exploited for fault accommodation. In this way, the controller can be designed on the faulty-free model. The key feature of the chapter is that the designed solutions were based on a differential geometry tool. With reference to the FDD module, the fault estimates were robust with respect to disturbance and other fault, and hence unbiased, thanks to the analytical decoupling provided by application of the NLGA. When the application to the aircraft model was considered, faults were decoupled from aerodynamic disturbances, such as vertical wind gusts. It was also shown that the performance of the autopilot in the transient phases in the presence of faults was improved by the added loop providing the asymptotic fault recovery. The simulations were based on a high-fidelity aircraft model, whose features were tested via extensive simulations. The results showed that the designed non-linear active fault-tolerant scheme was reliable.

References

[1] Mahmoud M, Jiang J, Zhang Y. Active Fault Tolerant Control Systems: Stochastic Analysis and Synthesis. Lecture Notes in Control and Information Sciences. Berlin, Germany: Springer-Verlag; 2003. ISBN: 3540003185.

[2] Blanke M, Kinnaert M, Lunze J, *et al.* Diagnosis and Fault-Tolerant Control. Berlin, Germany: Springer-Verlag; 2006.

[3] Zhang Y, Jiang J. Bibliographical review on reconfigurable fault-tolerant control systems. Annual Reviews in Control. 2008;32:229–252.

[4] Isermann R. Fault-Diagnosis Systems: An Introduction from Fault Detection to Fault Tolerance. 1st ed. Weinheim, Germany: Springer-Verlag; 2005. ISBN: 3540241124.

[5] Witczak M. Modelling and Estimation Strategies for Fault Diagnosis of Non-Linear Systems: From Analytical to Soft Computing Approaches. 1st ed. Lecture Notes in Control & Information Sciences. Berlin and Heidelberg: Springer-Verlag GmbH & Co.; 2007. ISBN: 978-3540711148.

[6] Ding SX. Model-based Fault Diagnosis Techniques: Design Schemes, Algorithms, and Tools. 1st ed. Berlin, Heidelberg: Springer; 2008. ISBN: 978-3540763031.

[7] Theilliol D, Join C, Zhang Y. Actuator fault tolerant control design based on a reconfigurable reference input. International Journal of Applied Mathematics and Computer Science. 2008;18(4):553–560. ISSN: 1641-876X. DOI: 10.2478/v10006-008-0048-1.

[8] Chen J, Patton RJ. Robust Model-Based Fault Diagnosis for Dynamic Systems. Boston, MA, USA: Kluwer Academic Publishers; 1999.

[9] De Persis C, Isidori A. A geometric approach to nonlinear fault detection and isolation. IEEE Transactions on Automatic Control. 2001;46(6):853–865.

[10] Bonfè M, Castaldi P, Geri W, *et al.* Design and performance evaluation of residual generators for the FDI of an aircraft. International Journal of Automation and Computing. 2007;4(2):156–163. Special Issue on FDD and FTC. ISSN: 1476-8186. DOI: 10.1007/s11633-007-0156-7.

[11] Fink MP, Freeman DCJ. Full-Scale Wind-Tunnel Investigation of Static Longitudinal and Lateral Characteristics of a Light Twin-Engine Airplane. N.A.S.A.; 1969. TN D-4983.

[12] Koziol J. Simulation Model for the Piper PA–30 Light Maneuverable Aircraft in the Final Approach. N.A.S.A.; 1971. DOT-TSC-FAA-71-11.

[13] Gray H. Wind-tunnel Tests of Eight-blade Single and Dual-rotating Propellers in the Tractor Position. Washington, DC, USA: National Advisory Committee for Aeronautics; 1943. NACA-WR-L-384.

[14] Lycoming. O–320 Operator's Manual. Williamsport, PA, USA: Lycoming; 2006.

[15] Bonfè M, Castaldi P, Geri W, *et al.* Fault detection and isolation for on-board sensors of a general aviation aircraft. International Journal of Adaptive Control and Signal Processing. 2006;20(8):381–408. ISSN: 0890-6327. DOI: 10.1002/acs.906.

[16] Stevens BL, Lewis FL. Aircraft Control and Simulation. 2nd ed. Hoboken, NJ: John Wiley and Son; 2003.

[17] Farrell JA, Sharma M, Polycarpou M. Longitudinal flight-path control using online function approximation. Journal of Guidance, Control, and Dynamics. 2003;26(6):885–897. DOI: 10.2514/2.6932.

[18] Castaldi P, Mimmo N, Simani S. Differential Geometry Based Active Fault Tolerant Control for Aircraft. Control Engineering Practice. 2014;32:227–235. Invited Paper. DOI:10.1016/j.conengprac.2013.12.011.

[19] Castaldi P, Geri W, Bonfè M, *et al.* Design of residual generators and adaptive filters for the FDI of aircraft model sensors. Control Engineering Practice. 2010;18(5):449–459. ACA'07 – 17th IFAC Symposium on Automatic Control in Aerospace Special Issue. Elsevier Science. ISSN: 0967-0661. DOI: 10.1016/j.conengprac.2008.11.006.

[20] Isidori A. Nonlinear Control Systems II. Berlin and Heidelberg: Springer-Verlag; 1999.

Chapter 10

Fault-tolerant trajectory tracking control of in-wheel motor vehicles with energy-efficient steering and torque distribution

Péter Gáspár[1], *András Mihály*[1], *and Hakan Basargan*[2]

As conventional gasoline- and diesel-propelled vehicles are giving way to the less polluting and more energy-efficient hybrid and fully electric vehicles, in-wheel motor vehicles are in the scope of research of automotive companies owing to their several advantages. The dynamical benefits of the in-wheel vehicle architecture and the inherent precise torque vectoring ability have already been broadly studied, see [1,2]. Especially, four-wheel independently actuated (4WIA) vehicles enable the usage of state-of-the-art techniques to further improve the dynamical behavior, safety, or economy of the vehicle, see [3,4].

As for all battery electric vehicles, limited range is one of the most crucial aspects for in-wheel vehicle operation. Hence, several methods have been presented for the range extension of 4WIA vehicles. Most of these methods are based on more effective regenerative braking of in-wheel vehicles [5], multi-objective optimization for wheel torque distribution [6], or methods to integrate supplementary energy sources for electric vehicles [7]. One of the biggest advantages of the presented method in this chapter is that it enables energy optimal actuator selection based on results of extensive prior analysis; thus the amount of online calculation can be reduced. Moreover, it only requires easily accessible road and vehicle velocity data.

Hence, the objective of the chapter is to design a novel fault-tolerant and energy-efficient torque distribution method for an autonomous 4WIA electric vehicle with four in-wheel independently controlled hub motors. In normal operating conditions, the presented high-level reconfigurable controller and the low-level wheel torque optimization method enables the in-wheel vehicle to establish better energy efficiency. Hence, in normal operating conditions battery, state-of-charge (SOC) values increase compared to the conventional control methods; thus a bigger range can be achieved for the electric vehicle. The designed integrated vehicle control strategy allows one to combine and supervise all control components affecting vehicle dynamics responses.

[1]Systems and Control Laboratory, Institute for Computer Science and Control, Budapest, Hungary
[2]Department of Control for Transportation and Vehicle Systems, Budapest University of Technology and Economics, Budapest, Hungary

Hence, if the demanded yawing torque cannot be realized with the four in-wheel motors due to actuator failure or performance degradation, the high-level controller reconfigures to apply more steering in order to stabilize the vehicle on the predefined trajectory. On the other hand, during a fault event in the steer-by-wire steering system, the high-level controller also reconfigures to apply more yawing torque for the vehicle.

The main goal of the design is to maximize the battery SOC to enhance the range of the autonomous electric vehicle, while preserving safe motion in the case of a fault event. Previous studies have already shown that significant amount of energy can be saved by appropriate actuator selection, see [8,9]. Here, the method is based on a lane-tracking reconfigurable linear parameter varying (LPV) controller, where the high-level control allocation between steering intervention and yaw moment generation aims to maximize battery SOC. A lithium-ion battery model and a permanent magnet synchronous motor (PMSM) model with regenerative braking ability have been implemented in order to simulate the operation of the 4WIA electric vehicle and to calculate battery SOC. The control reconfiguration is based on the preliminary analysis of the electric motor and the battery system for different road conditions and vehicle states, with the consideration of regenerative braking. Also, a multi-objective low-level in-wheel motor torque optimization has been designed to consider electric motor efficiency characteristics along with safety aspects of the vehicle motion. Finally, the reconfigurable control approach has been validated in a real-data driving scenario in the CarSim simulation environment.

10.1 Trajectory-tracking controller design

10.1.1 Vehicle modeling

For the design of an effective trajectory tracking controller, lateral and longitudinal dynamics of the vehicle are necessary to be formulated. Here, a bicycle model is used for vehicle modeling, see Figure 10.1. The motion equations of the longitudinal and lateral dynamics are given as

$$J\ddot{\psi} = c_1 l_1(\delta - \beta - \dot{\psi} l_1/\dot{\xi}) - c_2 l_2(-\beta + \dot{\psi} l_2/\dot{\xi}) + M_z \tag{10.1a}$$

$$m\ddot{y}_v = c_1(\delta - \beta - \dot{\psi} l_1/\dot{\xi}) + c_2(-\beta + \dot{\psi} l_2/\dot{\xi}) \tag{10.1b}$$

$$m\ddot{\xi} = F_l - F_d \tag{10.1c}$$

where J is the yaw inertia of the vehicle, m is the mass, l_1 and l_2 are geometric parameters, c_1 and c_2 are cornering stiffnesses, which can be identified [10]. While the yaw of the vehicle is expressed with ψ, β is the side-slip angle, $\ddot{y}_v = \dot{\xi}(\dot{\psi} + \dot{\beta})$ is the lateral and $\ddot{\xi}$ is the longitudinal acceleration. Due to the longitudinal velocity $\dot{\xi}$, the system is nonlinear. The inputs of the system are the longitudinal force (F_l), the brake yaw moment (M_z) and the front steering angle (δ).

In (10.1a)–(10.1c) the longitudinal disturbance force F_d contains the road slope, the rolling resistance of the vehicle and the aerodynamic drag. Thus, disturbance is described by the following nonlinear equation: $F_d = 0.5C_d\sigma A_F\dot{\xi}^2 + mg\sin\alpha_s + mgf\cos\alpha_s$, where C_d is the drag coefficient, A_F is the frontal surface, assumed to be

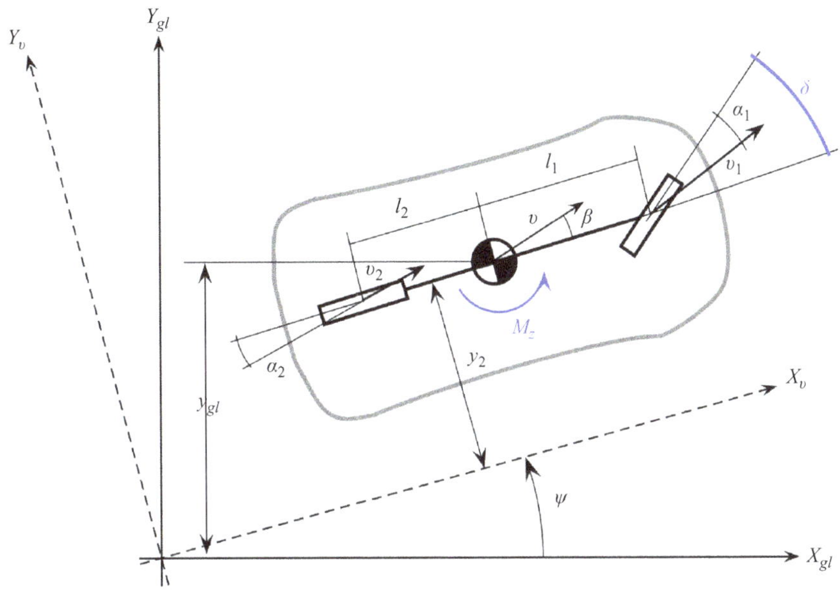

Figure 10.1 Single-track bicycle model

known for the vehicle, σ stands for air mass density, α_s denotes road slope angle, f represents road friction and g is the gravitational constant.

Lateral and longitudinal position tracking must be guaranteed in the design of the high-level controller. The error between the desired road curve and the lateral position controls the required lateral motion while the required longitudinal motion is made certain by velocity tracking. The world coordinate system (X_{gl} and Y_{gl}), in which the coordinate system of the vehicle rotates together with the vehicle, is used to define the road geometry's reference.

The calculated lateral positions of the vehicle in the two coordinate systems are shown in Figure 10.1. Consequently, the rotation of the vehicle in the reference road geometry calculation is considered as $y_{v,r} = -\sin(\psi)\, x_{gl,r} + \cos(\psi)\, y_{gl,r}$, where $x_{gl,r}$ and $y_{gl,r}$ are the longitudinal and lateral coordinates of the reference road geometry in the world coordinate system, $y_{v,r}$ is the lateral position of the reference road geometry in the coordinate system of the vehicle. It is assumed that $x_{gl,r}$ and $y_{gl,r}$ are given in look-up tables for different road courses and can be calculated from GPS data of the selected road. Here, GPS geodetic data are transformed to a local east-north-up coordinate system using coordinate transformation detailed in the literature. At the same time, based on the current position of the vehicle x_{gl} and y_{gl} (assumed to be measured by high precision GPS positioning system), the lateral position of the vehicle in its own coordinate system is also calculated as $y_v = -\sin(\psi)\, x_{gl} + \cos(\psi)\, y_{gl}$. Hence, lateral error $e_y = |y_{v,r} - y_v|$ can be calculated and defined as a performance criterion for the LPV controller design detailed in the following section.

10.1.2 Reconfigurable LPV controller design

From (10.1a)–(10.1c), the state-space representation form of the vehicle motion equation can be expressed as follows:

$$\dot{x} = A(\rho)x + B_1 w + B_2(\rho)u \tag{10.2a}$$

$$z = Cx + Dw \tag{10.2b}$$

$$y = Cx + Dw \tag{10.2c}$$

The state vector of the system can be written as

$$x = \left[\dot{\xi}\ \xi\ \dot{\psi}\ \beta\ \dot{y}_v\ y_v\right]^T, \tag{10.3}$$

containing the displacement of the vehicle, the longitudinal velocity, the side-slip angle, the yaw-rate, the lateral velocity and the position of the vehicle. The control inputs of the system are the longitudinal force, the front steering angle and the brake yaw-moment, which are given in an input vector as

$$u = \left[F_l\ \delta\ M_z\right]^T. \tag{10.4}$$

Both lateral and longitudinal dynamics must be considered in the trajectory tracking, i.e. the vehicle must track two reference signals at the same time. First, accurate velocity tracking must be provided in the longitudinal direction: $z_\xi = |\dot{\xi}_{ref} - \dot{\xi}|$. This requirement can be formulated as an optimization criterion: $z_\xi \to 0$. Second, minimization is needed in the difference between the lateral position and reference position $z_y = |y_{v,r} - y_v|$ of the vehicle, which can be formulated as the following optimization criterion $z_y \to 0$. These performances are built in a performance vector:

$$z_1 = \left[z_\xi\ z_y\right]^T. \tag{10.5}$$

Concurrently, actuator saturations must be handled. Physical construction limits and the tire-road adhesion conditions determine the maximal forces of braking systems and the driveline. In the control design, they are formulated as performance criteria:

$$z_2 = \left[\delta\ \ M_z\right]^T. \tag{10.6}$$

The velocity $\dot{\xi}$ of the vehicle affects the system matrices nonlinearly. Velocity is assumed to be estimated or measured, see, e.g. [11]. Hence, the nonlinear model can be transformed into an LPV model using a scheduling variable $\rho_1 = \dot{\xi}$. Velocity and the lateral position are measured as the output of the system, i.e. $y = \left[\dot{\xi}\ y_v\right]^T$.

The design of the controller is based on a weighting strategy, which is formulated via a closed-loop interconnection structure, see in Figure 10.2. Generally, input and output weighting functions are chosen according to the inverse of the specifications on the outputs and the specifications of disturbances. The purpose of the weighting function W_p is to define the performance specifications in a way that an appropriate balance is guaranteed between them. The objective of the weighting functions W_w and W_n is to consider the sensor and disturbance noises.

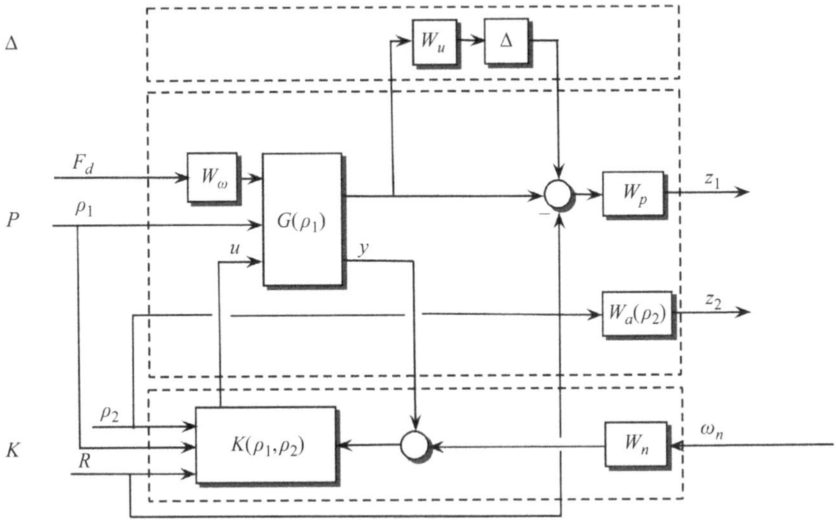

Figure 10.2 Closed-loop interconnection structure

Uncertainties of the system, such as unmodeled dynamics and parameter uncertainty, are located in a Δ block. Weighting function W_u handles the magnitude of the ignored dynamics.

Two performance signals are expressed to guarantee the tracking of the reference velocity and the lateral displacement with an admissibly small error. The weighting function is chosen in a second-order proportional form:

$$W_p = \gamma \frac{\alpha_2 s^2 + \alpha_1 s + 1}{T_1 s^2 + T_2 s + 1}, \tag{10.7}$$

where $T_{1,2}$, $\alpha_{1,2}$ and γ are designed parameters.

Note that although weighting functions are expressed in the frequency domain, in the control design their state-space representation forms are used. In the design of the weighting functions noted with W_a in Figure 10.2, the distribution among steering and differential torque generation can be changed according to the vehicle state. Hence, a weighting function for the steering angle input $W_{a\delta} = (\delta_{\max}\chi_1)/(\rho_2)$ and for the yaw moment input $W_{aMz} = (\rho_2)/(M_{z\max}\chi_2)$ is given, where χ_1 and χ_2 are design parameters selected to realize control redistribution, $\delta_{\max}$ is the maximum value of the steering angle, $M_{z\max}$ is the maximum yaw moment, while $\rho_2 \in [0.01, 1]$ is a monitoring variable representing the split between steering intervention and yaw moment generation. Therefore, by choosing $\rho_2 = 1$ only steering is applied throughout a cornering maneuver, while when selecting $\rho_2 = 0.01$ only differential torque is generated by the in-wheel motors of the vehicle. Defining the value of ρ_2 is based on an optimization process minimizing cornering energy. The control design is based on the LPV method that uses parameter-dependent Lyapunov functions, see [12,13].

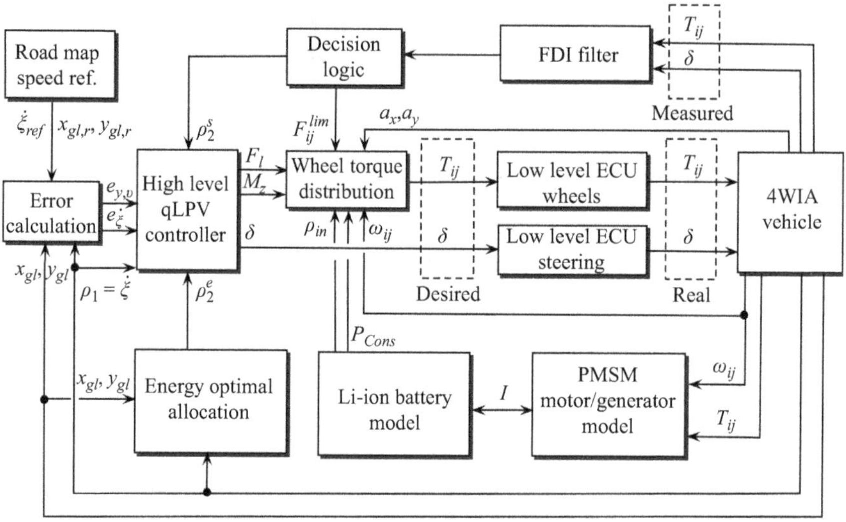

Figure 10.3 Architecture of the control system

The quadratic LPV performance problem is choosing the parameter-varying controller in such a way that the resulting closed-loop system is quadratically stable and the induced $\mathscr{L}_2$ norm from the disturbance and the performances is less than the value γ. The minimization statement is the following:

$$\inf_{K} \sup_{\rho \in \mathcal{F}_P} \sup_{\|w\|_2 \neq 0, w \in \mathcal{L}_2} \frac{\|z\|_2}{\|w\|_2}. \tag{10.8}$$

where w is the disturbance. The presence of a controller that solves the quadratic LPV γ-performance problem can be predicated as the applicability of a numerically identifiable set of linear matrix inequalities. Finally, the state-space representation of the LPV control $K(\rho)$ is established, see e.g. [13].

10.2 Fault-tolerant and energy optimal control synthesis

10.2.1 Control architecture

The architecture of the multilayer reconfiguration control method considering energy optimality and fault-tolerant capabilities is shown in Figure 10.3.

In the first layer of the hierarchical structure, the high-level LPV controller introduced in Section 10.1.2 calculates the inputs of the 4WIA vehicle based on multiple signals. First, the path of the vehicle is given by the predefined trajectory of the road map, while the reference velocities are also set in coherence with the speed limits for the given road. Next, velocity error $e_{\dot{\xi}} = \dot{\xi}_{ref} - \dot{\xi}$ and lateral deviation

$e_{y,v} = y_{v,r} - y_v$ from the reference values are calculated based on the vehicle current position $(x_{gl,v}; y_{gl,v})$ and velocity:

$$\rho_1 = \dot{\xi}. \tag{10.9}$$

These error signals $e_{\dot{\xi}}$ and $e_{y,v}$ along with scheduling variable $\rho_1 = \dot{\xi}$ and ρ_2 are fed to the LPV controller. Note that the value of ρ_2 is determined by the results of fault-tolerant reconfiguration and the energy optimal introduced in Sections 10.2.2 and 10.2.3.

For this purpose, a decision logic has been created by giving priority to the safety of the in-wheel vehicle. Hence, ρ_2 is defined as follows:

$$\rho_2 = \begin{Bmatrix} \rho_2^e, & \text{if } \rho_2^s = 0 \\ \rho_2^s, & \text{if } \rho_2^s \neq 0 \end{Bmatrix} \tag{10.10}$$

The second layer is responsible for the allocation of the high-level control signals, i.e. the longitudinal force, yaw moment and steering angle. The distribution method is based on current vehicle dynamics in order to avoid wheel slip, the characteristics of the in-wheel motors to maximize efficiency, and the detected faults are also considered to guarantee the reference tracking of the vehicle in the case of critical performance degradation of an electric motor. The detailed description of the wheel torque distribution method is given in Section 10.2.4.

The third layer models the operation of the low-level controllers, i.e. it represents the tracking of the distributed electric motor torques and the prescribed steering angle. Here, the dynamics of the steer-by-wire steering system is modeled as a first-order system as proposed in [14]. The current control of the in-wheel hub motors is simplified in this layer as proposed in [15], thus the following equation describes the relation between the prescribed wheel torque and the generated torque:

$$T_{motor}(s) = \frac{T(s)(1+\eta)}{1+2\zeta+2\zeta^2} \tag{10.11}$$

where T_{motor} is the electric motor torque, T is the prescribed torque, while ζ and η represent the dynamic response and steady state error of the electric hub motors.

10.2.2 Fault-tolerant reconfiguration

The performance of in-wheel motors may be degraded due to mechanical failures, the overheating of the engine or faults connected to the motor control system. The complexity of in-wheel vehicles and the increased potential dangers of a fault event make a fault tolerant control design necessary. In-wheel motors were studied by [16], which focused on achieving a high-torque density and the capability to sustain an adequate level of performance following a failure. Reference [17] proposed a fault-tolerant control system designed to accommodate hub motor faults by automatically reallocating the control effort among other healthy wheels. Reference [18] designed a fault-tolerant control based on the estimation of the maximum transmissible torque for the wheels. Recently, a sliding mode control has been developed by [19] to deal with a faulty in-wheel motor by rearranging steering geometry depending on the location of the fault.

Although faulty steering occurs rarely compared to the performance degradation or faults of in-wheel motors, it is also important to deal with such event to guarantee vehicle stability. The fault-tolerant control design of steer-by-wire steering systems has been studied by several authors, see [20,21]. Also, the steer-by-wire steering system failures can be handled more effectively in a 4WIA vehicle by using differential drive assisted steering, as presented in [22]. The performance degradations of both the steering system and in-wheel motors were addressed in a more general manner in [23].

Here, in order to eliminate or at least mitigate the effect of actuator failures connected both to the steer-by-wire steering system and the in-wheel motors, a high-level control reallocation method is presented. In the case of a faulty steering system, the aim is to substitute for the effect of the steering by reconfiguring the high-level controller to prescribe additional differential torque for the vehicle. The fault of the steering system is assumed to be detected by FDI filters, as proposed by [24]. When the fault is detected, the scheduling variable $\rho_2^s = 0.01$ is applied to override the value of the actual ρ_2. Thus, the high-level LPV controller is modified in such way that it prescribes solely yaw moment signal for the 4WIA vehicle. Hence in the case of a steering system failure, the cornering maneuvers are evaluated by using only the precise torque vectoring ability of the vehicle.

The fault of one or more in-wheel motors is handled in a more complex manner. The detection of in-wheel motor failures is also assumed to be done by FDI filters, such as the Kalman filtering-based algorithm presented in [25]. In the case of an in-wheel motor failure, scheduling variable $\rho_2^s = 1$ is set to substitute for the actual ρ_2 value. Hence, weighting function W_a of the LPV controller is modified in such way that the high-level controller prescribes solely steering intervention for the 4WIA vehicle to evaluate the cornering maneuver. Note that in order to avoid unintentional yaw moment generation, the wheel torque distribution algorithm discussed in Section 10.2.4 also needs to be modified to handle the fault of an in-wheel motor. Hence, the upper and lower bounds given as constraints for the wheel torque optimization method are modified in (10.14) for the faulty in-wheel motor.

The novelty of the presented approach lies in the high-level LPV control reconfiguration strategy based on the specific design of weighting functions handling actuator selection. The proposed method enables the vehicle to dynamically modify the prescribed control values best suited for the actual vehicle state and the corresponding priority regarding safety.

10.2.3 Energy optimal reconfiguration

The aim of energy optimal reconfiguration is to extend the range of the 4WIA electric vehicle. For this purpose, different methods have already been introduced. In [8, 9] the high-level control reconfiguration is based on the minimization of cornering energy, not considering the battery SOC and electric motor characteristics. Hence, a lithium-ion battery model and a PMSM model with regenerative braking ability have been implemented in order to consider battery-SOC in [26]. The simulation results have shown that although both methods provide effective solutions to enhance

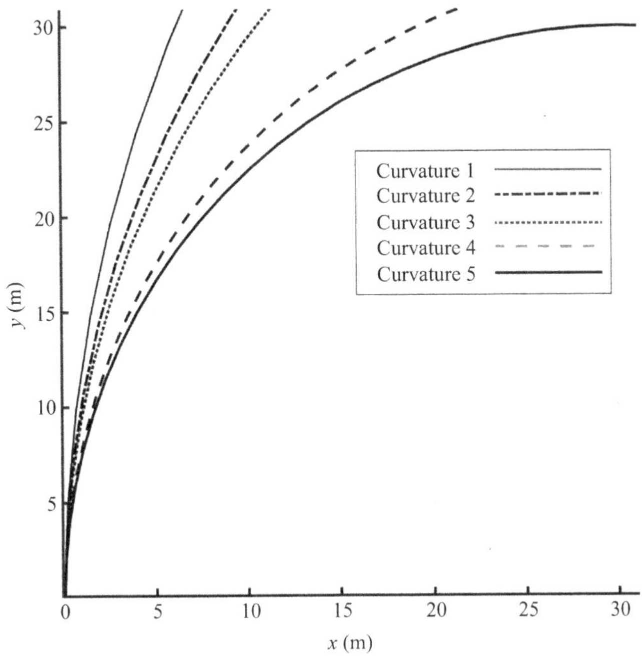

Figure 10.4 Selected curves for analysis

efficiency, they should be coupled to maximize the range of the vehicle. For this purpose, a novel approach is introduced, in which the scheduling variable ρ_2^s for the high-level control reconfiguration is designed in a way to consider battery SOC variations for different bends and vehicle velocities. Hence, the goal has been to create an energy efficiency map showing the optimal scheduling variable selection for a given curve radius and vehicle velocity. The energy map has been identified by simulation studies in a CarSim simulation environment, considering the battery and motor model introduced in Section 10.3.

For the analysis, five different types of curves have been selected on the Waterford Michigan racetrack with different curvature characteristics, see Figure 10.4. Sharper bends and milder curves have also been considered in the analysis in order to create a comprehensive energy efficiency map. Note that for the sake of simplicity, the elevation of the road has not been considered in the analysis, and the road adhesion has been assumed to be constant, representing that of dry asphalt. The simulated 4WIA vehicle must perform trajectory tracking on each curve with velocities between 10 and 60 km/h, while scheduling variable $\rho_2 \in [0.01, 1]$ is set to different values representing different allocation between steering and yaw moment generation. Meanwhile, battery characteristics, including battery SOC is monitored. For each simulation case, reduction of SOC values are compared, see Figure 10.5. Here, Curve 1 in Figure 10.5(a)

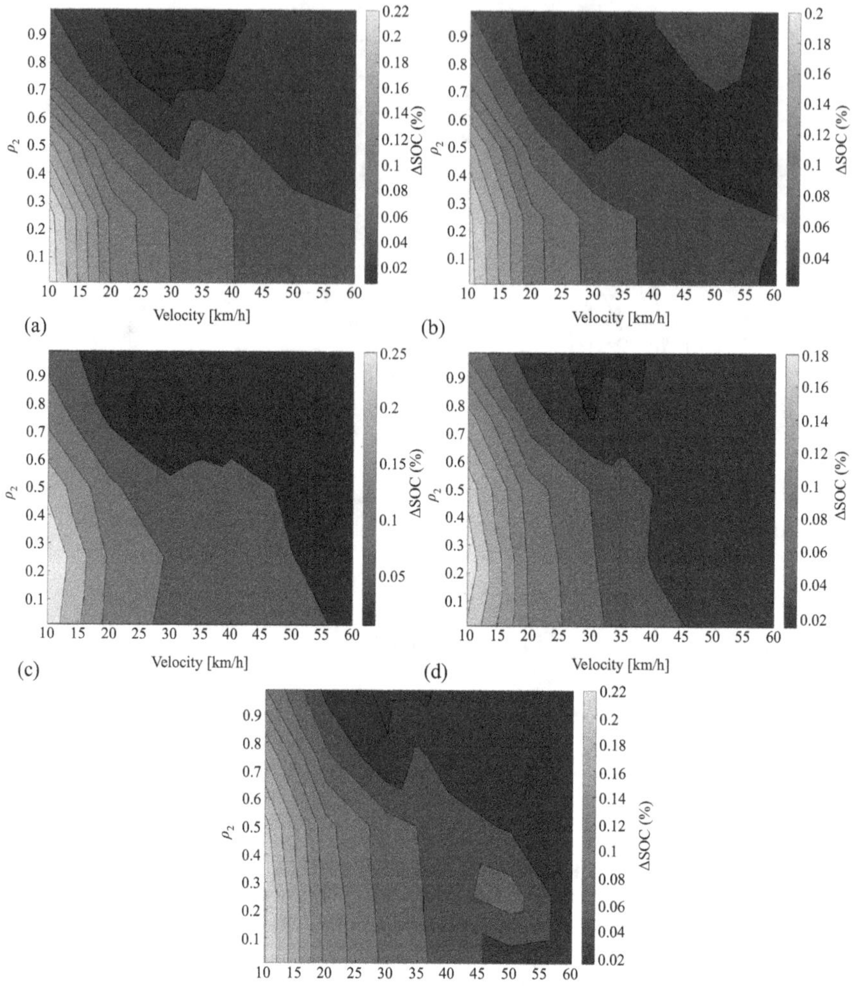

Figure 10.5 Energy efficiency map: (a) curve 1, (b) curve 2, (c) curve 3, (d) curve 4 and (e) curve 5

represents the mildest curve, while Curve 5 in Figure 10.5(e) stands for the sharpest bend. From the simulation results, the following conclusions can be drawn:

- While increasing vehicle velocity in the selected range, the battery SOC decreases less in the selected curves. This phenomenon is due to multiple causes: the electric in-wheel motors operate in a more efficient revolution range, friction resistance is more optimal, etc.

- As the vehicle velocity increases, the selection of $\rho_2 \in [0.01, 1]$ has less effect on the decrement of battery SOC. This phenomenon is more pronounced in sharper curves.
- For smaller velocities under 40 km/h, choosing a larger $\rho_2 \approx 1$ value (selecting steering over yaw-moment generation) results in better battery SOC, thus it is more energy optimal for the vehicle. Note that as the curves get sharper, this phenomenon becomes less pronounced.
- For velocities between 45 and 55 km/h, depending on the curvature, the default value of $\rho_2 = 1$ must be reduced to $0.5 < \rho_2 < 0.75$ in order to enhance battery SOC.
- For higher velocities between 55 and 65 km/h, as the curvature of the road becomes greater (turns sharper), the value of ρ_2 must be reduced even more to realize maximal efficiency. Thus, for sharper curves in this velocity range, larger yaw-moment generation results in more energy optimal motion for the 4WIA vehicle.
- The analysis also suggests that controlling the vehicle with solely torque vectoring ($\rho_2 \approx 0.01$) can adversely affect vehicle stability due to the large side-slip values of the wheels and the corresponding nonlinear effects ignored in vehicle modeling detailed in Section 10.1.1. Thus, in the reconfiguration design $\rho_2^s = 0.25$ is set as a minimal value.

The results of the simulation studies are used to create a look-up table for the online determination of ρ_2^s during the operation of the 4WIA vehicle. Hence, the value of ρ_2^s is selected from the predefined energy efficiency map based on the current vehicle velocity and road curvature. Note that road curvature can be easily calculated based on the coordinates of the predefined path, as detailed in [27]. By this means, the high-level controller can be reconfigured to adapt to road and vehicle dynamic conditions, resulting in better overall efficiency and enhanced range for the 4WIA vehicle.

10.2.4 Efficient wheel torque distribution

An in-wheel electrical vehicle can be directly driven and hence, the clutch, the transmission or other drive systems are not necessary. Therefore, power train efficiency is improved and control for each wheel is available [28]. Each in-wheel motor drives the wheel independently; hence, the degrees of freedom of the driving force distribution between the wheels increase [29]. The main aim of force distribution of the in-wheel motor drive system is to improve the energy efficiency and performance of the vehicle, such as dynamic performance, security, economy, etc. The pattern search algorithm is used to minimize the power consumption of motors while satisfying the yaw moment and longitudinal force. The power of in-wheel motors are directly related to their torque. Hence, the total power consumption of the in-wheel motors P_{total} is calculated as [6]

$$P_{total} = \sum_{\substack{i=f,r \\ j=L,R}} \frac{F_{ij} w_{ij} P_{in}}{P_{ij}} \tag{10.12}$$

where w_{ij}, $i \in [f = \text{front}, r = \text{rear}]$, $j \in [L = \text{left}, R = \text{right}]$ are the rotational speeds of the motors and P_{in} is equalized input power, which is defined as the allocated equalized power of one motor. It is calculated as total power consumption of motors per motor equally.

The maximum vertical wheel loads that can be transmitted to $F_{z,ij}$, $i \in [f = \text{front}, r = \text{rear}]$, $j \in [L = \text{left}, R = \text{right}]$ are calculated [30]:

$$F_{z,ij} = m\left(\frac{l_{[1;2]}g}{L} \pm \frac{ha_x}{L}\right)\left(0.5 \pm \frac{ha_y}{b_{[f;r]}g}\right) \tag{10.13}$$

where $L = l_1 + l_2$, h is the height of the mass center, a_y and a_x are lateral and longitudinal accelerations measured by inertial sensors and g is the gravitational constant. Note that the front wheel loads ($i = f$) are calculated by using l_2 and b_f with a negative sign in the first bracket, whereas the rear wheel loads ($i = r$) are calculated with l_1 and b_r with positive sign in the first bracket. Moreover, left wheel loads ($j = L$) are defined by a negative sign in the second bracket, while right wheel loads ($j = R$) with a positive sign.

The maximum and minimum longitudinal traction wheel forces that are transmissible for each wheel are given as $|F_{ij}^{\max}| = \mu F_{z,ij}$ and $|F_{ij}^{\min}| = -\mu F_{z,ij}$. The upper and lower bound can be expressed as

$$-\mu F_{z,ij} \leq F_{z,ij} \leq \mu F_{z,ij} \tag{10.14}$$

Note that in the case of an in-wheel motor fault or performance degradation, the upper and lower bounds are modified for the faulty motor to a limited value $F_{ij}^{\lim}$, in coherence with the fault detected by FDI filters.

At the same time, M_z is the yaw moment that is given by the high-level controller achieved by wheel forces generated by in-wheel motors. Therefore, the yaw moment M_z^{veh} generated by the longitudinal wheel force F_{ij} can be calculated as

$$M_z^{veh} = (-F_{fL} + F_{fR})\frac{b_f}{2} + (-F_{rL} + F_{rR})\frac{b_r}{2} \tag{10.15}$$

Two equalities need to satisfy the demands of the high-level controller, including the desired longitudinal force reference and the yaw moment reference. Satisfying the yaw moment requirement is expressed as

$$M_z^{veh} = M_z \tag{10.16}$$

Meanwhile, a nonlinear constraint is presented in order to ensure the performance of velocity tracking. Therefore, the sum of the wheel forces must be equal to the longitudinal force F_l which is given by the high-level controller:

$$\sum F_{ij} - F_l = 0 \tag{10.17}$$

The objective function is

$$J = \min\left(P_{cons} - P_{total}\right) \tag{10.18}$$

where P_{cons} is the total power consumption given by the battery model. Therefore, the wheel force distribution optimization problem is solved, where the objective function

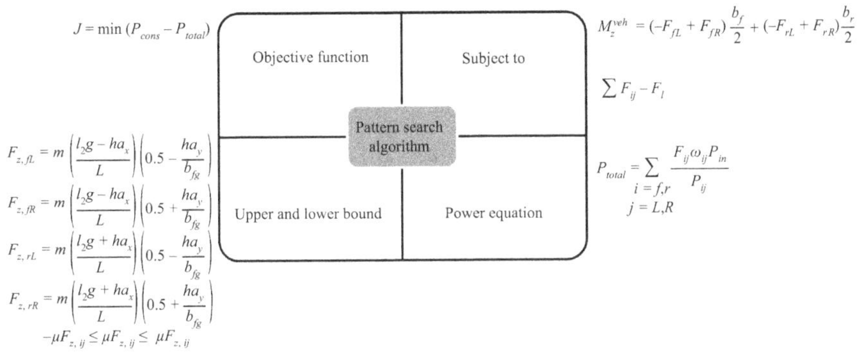

Figure 10.6 Pattern search algorithm methodology

given in (10.18) is minimized by the lower and upper bounds calculated by (10.14) and (10.16) and (10.17) are satisfied. The methodology of the pattern search algorithm is given in Figure 10.6 with formulas. Finally, the torques for the in-wheel motors can be expressed as

$$T_{ij} = R_{eff} F_{ij} \tag{10.19}$$

10.3 Electric motor and battery model

10.3.1 Lithium-ion battery

A battery, which is the main element of electric vehicles, stores a great amount of energy to be released when it is necessary. The battery allows regenerative braking and supplements a slow dynamic energy source [31]. A lithium-ion battery (Li-ion) is a rechargeable battery in which lithium ions move from the negative to the positive electrode. Li-ion batteries use intercalated lithium compound as electrode materials. The Lithium-ion battery has a number of advantages compared to the other battery types [32]:

- increased nominal voltage,
- high efficiency and energy density,
- high resistance for external conditions,
- small size and specific weight,
- maintenance is not required,
- fast, more efficient charging and increased lifetime.

Li-ion batteries must not be completely discharged and this is their most important disadvantage. Completely discharging reduces the lifetime of the battery. Hence, discharging the battery with high current is harmful and dangerous. The main reason for the preference of Li-ion batteries in the automotive market is energy density. Currently, they are preponderant for portable electric devices, cell phones and laptops as mobile power sources [33].

10.3.2 Battery pack

Several widespread battery models can be found in the literature, including detailed models containing chemical reactions and simplified equivalent circuits and polynomial models. In [34] the detailed chemical reactions are collected in a state-space model for NiMH batteries. In [35] a detailed algorithm for the estimation of NiMH battery SOC can be found, which method can also be adapted to lead–acid and Li-ion batteries as well. A generic battery model of Simulink, which is based on the model of Shepherd [36] under the Simscape library is used for the Li-ion battery pack model. The SimPowerSystems battery model provides current, voltage and SOC data from *m* output and possibly modifies many preferences easily such as battery type, nominal voltage, rated capacity, battery response time, maximum capacity, nominal discharge voltage, etc. Note that temperature and aging effects are not considered for the sake of simplicity.

The battery charge and discharge are based on the Shepherd model, which can be formulated as described in (10.21) for charge, and (10.20) for discharge.

$$f_1(i_t, i^*, i) = E_0 - K \cdot \frac{Q}{Q - i_t} \cdot i^* - K \cdot \frac{Q}{Q - i_t} \cdot i_t + A \cdot \exp(-B \cdot i_t) \quad (10.20)$$

$$f_2(i_t, i^*, i) = E_0 - K \cdot \frac{Q}{i_t + 0.1 \cdot Q} \cdot i^* - K \cdot \frac{Q}{Q - i_t} \cdot i_t + A \cdot \exp(-B \cdot i_t) \quad (10.21)$$

where i is the battery current, i_t is the extracted capacity, i^* is the low-frequency current dynamics, E_0 is the constant voltage, K is the polarization constant, Q is the maximum battery capacity, A is the exponential voltage and B is the exponential capacity. In the battery cell model, the nominal voltage of the battery is 400 V, rated capacity is 58.5 A h, initial SOC is 80% and battery response time is 30 s.

DC–DC converters are circuits that use high-frequency switching and indicators, capacitors and transformers for smoothing out switching noise into regulated DC voltages. Buck converter type is implemented to present study between battery model and motor model to convert 400–500 V, which is the working voltage of the in-wheel motor model.

10.3.3 Motor model

A PMSM is an electrical machine that has the stator phase windings and rotor permanent magnets. Three-phase stator windings create a turning magnetic field through the three-phase AC [37]. The rotor is furnished with a superior permanent magnet on the surface and with ferromagnetic materials within, for example, neodymium iron, boron or uncommon earth magnetic materials to acquire a solid magnetic field [37].

A permanent magnet synchronous machine block from the Simulink Simscape library works in either generator or motor mode. The operating mode is determined by the mechanical torque sign (positive for engine mode, negative for generator mode). In this study regeneration energy can be applied because of this feature.

The equations, which are located below, are expressed in the rotor reference frame. All of the values in the rotor reference frame are referred to the stator:

$$\frac{d}{dt}i_d = \frac{1}{L_d}V_d - \frac{R}{L_d}i_d + \frac{L_q}{L_d}pw_m i_q \tag{10.22a}$$

$$\frac{d}{dt}i_d = \frac{1}{L_d}V_d - \frac{R}{L_d}i_d + \frac{L_q}{L_d}pw_m i_q - \frac{\lambda p w_d}{L_q} \tag{10.22b}$$

$$T_e = 1.5p[\lambda i_q + (L_d - L_q)i_d i_q] \tag{10.22c}$$

where L_q and L_d express inductances of the q and d axes, R signifies the resistance of stator windings, i_q and i_d are the currents of the axes of q and d, V_q and V_d are the voltages of the q and d axes, w_m is the angular velocity of the rotor, λ is the flux-induced amplitude by the permanent magnets of the rotor in the stator phases p number of pole pairs and T_e expresses electromagnetic torque.

There is no variation in the phase inductance for a round motor. The inductances of d and q are given by formulas below for a salient round motor:

$$L_q = L_d = \frac{L_a b}{2} \tag{10.23a}$$

$$L_q = \frac{\min(L_a b)}{2} \tag{10.23b}$$

$$L_d = \frac{\max(L_a b)}{2} \tag{10.23c}$$

For the mechanical system of three-phase permanent magnet synchronous machine, the angular velocity of the rotor is expressed as follows:

$$\frac{d}{dt}w_m = \frac{1}{J}(T_e - T_f - Fw_m - T_m) \tag{10.24a}$$

$$\frac{d\theta}{dt} = w_m \tag{10.24b}$$

where J combines the inertia of the rotor and the load, F combines the viscous friction of the rotor and the load, θ is the rotor angular position, T_m is the shaft mechanical torque, T_f is the shaft static friction torque and ω_m is the rotor's angular velocity.

Complete motor model consists of eight motors, four of them are for recharging process and other four motors are for charging process in the case of negative torque. Charging motors enable one to increase SOC of the battery. At the same time, Simscape library is used for the inverter, speed and vector controller, which are located in the motor model.

10.4 Simulation results

The selected vehicle for the simulation is a small 4WIA vehicle, with the main physical and dynamic parameters listed in Table 10.1.

Table 10.1 Parameters of the 4WIA vehicle

Parameter	**Value**	**Unit heads**
Vehicle mass (m)	830	kg
Yaw moment of inertia (J)	1, 110.9	kg m^2
Distance from COG to front axle (l_1)	1.103	m
Distance from COG to rear axle (l_2)	1.244	m
Tread front (b_f)	1.416	m
Tread rear (b_r)	1.375	m
Height of COG (h_{COG})	0.54	m
Cornering stiffness front (c_1)	22	kN/rad
Cornering stiffness rear (c_2)	85	kN/rad
Aerodynamic drag coefficient (c_w)	0.343	–
Front contact surface (A)	1.6	m^2

The simulation task for the autonomous 4WIA vehicle is to follow the Waterford Michigan racetrack. The geometry of the racetrack is shown in Figure 10.7. As it can be seen in Figure 10.7(a) and (b), the path of the vehicle contains different types of bends and road slopes. Note that the reference velocity for the autonomous 4WIA vehicle is also altered during the simulation, see Figure 10.7(c).

The goal of the simulation is to demonstrate the effectiveness of the proposed fault-tolerant and energy optimal reconfiguration control method. Thus, two simulations are performed: in the first simulation the vehicle is controlled by a fixed high-level controller setup of $\rho_2 = 0.5$ and by an analytical wheel torque allocation method presented in [38] (normal case), while in the second simulation the control is reconfigured based on the results of the battery SOC-based analysis and an energy optimal torque allocation method detailed in the previous chapter (energy optimal case).

Several vehicle dynamic signals can be measured in real vehicles using widespread wheel speed sensors and a gyroscopes, thus vehicle velocity, yaw rate, longitudinal and lateral accelerations are also measured in CarSim. As it can be seen in Figure 10.8(a) the autonomous 4WIA vehicle is subject to high lateral accelerations, as well as high yaw rate values, see Figure 10.8(b).

The scheduling variables of the LPV controller are shown in Figure 10.9. Scheduling variable ρ_1 is the velocity of the 4WIA vehicle and is similar in both cases, see Figure 10.9(a). However, significant differences can be seen in choosing the scheduling variable ρ_2 as depicted in Figure 10.9(b), which is responsible for the split between steering intervention and yaw-moment generation. In the default case (normal), a fix $\rho_2 = 0.5$ is selected, which represents steering-oriented cornering supplemented with remarkable torque vectoring. With the proposed reconfiguration considering battery SOC based on the road curvature and vehicle velocity, ρ_2 is altered in coherence with the vehicle state and the current curvature of the road.

The high-level control signals of the autonomous 4WIA vehicle are shown in Figure 10.10 for both cases. Note that the longitudinal forces depicted in

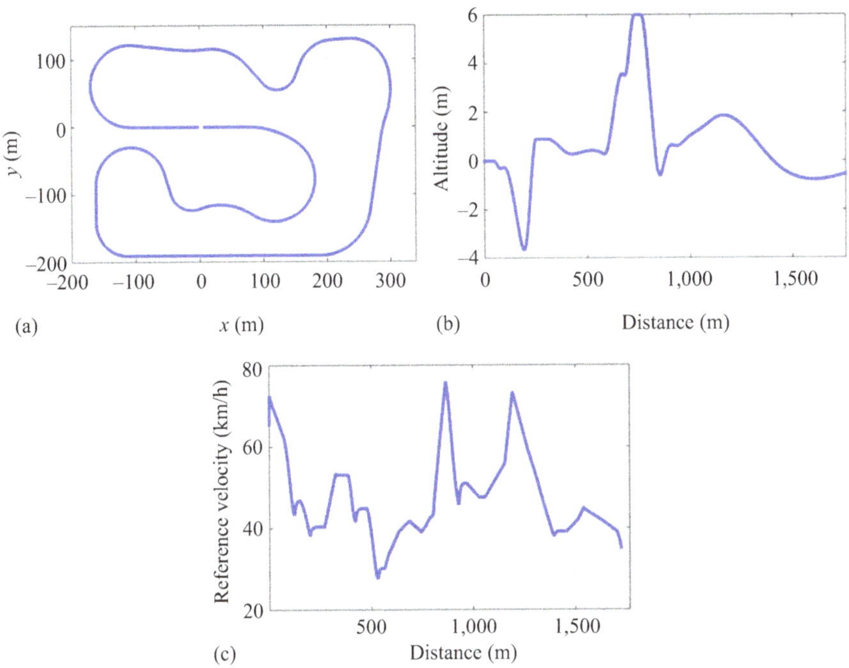

Figure 10.7 Reference velocity and trajectory: (a) racetrack X–Y plane, (b) altitude and (c) reference velocity

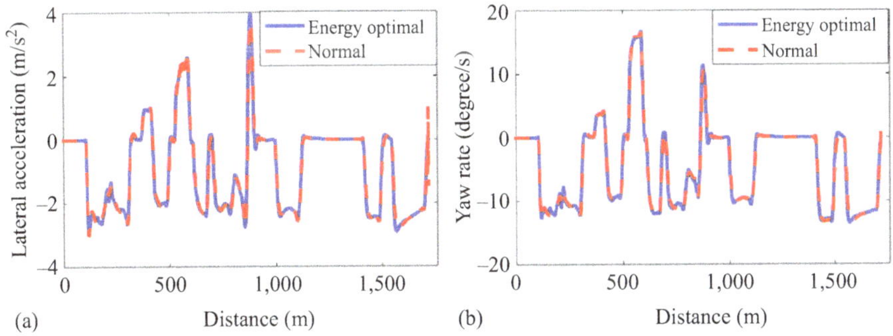

Figure 10.8 Dynamics of the 4WIA vehicle: (a) lateral acceleration and (b) yaw rate

Figure 10.10(a) in the normal and the energy optimal cases are similar to each other. On the other hand, with the proposed energy optimal reconfiguration, steering is slightly more pronounced in most corners, see Figure 10.10(b). Moreover, yaw-moment generation has been significantly reduced, see Figure 10.10(c). Thus, it can be stated that for the prescribed velocities on the selected road, the reconfiguration algorithm designs a more steering-oriented actuator selection.

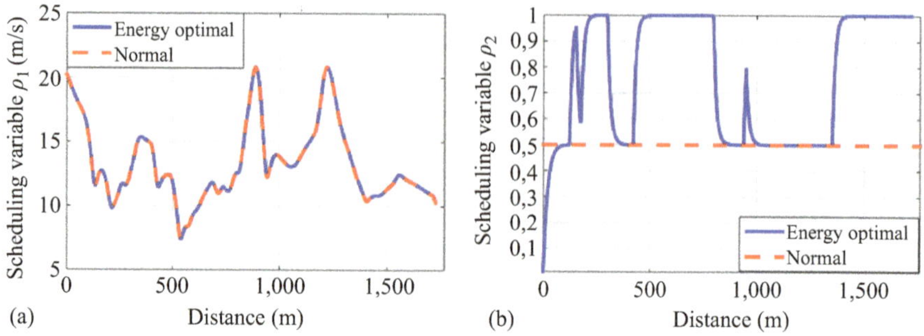

Figure 10.9 Scheduling variables of the LPV controller: (a) ρ_1 and (b) ρ_2

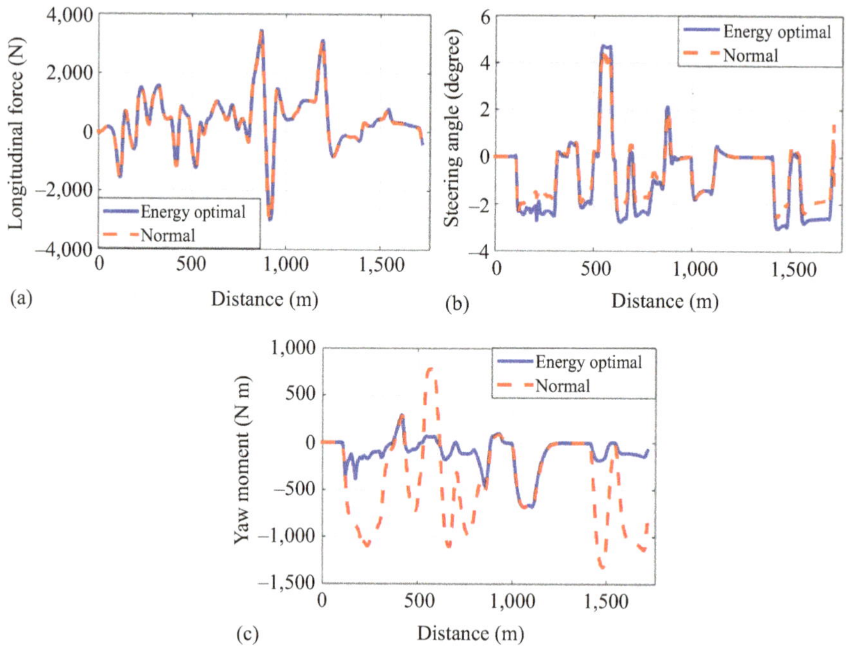

Figure 10.10 High-level control signals: (a) longitudinal force, (b) steering angle, (c) yaw moment

Correspondingly, the in-wheel brake and drive torques depicted in Figure 10.11(b) in the energy optimal control are smaller than those in the normal control shown in Figure 10.11(a).

The performances of the proposed method are shown in Figure 10.12. It is well demonstrated that the 4WIA autonomous vehicle follows the prescribed altering velocity profile with small deviations for both cases, see Figure 10.12(a). Despite the challenging road environment of the racetrack with several different types of

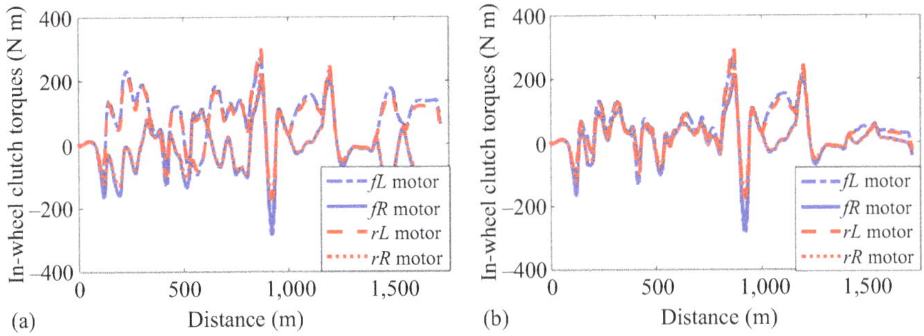

Figure 10.11 In-wheel clutch torques: (a) without reconfiguration and (b) with energy optimal reconfiguration

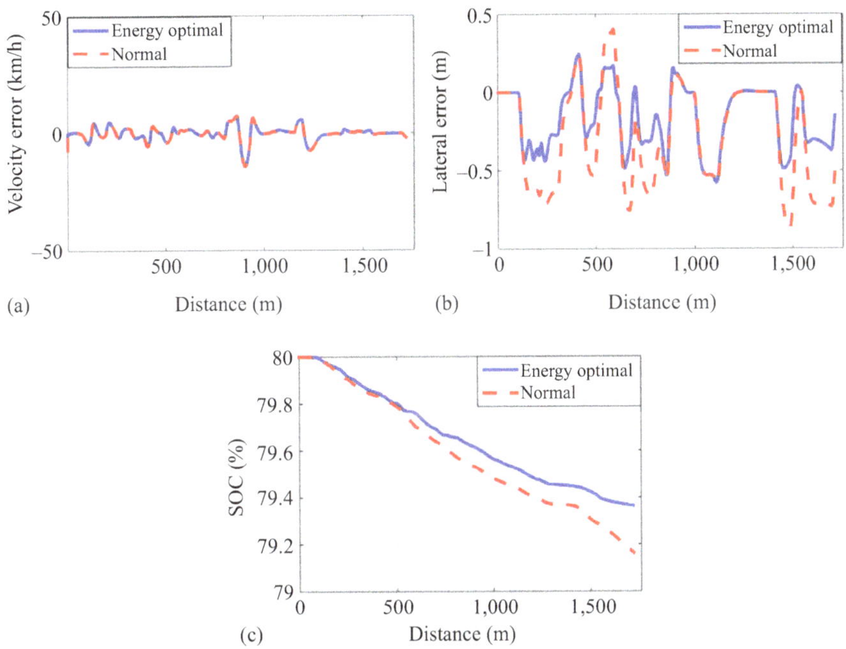

Figure 10.12 Performances of different methods: (a) velocity error, (b) lateral error, (c) battery SOC

bends, the 4WIA vehicle is able to follow the path with small lateral errors, see Figure 10.12(b). It is well demonstrated that the lateral error is also decreased by applying the energy optimal reconfiguration control method and the wheel torque distribution method considering roll and pitch dynamics of the vehicle. Finally, the efficiency of the proposed method is demonstrated by comparing battery SOC values, as depicted in Figure 10.12(c). With the proposed energy optimal reconfiguration method, the initial SOC value of 80% has dropped down to only around 79.4%, while in the normal case battery SOC value decreased to around 79.2%, which shows a significant difference.

10.5 Conclusion

In this chapter a reconfigurable trajectory-tracking control design method has been presented for autonomous in-wheel electric vehicles with independently controlled hub motors and the steer-by-wire steering system. The high-level control reconfiguration has been implemented through the design of a scheduling variable using the LPV framework in order to deal with fault events, while in normal operating conditions the objective of the reconfiguration is to maximize battery SOC, thus enhancing the range of the in-wheel electric vehicle. The energy optimal control reconfiguration has been designed based on the results of preliminary simulations with a high-fidelity vehicle and electrical models on different road conditions. Finally, the efficiency of the proposed method has been demonstrated in a real-data CarSim simulation, showing significant energy saving by the proposed method.

References

[1] Wu FK, Yeh TJ, Huang CF. Motor control and torque coordination of an electric vehicle actuated by two in-wheel motors. Mechatronics. 2013;23:46–60.

[2] Castro R, Araújo RE, Tanelli M, *et al.* Torque blending and wheel slip control in EVs with in-wheel motors. Vehicle System Dynamics. 2012;50:71–94.

[3] Xiong L, Yu Z, Wang Y, *et al.* Vehicle dynamics control of four in-wheel motor drive electric vehicle using gain scheduling based on tyre cornering stiffness estimation. Vehicle System Dynamics. 2012;50:831–846.

[4] Shuai Z, Zhang H, Wang J, *et al.* Lateral motion control for four-wheel-independent-drive electric vehicles using optimal torque allocation and dynamic message priority scheduling. Control Engineering Practice. 2013;24:55–66.

[5] Wang R, Chen Y, Feng D, *et al.* Development and performance characterization of an electric ground vehicle with independently actuated in-wheel motors. Journal of Power Sources. 2011;196:3962–3971.

[6] Cheng CL, Xu Z. Wheel torque distribution of four-wheel-drive electric vehicles based on multi-objective optimization. Energies. 2015;8:3815–3831.

[7] Ocran TA, Cao J, Cao B, *et al.* Artificial neural network maximum power point tracker for solar electric vehicle. Tsinghua Science and Technology. 2005;10:204–208.

[8] Mihály A, Gáspár P. Robust and fault-tolerant control of in-wheel vehicles with cornering resistance minimization. European Control Conference (ECC). 2016;p. 2590–2595.

[9] Mihály A, Gáspár P, Németh B. Robust fault-tolerant control of in-wheel driven bus with cornering energy minimization. Strojniški Vestnik – Journal of Mechanical Engineering. 2017;63(1):35–44.

[10] Szabó B. Vehicle test based validation of a tire brush model using an optical velocity sensor. Periodica Polytechnica. 2012;40:33–38.

[11] Song CK, Uchanski M, Hedrick JK. Vehicle speed estimation using accelerometer and wheel speed measurements. Proc of the SAE Automotive Transportation Technology, Paris. 2002.

[12] Bokor J, Balas G. Linear parameter varying systems: A geometric theory and applications. 16th IFAC World Congress, Prague. 2005.

[13] Wu F, Yang XH, Packard A, *et al.* Induced l^2-norm control for LPV systems with bounded parameter variation rates. International Journal of Nonlinear and Robust Control. 1996;6:983–998.

[14] Takanori F, Shogo M, Kenji M, *et al.* Active steering systems based on model reference adaptive nonlinear control. Vehicle System Dynamics: International Journal of Vehicle Mechanics and Mobility. 2004;42:301–318.

[15] Tahami F, Kazemi R, Farhanghi S. A novel driver assist stability system for all-wheel-drive electric vehicles. IEEE Transactions on Vehicular Technology. 2003;52:683–692.

[16] Ifedi CJ, Mecrow BC, Brockway STM, *et al.* Fault tolerant in-wheel motor topologies for high performance electric vehicles. IEEE Transactions on Industry Applications. 2013;49:1249–1257.

[17] Wang R, Wang J. Fault-tolerant control for electric ground vehicles with independently-actuated in-wheel motors. Journal of Dynamic Systems, Measurement, and Control. 2012;134(2):021014.

[18] Hu JS, Yin D, Hori Y. Fault-tolerant traction control of electric vehicles. Control Engineering Practice. 2011;19(2):204–213.

[19] Li B, Du H, Li W. Fault-tolerant control of electric vehicles with in-wheel motors using actuator-grouping sliding mode controllers. Mechanical Systems and Signal Processing. 2016;19:72–73:462–485.

[20] Tian C, Zong C, He L, *et al.* Fault tolerant control method for steer-by-wire system. Proceedings of the 2009 IEEE International Conference on Mechatronics and Automation. 2009;p. 291–295.

[21] Zheng B, Altemare C, Anwar S. Fault tolerant steer-by-wire road wheel control system. American Control Conference. 2005;p. 1619–1624.

[22] Hu C, Jing H, Wang R, *et al.* Fault-tolerant control of FWIA electric ground vehicles with differential drive assisted steering. 9th IFAC Symposium on Fault Detection, Supervision and Safety for Technical Processes (Safeprocess'15). 2015;48;p. 1180–1185.

[23] Jing H, Wang R, Chadli M, *et al.* Fault-tolerant control of four-wheel independently actuated electric vehicles with active steering systems. 9th IFAC Symp Fault Detection, Supervision and Safety for Technical Processes. 2015;48; p. 1165–1172.

[24] Im JS, Ozaki F, Yeu TK, *et al.* Model-based fault detection and isolation in steer-by-wire vehicle using sliding mode observer. Journal of Mechanical Science and Technology. 2009;23:1991–1999.

[25] Nguyen TH, Chen BC, Yin D, *et al.* Active fault tolerant torque distribution control of 4 in-wheel motors electric vehicles based on Kalman filter

approach. International Conference on System Science and Engineering (ICSSE). 2017;p. 1619–1624.

[26] Mihály A, Gáspár P. Reconfiguration control of in-wheel electric vehicle based on battery state of charge. European Control Conference (ECC). 2018;p. 243–248.

[27] Mihály A, Németh B, Gáspár P. Look-ahead control of road vehicles for safety and economy purposes. Control Conference (ECC), 2014 European. 2014; p. 714–719.

[28] Al Emran Hasan MM, Ektesabi M, Kapoor A. Pollution control and sustainable urban transport system-electric vehicle. International Journal of Transport and Vehicle Engineering. 2011;5(6):374–379.

[29] Papelis YE, Watson GS, Brown TL. An empirical study of the effectiveness of electronic stability control system in reducing loss of vehicle control. Accident Analysis & Prevention. 2010;42(3):929–934.

[30] Kiencke U, Nielsen L. Automotive Control Systems. Berlin, Heidelberg: Springer Verlag; 2005.

[31] Tremblay O, Dessaint LA. Experimental validation of a battery dynamic model for EV applications. World Electric Vehicle Journal. 2009;3(2):289–298.

[32] Han X, Ouyang M, Lu L, *et al.* Cycle life of commercial lithium-ion batteries with lithium titanium oxide anodes in electric vehicles. Energies. 2014;7(8):4895–4909.

[33] Deng D, Kim MG, Lee JY, *et al.* Green energy storage materials: Nanostructured TiO_2 and Sn-based anodes for lithium-ion batteries. Energy & Environmental Science. 2009;2(8):818–837.

[34] Barbarisi O, Vasca F, Glielmo L. State of charge Kalman filter estimator for automotive batteries. Control Engineering Practice. 2006;14:264–275.

[35] Verbrugge M, Tate E. Adaptive state of charge algorithm for nickel metal hydride including hysteresis phenomena. Journal of Power Sources. 2004;126:236–249.

[36] Shepherd CM. Design of primary and secondary cells II. An equation describing battery discharge. Journal of the Electrochemical Society. 1965;112(7):657–664.

[37] Boby K, Kottalil AM, Ananthamoorthy N. Mathematical modelling of PMSM vector control system based on SVPWM with PI controller using Matlab. International Journal of Advanced Research in Electrical, Electronics and Instrumentation Engineering. 2013;2(1):689–695.

[38] Gáspár P, Bokor J, Mihály A, *et al.* Robust reconfigurable control for in-wheel electric vehicles. 9th IFAC Symposium on Fault Detection, Supervision and Safety for Technical Processes (Safeprocess'15). 2015;p. 36–41.

Chapter 11

Nullspace-based input reconfiguration architecture for over-actuated aerial vehicles

Tamás Péni[1], Bálint Vanek[1], György Lipták[1], Zoltán Szabó[1], and József Bokor[1]

A dynamic input reconfiguration architecture is proposed for over-actuated vehicles to accommodate actuator failures. The method is based on the dynamic nullspace computed from the linear parameter-varying model of the plant dynamics. If there is no uncertainty in the system, then any signal filtered through the nullspace will have no effect on the plant outputs. This makes it possible to reconfigure the inputs without influencing the nominal control loop and thus the nominal control performance. Since the input allocation mechanism is independent of the structure of the baseline controller, it can be applied even if the baseline controller is not available in analytic form. The applicability of the proposed algorithm is demonstrated via the case study of designing a fault-tolerant controller for the Rockwell B-1 Lancer aircraft.

The most stringent requirements for improved reliability and environmental sustainability of safety critical flight control systems could only be satisfied with the most advanced fault-tolerant control (FTC) techniques [1,2]. The FTC system is required to detect and identify the failure and then to compensate its effect by reconfiguring the control system [3]. Focus on the environmental impact of the aircraft triggers the need for higher performance flight control systems, which leads to a paradigm shift from robust passive FTC towards active methods relying on switching, gain scheduled or linear parameter-varying (LPV) methods with certifiable algorithms [4]. In the past few years, a wide variety of FTC design approaches have been proposed [5,6].

This chapter focuses on the reconfiguration task in the case of actuator failures. The approach considered here is the control input reallocation [7], where the aim is to compensate the actuator failures by reconfiguring the remaining flight control surfaces such that the performance degradation caused by the failure is as small as possible. One possible approach to solve this problem is based on the nullspace (or kernel) of the aircraft dynamics. These algorithms can be applied only if control input redundancy is available in the system [8]. As in aerospace applications this is often the case [9], this approach is a promising method for fault-tolerant flight control design.

[1]Systems and Control Laboratory, Institute for Computer Science and Control, Budapest, Hungary

The classical kernel-based algorithms assume constant input direction matrices and thus use static matrix kernels [7,8]. This concept is extended in [10,11] by using dynamic nullspace generators. Though the concept presented in these papers is similar to that is discussed here, the algorithms in [10,11] highly depend on the linear time-invariant (LTI) framework and thus their extension to the parameter-varying case rises several theoretical problems. This chapter proposes a different approach, which is promising for LPV applications as well.

The main component of the proposed reconfiguration architecture is the nullspace of the controlled LPV plant model. Although the nullspace of a dynamical system bears significant importance on several other fields of control as well [12–14], its numerical computation has been solved only partially so far. In [15] an algorithm based on matrix pencils is proposed to compute the dynamic kernel of an LTI system. Although this approach is computationally efficient, it is based on frequency domain formulation, which prevents its extension to LPV systems. In [12] the nullspace of a parameter-dependent, memoryless matrix is addressed in connection with controller design. The paper uses linear-fractional representation (LFR), but it does not consider the case of dynamic kernels. Moreover, the method of [12] does not analyse and thus cannot guarantee the well-definedness of the kernel basis, which are also necessary to use the kernel in any further design process. The LFR-Toolbox [16] also provides a method for nullspace computation. The algorithm is similar to the one given in [12], and it can be applied for dynamical systems as well, but it does not work for the general case as it requires certain rank conditions to be satisfied. In this chapter, we revise the existing methods and assemble a complete tool for kernel computation that can be applied for parameter-dependent matrices and LTI/LPV dynamical systems as well.

The chapter is structured as follows: Sections 11.1 and 11.2 are devoted to the numerical computation of the parameter-varying nullspace. The proposed actuator reconfiguration architecture is discussed in Section 11.3. The simulation example, using the B-1 aircraft, is presented in Section 11.4, while the results derived in the chapter are concluded in Section 11.5.

11.1 Inversion-based nullspace computation for parameter-varying systems

In this section, the definition of the nullspace of a linear map is defined and the problem related to its numerical computation is addressed.

11.1.1 Nullspace of a linear map

Definition 11.1 (nullspace and its generator). *Let L denote a linear map assigning to the elements of a linear input space $\mathscr{U}$ the elements of a linear output space $\mathscr{Y}$. The (right) nullspace (kernel) of L, denoted by $\mathscr{N}(L)$, is a subset (subspace) of $\mathscr{U}$ that are mapped to 0, i.e. $\mathscr{N}(L) = \{u \in \mathscr{U} \mid Lu = 0\}$. If for some linear map $N : \mathscr{W} \to \mathscr{U}$, $\mathscr{N}(L) = \mathrm{Im}(N)$, then N is called the* generator *of the nullspace.*

The following lemma gives the basis of the nullspace construction algorithms derived afterwards.

Lemma 11.1. *Let* $L : \mathscr{U} \to \mathscr{Y}$ *and* $Q : \mathscr{U} \to \mathscr{Z}$ *be two linear maps such that* $\begin{bmatrix} L \\ Q \end{bmatrix} : \mathscr{U} \to \mathscr{Y} \times \mathscr{Z}$ *is invertible. Then* $\mathscr{N}(L) = \operatorname{Im}\left(\begin{bmatrix} L \\ Q \end{bmatrix}^{-1}\begin{bmatrix} 0 \\ I \end{bmatrix}\right)$.

Proof. $\Leftarrow$) Let $z \in \operatorname{Im}\left(\begin{bmatrix} L \\ Q \end{bmatrix}^{-1}\begin{bmatrix} 0 \\ I \end{bmatrix}\right)$, i.e. $z = \begin{bmatrix} L \\ Q \end{bmatrix}^{-1}\begin{bmatrix} 0 \\ I \end{bmatrix} r$, with some r. If $[M\ N] = \begin{bmatrix} L \\ Q \end{bmatrix}^{-1}$, then $Lz = L[M\ N]\begin{bmatrix} 0 \\ I \end{bmatrix} r = [I\ 0]\begin{bmatrix} 0 \\ I \end{bmatrix} r = 0$.

$\Rightarrow$) Let z s.t. $Lz = 0$. Then with $r = Qz$, $\begin{bmatrix} L \\ Q \end{bmatrix} z = \begin{bmatrix} 0 \\ Qz \end{bmatrix} = \begin{bmatrix} 0 \\ I \end{bmatrix} r$, which implies $z \in \operatorname{Im}\left(\begin{bmatrix} L \\ Q \end{bmatrix}^{-1}\begin{bmatrix} 0 \\ I \end{bmatrix}\right)$.

□

Since the nullspace in Lemma 11.1 is generated by using the inverse of the extended map $\begin{bmatrix} L \\ Q \end{bmatrix}$, therefore the algorithms based on this formula are called *inversion based*. We consider now two special classes of linear maps – the parameter-varying memoryless matrices and LPV systems – and construct their nullspace by using Lemma 11.1.

11.1.2 Memoryless matrices

Let $M(\rho)$ be a parameter-dependent memoryless matrix such that $M : \Omega \to \mathbb{R}^{n_y \times n_u}$ where ρ collects the n_ρ time-varying parameters taking values from the set $\Omega \subseteq \mathbb{R}^{n_\rho}$. For concise notation we make the following convention: ρ denotes in general the time-varying parameters, but when we refer to ρ with $\rho \in \Omega$, we mean all possible values of the parameter vector. Assume $0 \in \Omega$, $n_y < n_u$. Assume also that $M(\rho)$ has full row rank for all $\rho \in \Omega$. (We focus only on this case, because it is enough to construct the nullspace of an LPV system in the next section.) If the parameter dependence is linear fractional, then $M(\rho)$ can be given in LFR as follows:

$$M(\rho) = M_{21}\Delta(\rho)(I - M_{11}\Delta(\rho))^{-1}M_{12} + M_{22} = \mathscr{F}_u\left(\begin{bmatrix} M_{11} & M_{12} \\ M_{21} & M_{22} \end{bmatrix}, \Delta\right) \tag{11.1}$$

where M_{ij} are constant matrices, $\Delta : \Omega \to \Delta_{set} \subset \mathbb{R}^{r_\Delta \times c_\Delta}$ is a parameter-dependent matrix whose entries are linear in ρ and M_{22} is of full row rank. The representation is *well defined* if $I - M_{11}\Delta(\rho)$ is invertible for all $\rho \in \Omega$.

According to Definition 11.1 the kernel of $M(\rho)$ is the set of all input vectors $u \in \mathbb{R}^{n_u}$ satisfying $M(\rho)u = 0$, $\forall \rho \in \Omega$. Based on Lemma 11.1 the generator of the nullspace can be constructed by finding a (generally) parameter-dependent matrix $Q(\rho)$ such that $[M(\rho)^T\ Q(\rho)^T]^T$ is invertible for all $\rho \in \Omega$. Since some additional smoothness properties are also prescribed in general for the parameter dependence of the nullspace generator, thus finding that an admissible $Q(\rho)$ is difficult in practice. Therefore, based on the idea in [16], the following algorithm can be applied: choose some parameter value ρ^* and find (a parameter independent) Q such that $[M(\rho^*)^T \ Q^T]^T$ is invertible. (Without loss of generality, we can choose $\rho^* = 0$, i.e. $M(\rho^*) = M_{22}$ and Q can be chosen to be a basis of the nullspace of M_{22}.) Then use Q over the entire parameter domain, i.e. consider the matrix $[M(\rho)^T \ Q^T]^T$. Based on [16], the *formal* inverse of $[M(\rho)^T \ Q^T]^T$ and thus by Lemma 11.1 the *formal* generator of $\mathcal{N}(M(\rho))$ can be obtained in LFR form as follows:

$$N(\rho) = \mathscr{F}_u\left(\begin{bmatrix} M_{11} - M_{12}XM_{21} & M_{12}Y \\ -XM_{21} & Y \end{bmatrix}, \Delta\right) \tag{11.2}$$

where $\begin{bmatrix} M_{22} \\ Q \end{bmatrix}^{-1} = [X \ Y]$ such that $M_{22}X = I$ and $M_{22}Y = 0$.

We called the inverse and the generator *formal*, because if $[M(\rho)^T \ Q^T]^T$ is not invertible for all $\rho \in \Omega$, then (11.2) is not well defined. Well-definedness and thus the invertibility can be achieved by finding a factorisation $N(\rho) = N_r(\rho)M_r(\rho)^{-1}$ such that $N_r(\rho)$ is well defined. It is clear that $N_r(\rho)$ spans the same nullspace as $N(\rho)$, so $N(\rho)$ can be replaced by $N_r(\rho)$. Unfortunately, there is no general algorithm that finds the factorisation above in all cases. Instead, there are approaches that generally work in the practical situations.

M1. Before trying any factorisation method, it is advisable to minimise the realisation of $N(\rho)$ first. This means finding an equivalent representation for N with smaller Δ. The observability and reachability decomposition proposed in [17], [18] and implemented by the `minlfr` function of the LFR Toolbox [16] is a possible algorithm to minimise the realisation. Although the algorithm is efficient in several cases, it has non-negligible limits: it can work only with diagonal Δ and seeks only transformations, which commute with the parameter block.

M2. Find a matrix F such that $N_r(\rho)$ and $M_r(\rho)$ defined by

$$\begin{bmatrix} M_r(\rho) \\ N_r(\rho) \end{bmatrix} = \mathscr{F}_u\left(\left[\begin{array}{c|c} N_{11} + N_{12}F & N_{12} \\ \hline F & I \\ \hline N_{21} + N_{22}F & N_{22} \end{array}\right], \Delta\right) \tag{11.3}$$

are well defined. It follows from the construction that $N(\rho) = N_r(\rho)M_r(\rho)^{-1}$, so (11.3) is a suitable factorisation of the kernel. Unfortunately, this approach cannot be applied in all cases: there are ill-defined systems (even in minimal realisation), for which there does not exist a suitable F, but a well-defined left–right factorisation can still be found. See the example in [19].

M3. It is clear that the entries of $N(\rho)$ are rational functions of δ_i-s. If $p(\rho)$ denotes the product of terms, which are responsible for the poles of $N(\rho)$, a possible factorisation can be obtained by choosing $N_r(\rho) = N(\rho)(p(\rho)/q(\rho))$ and $M_r(\rho) = (p(\rho)/q(\rho))$, where $q(\rho)$ is an arbitrary polynomial, which does not have zero in Ω. In this way, the ill-definedness can be fully eliminated. The only weak point of this algorithm is the difficulty of finding the poles of $N(\rho)$ in larger dimensional cases.

11.1.3 LPV systems

Consider now an LPV system in the usual state-space form:

$$G: \begin{array}{l} \dot{x} = A(\rho)x + B(\rho)u \\ y = C(\rho)x + D(\rho)u' \end{array} \tag{11.4}$$

where ρ collects the time-varying scheduling parameters taking values from $\Omega \subseteq \mathbb{R}^{n_\rho}$; $u : \mathbb{R}_+ \to \mathbb{R}^{n_u}, y : \mathbb{R}_+ \to \mathbb{R}^{n_y}, x : \mathbb{R}_+ \to \mathbb{R}^{n_x}$ are the input, output and the state vector, respectively. If $x(0) = 0$, (11.4) defines a linear map that uniquely assigns to any input signal $u(t)$ an output signal $y(t)$. Moreover, this map, denoted by $G(\rho, \mathscr{I})$, can be given in LFR as follows:

$$G(\rho, \mathscr{I}) = \mathscr{F}_u\left(\begin{bmatrix} A(\rho) & B(\rho) \\ C(\rho) & D(\rho) \end{bmatrix}, \mathscr{I}\right), \tag{11.5}$$

where $\mathscr{I}$ is used to denote an n_x-dimensional diagonal matrix of integral operators such that $v = \mathscr{I}z$ means that $v_i(t) = \int_0^t z_i(t)dt,\ i = 1, \dots, n_x$ with $v(0) = x(0) = 0$. Using Definition 11.1, we can now define the nullspace of $G(\rho, \mathscr{I})$ as a set of all input signals that are mapped to the constant 0 output. Due to the zero initial condition the definition of the nullspace makes sense even if G is an unstable system. In the remaining part of the section, we give an algorithm to construct the generator of the nullspace.

In order to proceed, two additional assumptions are made on G: (i) the entries of $D(\rho)$ are linear fractional functions of the parameters and (ii) $D(\rho)$ is a full row rank matrix for all $\rho \in \Omega$. Assumption (i) is not restrictive and necessary to use LFR formulation. Assumption (ii) facilitates the discussion and will be relaxed at the end of the section. The two assumptions imply that there exists a well-defined generator $N_D(\rho)$ for the nullspace of $D(\rho)$ such that $D'(\rho) := \begin{bmatrix} D(\rho) \\ N_D(\rho) \end{bmatrix} = \mathscr{F}_u\left(\begin{bmatrix} D'_{11} & D'_{12} \\ D'_{21} & D'_{22} \end{bmatrix}, \Delta'\right)$ is invertible for all $\rho \in \Omega$. The inverse can be determined by applying the following formula [16] with similar considerations as in the previous section:

$$\begin{aligned} D'(\rho)^{-1} &= \mathscr{F}_u\left(\begin{bmatrix} D''_{11} & D''_{12} \\ D''_{21} & D''_{22} \end{bmatrix}, \Delta'\right) \\ D''_{11} &= D'_{11} - D'_{12}(D'_{22})^{-1}D'_{21}, \qquad D''_{12} = D'_{12}(D'_{22})^{-1} \\ D''_{21} &= -(D'_{22})^{-1}D'_{21}, \qquad D''_{22} = (D'_{22})^{-1} \end{aligned}$$

The invertibility of D'_{22} is guaranteed by the fact that $D'(\rho)$ is invertible for all $\rho \in \Omega$ and by assumption, $0 \in \Omega$. Let $D'(\rho)^{-1}$ be partitioned as $[X(\rho)\ Y(\rho)]$ such that $D(\rho)X(\rho) = I$ and $D(\rho)Y(\rho) = 0$. Then, by using (11.2) the generator of $\mathscr{N}(G(\rho, \mathscr{I}))$ can be computed as

$$N(\rho, \mathscr{I}) = \mathscr{F}_u\left(\begin{bmatrix} A_o(\rho) & B_o(\rho) \\ C_o(\rho) & D_o(\rho) \end{bmatrix}, \mathscr{I}\right) \tag{11.6}$$

$$\begin{aligned} A_o(\rho) &= A(\rho) - B(\rho)X(\rho)C(\rho), \qquad B_o(\rho) = B(\rho)Y(\rho) \\ C_o(\rho) &= -X(\rho)C(\rho), \qquad D_o(\rho) = Y(\rho) \end{aligned}$$

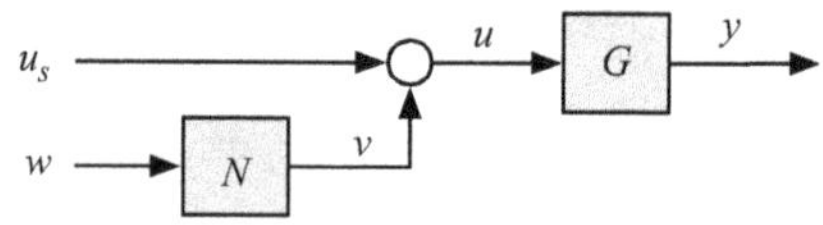

Figure 11.1 A typical interconnection of G with its nullspace generator G_o in a reconfiguration architecture

Note that $N(\rho, \mathscr{I})$ is an LFR of integrators; thus it is always well defined [19]. By construction, the linear mapping $v = N(\rho, \mathscr{I})w$ assigns to the input signal $w(t)$ the output response $v(t)$ of the dynamical (LPV) system:

$$N: \begin{array}{l} \dot{x}_o = A_o(\rho)x_o + B_o(\rho)w \\ v = C_o(\rho)x_o + D_o(\rho)w \end{array}, \tag{11.7}$$

where $x_o(0) = 0$. Due to practical applicability we are interested in stable nullspace generator; therefore, if N is not stable, it has to be stabilised. For this, we can use a similar factorisation as (11.3). To proceed, we make assumption (iii): there exists a stabilising state feedback gain $F(\rho)$ such that $A_o(\rho) + B_o(\rho)F(\rho)$ is asymptotically stable. Then with

$$\begin{bmatrix} M_r(\rho, \mathscr{I}) \\ N_r(\rho, \mathscr{I}) \end{bmatrix} = \mathscr{F}_u\left(\left[\begin{array}{c|c} A_o(\rho) + B_o(\rho)F(\rho) & B_o(\rho) \\ \hline F(\rho) & I \\ C_o(\rho) + D_o(\rho)F(\rho) & D_o(\rho) \end{array}\right], \mathscr{I}\right)$$

the equality $N(\rho, \mathscr{I}) = N_r(\rho, \mathscr{I})M_r(\rho, \mathscr{I})^{-1}$ holds, so $N_r(\rho, \mathscr{I})$ defines a stable generator system.

In the FTC framework, the role of N is to generate an input reconfiguration signal for G that does not influence its output. The general interconnection is depicted in Figure 11.1. G is assumed to be connected in a closed loop such that u_s is a stabilising input generated by some (baseline) controller. Let us examine the system behaviour in a practical situation, when the initial state of G differs from zero and is not known either. Assume that N is stable and starts from zero initial state. Since the systems are linear, G is stabilised via $u_s(t)$ and N is stable, the state transients generated due to the nonzero initial condition die out and the output of G converges to the response corresponding to $u_s(t)$. Consequently, $w(t)$ does not influence $y(t)$.

Finally, assume that condition (i) does not hold, i.e. $D(\rho)$ is not a full row rank matrix. We show that in this case the full row rank property can still be achieved by iteratively replacing the outputs causing rank deficiency by their time derivatives until a full row rank D' matrix is obtained. This idea is the same as that is used in [20] for constructing the inverse of a linear system. The main steps of the procedure are summarised in the following algorithm. For simplicity, we present the procedure for LTI systems. The parameter-varying case can be similarly handled.

If the algorithm terminates in finite steps, then G' can be constructed. Since only derivations and invertible transformations are applied in each step, G' has the same kernel as of G and, by construction, D_k has full row rank. Consequently, (11.6) can be

Algorithm 1: Assume for simplicity that $[C\ D]$ has full row rank, i.e. the trivial output redundancy is excluded. Let $C_1 = C, D_1 = D, y_1 = y$ and $k = 1$.

1. If $D_k \neq 0$ compute its SVD as $D_k = [U_{k,1}\ U_{k,2}]\begin{bmatrix}\Sigma_k & 0\\ 0 & 0\end{bmatrix}\begin{bmatrix}V_{k,1}^*\\ V_{k,2}^*\end{bmatrix}$ and perform a linear transformation using the invertible matrix U_k^*: $\tilde{y}_k = U_k^* y_k = \begin{bmatrix}U_{k,1}^* C_k x + \Sigma_k V_{k,1}^* u\\ U_{k,2}^* C_k x\end{bmatrix} = \begin{bmatrix}C_{k,1}x + D_{k,1}u\\ C_{k,2}x\end{bmatrix} = \begin{bmatrix}\tilde{y}_{k,1}\\ \tilde{y}_{k,2}\end{bmatrix}$. Trivially, if $D_k = 0$, then $\tilde{y}_{k,2} = y_k$ and there is no $\tilde{y}_{k,1}$.
2. Replace $\tilde{y}_{k,2}$ by its time derivative, i.e. let the new output equation be defined as follows:
$y_{k+1} = \begin{bmatrix}C_{k,1}\\ C_{k,2}A\end{bmatrix}x + \begin{bmatrix}D_{k,1}\\ C_{k,2}B\end{bmatrix}u = C_{k+1}x + D_{k+1}u.$
3. Set $k := k + 1$. If D_k is of full row rank, then terminate and go to step 4. Otherwise, go to step 1.
4. Let the transformed system be defined as follows:

$$G': \begin{array}{l}\dot{x} = Ax + Bu\\ y = C_k x + D_k u\end{array} \tag{11.8}$$

applied. The main concept of the algorithm can be extended to LPV systems, but in that case the transformed system will depend on the time derivatives of the scheduling parameters as well. As the generator computed from G' inherits this dependence, the time derivatives of the parameters have to be known (measured or precisely computed) to evaluate the generator system.

11.2 Geometry-based nullspace construction

The inversion-based approach presented in the previous section is not the only way of computing the nullspace of a parameter-varying system. A fundamentally different method can be derived by using the geometric view of system theory. This section gives a brief overview of the geometry-based algorithms and points out their advantageous properties and main limitations.

Embedding the basic concepts of control in the system of geometry and the interpretation (and re-interpretation) of the results of mathematical system theory by using a geometric approach was initiated in the beginning of the 1970s by Basile, Marro and Wonham. By now, the approach has proved to be an effective means to the analysis and design of control systems and the idea gained some popularity that was followed by many authors successfully (e.g. [10,11]). Good summaries of the subject can be found in the classical books of [21,22].

The linear geometric systems theory was extended to non-linear systems in the 1980s, see, e.g. [23,24]. In the non-linear theory, the underlying fundamental concepts are almost the same, but the mathematics is different. For non-linear systems the tools from differential geometry and Lie theory are primarily used. Due to the computational complexity involved, these general nonlinear methods have limited applicability in practice.

The concept of invariant subspace known from the geometric theory of LTI systems were extended to LPV dynamics by introducing the notion of parameter-varying invariant subspace, see [25]. We emphasise that despite its names, these subspaces are not parameter dependent! In introducing the various parameter-varying invariant subspaces, an important goal was to set notions that lead to computationally tractable algorithms for the case when the parameter dependency of the system matrices is affine. These invariant subspaces play the same role in the solution of the fundamental problems, such as disturbance decoupling, unknown input observer design, fault detection, as their counterparts in the time-invariant context, see [26,27].

11.2.1 Parameter-varying invariant subspaces

The invariant subspace concept, which is a cornerstone of the classical LTI geometric framework, can be extended to LPV systems as follows:

Definition 11.2. *A subspace $\mathscr{V}$ is called parameter-varying invariant subspace for the family of the linear maps $A(\rho)$ (or shortly $\mathscr{A}$-invariant subspace) if*

$$A(\rho)\mathscr{V} \subset \mathscr{V} \quad \text{for all } \rho \in \Omega. \tag{11.9}$$

Definition 11.3. *Let $\mathscr{B}(\rho)$ denote Im $B(\rho)$. Then a subspace $\mathscr{V}$ is called a parameter-varying (A,B)-invariant subspace (or shortly $(\mathscr{A}, \mathscr{B})$-invariant subspace) if for all $\rho \in \Omega$:*

$$A(\rho)\mathscr{V} \subset \mathscr{V} + \mathscr{B}(\rho). \tag{11.10}$$

As in the classical case this definition is equivalent to the existence of a mapping $F \circ \rho : [0, T] \to \mathbb{R}^{m\times n}$ such that $(A(\rho) + B(\rho)F(\rho))\mathscr{V} \subset \mathscr{V}$. The set of $(\mathscr{A}, \mathscr{B})$-invariant subspaces contained in a given set is an upper semilattice with respect to the subspace addition; hence it has a maximal element. As in the LTI case, we denote by $\mathscr{V}^*$ the maximal $(\mathscr{A}, \mathscr{B})$-invariant subspace contained in ker C.

Let us denote the maximal $\mathscr{A}$-invariant subspace contained in a constant subspace $\mathscr{K}$ by $\langle \mathscr{K} | A(\rho)\rangle$. For the LPV case, one can get the following definition:

Definition 11.4. *A subspace $\mathscr{R}$ is called parameter-varying controllability subspace if there exists a constant matrix K and a parameter-varying matrix $F : [0, T] \to \mathbb{R}^{m\times n}$ such that*

$$\mathscr{R} = \langle \mathscr{A} + B\mathscr{F} | Im\ BK\rangle, \tag{11.11}$$

where $\mathscr{A} + B\mathscr{F}$ denotes the system $A(\rho) + BF(\rho)$.

As in the classical case, the family of controllability subspaces contained in a given subspace $\mathscr{K}$ has a maximal element. We denote by $\mathscr{R}^*$ the maximal controllability subspace that corresponds to $\mathscr{K} = \ker C$.

From a practical point of view, it is an important question to characterise these subspaces by a finite number of conditions. If the parameter dependence is affine, these subspaces can be computed by efficient algorithms in a finite number of steps, for details see [25,28].

11.2.2 Nullspace construction for LPV systems

In this section, we construct a nullspace generator for LPV systems given in the following form:

$$\begin{aligned} \dot{x}(t) &= A(\rho)x(t) + Bu(t) \\ y(t) &= Cx(t) + Du(t), \end{aligned}$$

where we assume that $A(\rho)$ depends affinely on the parameters. Note that the results can also be extended to the case where B is also parameter varying. Let us denote by $\mathscr{V}^*$ the weakly unobservable subspace, i.e. the set of initial conditions for which there exists an input function such that the ensuing output is identically zero. $\mathscr{V}^*$ is the largest subspace such that there exists a static feedback gain $F(\rho)$ ensuring

$$(A(\rho) + BF(\rho))\mathscr{V} \subset \mathscr{V}, \quad (C + DF(\rho))\mathscr{V} = 0, \quad \rho \in \Omega.$$

These gains F are called friends of $\mathscr{V}$. Recall that the controllable weakly unobservable subspace $\mathscr{R}^* \subset \mathscr{V}^*$, i.e. the set of initial conditions for which there exists an input function able to steer the state to zero in finite time while keeping the output identically zero, has the same friends as $\mathscr{V}^*$. Moreover, it is known that in the LTI case the spectrum on $\mathscr{R}^*$ is freely assignable. Let us choose the invertible input mixing map

$$T_u = \begin{bmatrix} V_1 & V_2 & V_3 \end{bmatrix}$$

such that

$$\mathrm{im} V_1 = B^{(-1)}\mathscr{R}^*, \quad \mathrm{im}\begin{bmatrix} V_1 & V_2 \end{bmatrix} = \ker D$$

and the invertible output mixing map $T_y = \begin{bmatrix} W_1 \\ W_2 \end{bmatrix}$ such that $W_1 D V_3 = 0$ and $W_2 D V_3 = I$, respectively. Chose a state transformation ($\xi = T^{-1}x$) matrix

$$T = \begin{bmatrix} T_1 & T_2 & T_3 \end{bmatrix}$$

such that

$$\mathrm{Im} T_1 = \mathscr{R}^*, \quad \mathrm{Im}\begin{bmatrix} T_1 & T_2 \end{bmatrix} = \mathscr{V}^*.$$

Note that these matrices do not depend on the parameters. Applying all these transformations on the LPV system, one has [22,29]

$$\dot{\xi} = \bar{A}(\rho)\xi + \bar{B}\bar{u}, \quad \bar{u} = T_u^{-1}u, \tag{11.12}$$

$$\bar{y} = \bar{C}\xi + \bar{D}\bar{u}, \quad \bar{y} = T_y y \tag{11.13}$$

with

$$\bar{A}(\rho) = \begin{bmatrix} \bar{A}_{11} & \bar{A}_{12} & \bar{A}_{13} \\ \bar{A}_{21} & \bar{A}_{22} & \bar{A}_{23} \\ \bar{A}_{31} & \bar{A}_{32} & \bar{A}_{33} \end{bmatrix}(\rho), \quad \bar{B} = \begin{bmatrix} \bar{B}_{11} & \bar{B}_{12} & \bar{B}_{13} \\ 0 & \bar{B}_{22} & \bar{B}_{23} \\ 0 & \bar{B}_{32} & \bar{B}_{33} \end{bmatrix}$$

$$\bar{C} = \begin{bmatrix} 0 & 0 & \bar{C}_{13} \\ \bar{C}_{21} & \bar{C}_{22} & \bar{C}_{23} \end{bmatrix}, \qquad \bar{D} = \begin{bmatrix} 0 & 0 & 0 \\ 0 & 0 & I \end{bmatrix}.$$

One should take an $\bar{F}_1(\rho)$ (e.g. by considering the first block column of a friend of $\mathscr{V}^*$) that fulfils conditions

$$\bar{C}_{21} + \bar{F}_{13}(\rho) = 0, \tag{11.14}$$

$$\bar{A}_{21}(\rho) + \bar{B}_{22}\bar{F}_{12}(\rho) + \bar{B}_{23}\bar{F}_{13}(\rho) = 0, \tag{11.15}$$

$$\bar{A}_{31}(\rho) + \bar{B}_{32}\bar{F}_{12}(\rho) + \bar{B}_{33}\bar{F}_{13}(\rho) = 0, \tag{11.16}$$

and renders the parameter-varying matrix $\bar{A}_0(\rho) = \bar{A}_{11}(\rho) + \bar{B}_{11}\bar{F}_{11}(\rho) + \bar{B}_{12}\bar{F}_{12}(\rho) + \bar{B}_{13}\bar{F}_{13}(\rho)$ stable (e.g. quadratically). Note that for strictly proper systems, condition (11.14) vanishes.

After performing all these preparatory steps, we are ready to design the nullspace generator Γ for the LPV system:

$$\dot{\zeta} = \bar{A}_0(\rho)\zeta + \bar{B}_{11}v, \quad \zeta(0) = 0 \tag{11.17}$$

$$u = T_u(\bar{F}_1(\rho)\zeta + Ev), \tag{11.18}$$

with $E = \begin{bmatrix} I & 0 & 0 \end{bmatrix}^T$ and $\bar{F}_1(\rho)$ the first block column of $\bar{F}(\rho)$. Note that while the dimensions of the matrices follow from the dimension of the invariant subspaces, one can figure out that $v(t) \in \mathbb{R}^{m-p}$, as for the inversion-based approach. One can check that Γ defined by (11.17) and (11.18) is indeed a nullspace generator. Plug in (11.18) into (11.12) and take the new state as $e = \begin{bmatrix} \xi_1^T - \zeta^T & \xi_2^T & \xi_3^T \end{bmatrix}^T$ to get

$$\dot{e} = \bar{A}e, \quad e(0) = 0,$$

$$\dot{\zeta} = \bar{A}_0\zeta + \bar{B}_{11}v$$

$$\bar{y} = \begin{bmatrix} 0 & 0 & \bar{C}_{13} \\ 0 & \bar{C}_{22} & \bar{C}_{23} \end{bmatrix} e,$$

i.e. $\bar{y} = 0$.

Remark 11.1. *In order to implement the algorithm above the weakly unobservability subspaces $\mathscr{R}^*$, $\mathscr{V}^*$ have to be determined first. References [25,28] provide efficient algorithms for the case when $D = 0$. If $D \neq 0$ one can compute $\mathscr{R}^*$ and $\mathscr{V}^*$ by using an equivalent strictly proper system [22,29]. Then $\mathscr{V}^*$ is the maximal $(\mathscr{A}, \mathscr{B})$-invariant subspace contained in* ker*C and $\mathscr{R}^*$ is the corresponding maximal controllability subspace.*

We conclude the section by making a short comparison of the two approaches: it is obvious that the geometry-based approach has serious limitations: first, it can be applied only for LPV systems with affine parameter dependence and second, the parameter-invariant subspaces are robust counterparts of the classical ones, which implies that the methods based on these concepts provide only sufficient conditions. Observe that the key element of the approach is that state transformation is defined by a constant matrix.

If the geometry-based design can be performed, we can obtain a very economic description of the nullspace by using a 'one shot' algorithm, avoiding the intricacies of the inversion-based approach.

There is a more subtle issue, however, revealed by these two approaches. The inversion-based method relies on techniques that consider the system from an external (input–output) perspective even if we represent the system by using a particular state-space description (this is indispensable for LPV systems!).

The problem appears at the point where this system is replaced with an 'equivalent' one and a reparameterisation is needed (the derivatives of the scheduling variables might occur). It is apparent that this parameterisation is not present in the geometry-based approach which always works on the same system and, which is very important in this context, uses only parameter-independent transformations. Thus, derivatives of the parameters cannot occur. Thus, the general question that needs more research and goes far beyond the scope of this chapter is that how the two LPV systems can be compared from an input–output perspective (e.g. when the two systems are equal).

11.3 Control input reconfiguration architecture for compensating actuator failures

In this section, an input reconfiguration method is proposed to compensate actuator failures. The concept extends the main idea of [8] to parameter-varying systems that do not necessarily have static nullspace. The architecture presented in Figure 11.2 is similar to that proposed in [10], but the concept we follow is basically different.

Before proceeding, we introduce the following notational convention: if S denotes a dynamical system, then its ℓth input and output will be denoted by $u_{S,\ell}$ and $y_{S,\ell}$, respectively. Let the n_u input–n_y output LPV plant be denoted by G. Assume the actuators driving the inputs of G are modelled by LTI systems collected in the n_u input–n_u output block A. Let K_B denote the baseline controller designed to guarantee the stability and control performance in fault-free case. For simplicity, consider now the case of one actuator fault affecting the ℓth control input such that $u_{G,\ell} = y_{A,\ell} + f = A_\ell(y_{K_B,\ell}) + f$, where f is an external fault signal, A_ℓ is the ℓth actuator dynamics. Suppose that $u_{G,\ell}$ (or equivalently f) is available for measurement or can be reconstructed by a suitable FDI filter, e.g. [30]. Let $\bar{A}$ be an n_u input–n_u output LTI shaping filter defined later and let N be the (stable) generator for the nullspace of $G \cdot A \cdot \bar{A}$. The procedure of input reconfiguration is the following. If controller K_N is designed such that $v_{3,\ell}$ (the ℓth component of v_3) tracks the fault signal f, then $u_{A,\ell} = y_{K_B,\ell} + v_{2,\ell}$, i.e. $y_{A,\ell} = A_\ell(y_{K_B,\ell}) + A_\ell(v_{2,\ell}) = A_\ell(y_{K_B,\ell}) + f = u_{G,\ell}$, i.e. the ℓth output of the actuator block equals the faulty input to the plant. Therefore, it looks as if the faulty input had been generated by the baseline controller. The input reconfiguration is performed via v_2, because it modifies the other control inputs such that they adapt to the faulty input. Since v_2 is generated from the nullspace, it does not have any effect on the plant's behaviour, i.e. the baseline control loop remains intact and the nominal control performance is preserved.

To complete the discussion, the role of the shaping filter $\bar{A}$ remains to be clarified. This dynamical block can be chosen by the designer to shape the actuator dynamics in order to simplify the nullspace computation. For example, if A is invertible then one can choose $\bar{A} = A^{-1}$ and thus the actuators are eliminated from the reconfiguration procedure. If A is not invertible, it may be possible to find an invertible $\hat{A}$, which approximates A over the frequency range of its bandwidth. Then $\hat{A} = \hat{A}^{-1}$ can be chosen. If the approximation does not work either, $\bar{A}$ can still be chosen to 'advantageously shape' the actuator dynamics. For example, [10] proposes to choose $\bar{A}$ such that $A \cdot \bar{A} = \kappa(s)I$, where $\kappa(s)$ is a user-defined transfer function. This unification simplifies the generator system and can improve the conditioning of the numerical computations related to the construction of N.

Finally, we add some comments and remarks to the architecture above.

1. It is clear that the method works perfectly only if the exact model of the plant is known, since the kernel and the plant should perfectly match in order that the effect of the input reconfiguration on the control performance could be perfectly cancelled. If there is modelling uncertainty in the system, then the robustness of the reconfigured system has to be checked either by simulations or by using a classical analysis tool based on $\mathscr{L}_2$ gain computation and integral quadratic constraints, see e.g. [31]. It is a possible option as well to improve the robustness properties of the reconfigured system by tuning the tracking controller K_N.
2. The design of the trajectory-tracking controller K_N is not difficult in general as the states of every dynamical model in the reconfiguration block (i.e. N, $\bar{A}$ and A) are fully available. This makes it possible to design a constrained (e.g. a model predictive [32], [33]) controller to respect hard input limitations. Also, when modelling uncertainty is present, the tracking controller can also be used to attenuate the performance degradation caused by the mismatch of the (nominal) nullspace and the (uncertain) plant. To handle this additional objective, this requirement has to be transformed to a performance target considered in the design of K_N.
3. The number and type of faults that can be compensated by the proposed method depend on the dimension and dynamical properties of the kernel space. If the dynamical properties are not appropriate (e.g. the kernel dynamics contain too fast or too slow dynamical components) the stabilising state feedback gain $F(\rho)$ or the tracking controller K_N can be used to shape these dynamics and set the required behaviour.
4. Note that the proposed reconfiguration framework can be applied if the system to be controlled is nonlinear but can be rewritten in quasi-LPV form as $\dot{x} = A(\rho(x))x + B(\rho(x))u$, $y = C(\rho(x))x + D(\rho(x))u$, where $\rho(x)$ is the magnitude

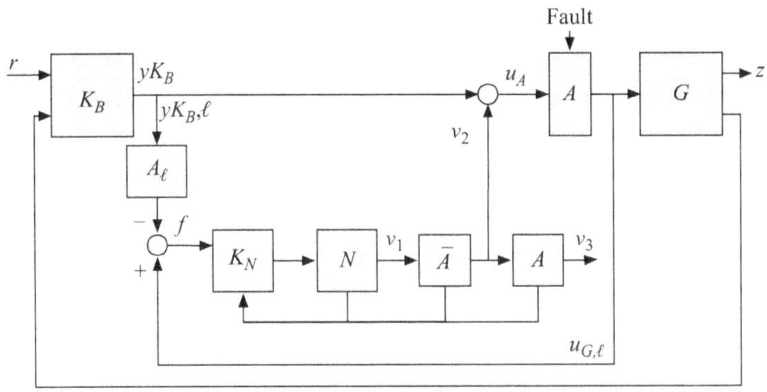

Figure 11.2 Input reconfiguration architecture

and rate bounded and available for measurement. In this case the scheduling variable is a function of the state, but for the reconfiguration block it is an external signal as it does not depend on the state of the nullspace dynamics. The derivations and the reconfiguration framework above thus remain applicable.

11.4 Reconfigurable fault-tolerant control of the B-1 aircraft

The Rockwell B-1 Lancer is a supersonic bomber that was introduced in the 1970s. It has four turbofan engines and variable wing sweep. In order to analyse the aeroelastic issues occurring at subsonic speeds, a high-fidelity simulator was developed in [34].

11.4.1 The non-linear flight simulator

To test the reconfigurable architecture in a realistic environment, the high-fidelity mathematical model is used. The simulator consists of the following main components: the nonlinear dynamics of the B-1 aircraft, the dynamics of the actuators and the stability augmentation system (SAS). In this case study, the flexible components were neglected; only the rigid body dynamics were used. The simplified structure of the simulator is presented in Figure 11.3.

The inputs to the non-linear model are the left (L) and right (R) horizontal tails (u_{HR}, u_{HL}), wing upper surface spoilers (u_{SPR}, u_{SPL}), split upper (RU) and lower (RL) rudders (u_{RU}, u_{RL}), and control vanes (u_{CVR}, u_{CVL}). All are measured in degree. Since the flexible part was neglected, the control vanes appear as free inputs that can be used in actuator reallocation. This configuration corresponds to the subsonic, clean configuration case. The model has ten states: $x = [\phi, \theta, \psi, p, q, r, u, v, w, h]$, where ϕ, θ, ψ are the roll-, pitch- and yaw angles in radian, p, q, r are the corresponding angular rates in radian per second, u, v, w denote the body axis velocities in ft/s and h is the height in foot. The model is scheduled with height and the Mach number M. The inner loop system has four outputs, $y = [p,\ q,\ r,\ a_y]^T$, where a_y is the side

Table 11.1 Properties of the actuators

Actuator	Poles	Lower limit (degree)	Upper limit (degree)
Horizontal tails	−10, −57.14	−20	10
Wing u/s spoilers	−10, −20	0	45
Lower rudder	−20, −80	−10	10
Upper rudder	−10	−10	10
Control vanes	−50	−20	20

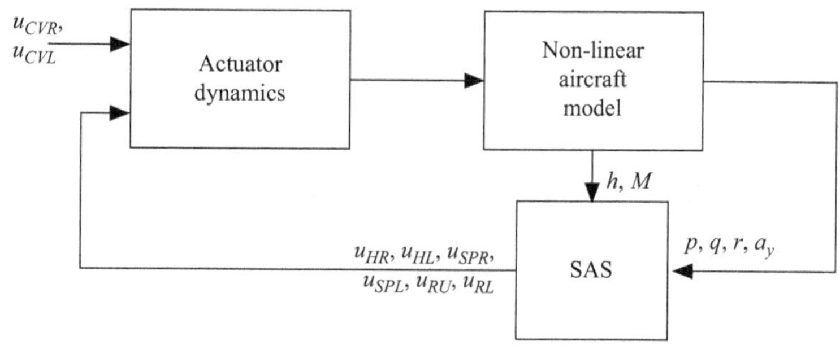

Figure 11.3 Basic configuration of the closed loop system

acceleration measured in gram, which is often referred to as side load factor n_y. The structure of the dynamics can be given as follows:

$$\dot{x} = F(\rho, x, u)$$
$$y = \begin{bmatrix} E_{4,5,6}x \\ h(\rho, x) \end{bmatrix} + \begin{bmatrix} 0 \\ \ell(\rho, x) \end{bmatrix} u$$

where matrix $E_{4,5,6}$ selects p, q, r from the state vector and $h(\rho, x)$, $\ell(\rho, x)$ are nonlinear functions defining a_y. The scheduling variable ρ is a two-dimensional vector comprising the height and the Mach number, i.e. $\rho = [h,\ M]$. We assume that the flight envelope is defined such that ρ takes values only from polytope Ω, defined by the following four vertices: $[10,000,\ 0.55]$, $[10,000,\ 0.7]$, $[20,000,\ 0.65]$, $[20,000,\ 0.8]$.

The actuators are modelled by first-order linear systems with output saturation. The poles of the LTI models and the saturation limits are collected in Table 11.1. The sensors dynamics are neglected from the present investigation due to their high bandwidth.

The SAS is responsible for the inner loop rigid body control of the aircraft, providing pitch- and roll-rate tracking and a yaw damper function using the measurements defined above and the reference stick commands from the pilot. In our framework, SAS is considered as the baseline controller K_B.

11.4.2 Construction of the LPV model

In order to construct the LPV model, the non-linear dynamics are trimmed and linearised at 12 equidistant points of the flight envelope characterised by Ω. After removing the unstable spiral mode from the model, least-squares interpolation is used to connect the 12 LTI systems into the following affine LPV model:

$$G(\rho, \mathscr{I}) = \mathscr{F}_u\left(\begin{bmatrix} A(\rho) & B(\rho) \\ C(\rho) & D(\rho) \end{bmatrix}, \mathscr{I}\right), \text{ where} \tag{11.19}$$

$$A(\rho) = A_0 + A_1 h + A_1 M \quad B(\rho) = B_0 + B_1 h + B_1 M$$

$$C(\rho) = C_0 + C_1 h + C_1 M \quad D(\rho) = D_0 + D_1 h + D_1 M$$

The results are verified by computing the gap- and ν-gap metrics [35,36] between (11.19) and the linearised models at the 14-grid points. Since none of the gap metrics exceed 0.025 among the points, the LPV model is accepted as a good approximation of the original dynamics.

11.4.3 Actuator inversion and nullspace computation

Since the actuator dynamics are stable, they can be augmented with suitably fast left half plane zeros to make them biproper, and hence invertible. To get good approximation and avoid high transients, the zeros are chosen to be four times faster than the poles of each actuator.

The next step is the computation of the nullspace of $G(\rho, \mathscr{I})$. For this, the first three outputs p, q, r have to be derived because $D(\rho)$ is not a full row rank matrix. After derivation, the rank condition is satisfied so no more derivations are necessary to determine the kernel. The output equations do not depend on ρ, nor does the kernel dynamics. On the other hand, the $N(\rho, \mathscr{I})$ obtained is unstable, so it has to be stabilised. For this, following the formal coprime factorisation method, a stabilising state feedback $F(\rho)$ is designed by solving the feasibility problem as follows:

$$A(\rho)Q + QA(\rho)^T + B(\rho)Y + Y^T B^T(\rho) + 2cQ \prec 0, \quad \begin{bmatrix} u_{\lim}^2 I & Y \\ Y^T & Q \end{bmatrix} \succ 0 \tag{11.20}$$

where $Q = P^{-1}$, $V(x) = x^T Px$ is the Lyapunov function and constants $c, u_{\lim}$ are used to control the location of the poles (at fixed ρ values) and the norm of the feedback gain matrix, respectively [37]. The free variables in (11.20) are Q and Y from which the feedback gain is computed as $F := YQ^{-1}$, which is now parameter independent. (Parameter-dependent feedback gain can also be constructed by choosing Q and Y parameter dependent.) To solve (11.20) the matrix inequalities are reduced to a finite set of linear matrix inequalities (LMIs) by evaluating them over a suitable dense grid over Ω [38].

We found in this particular example that the parameters c and $u_{\lim}$ can be chosen such that the kernel dynamics become significantly slower than the closed loop formed by the plant and the baseline controller. This makes it possible to truncate the states and substitute the dynamical kernel by a static, parameter-dependent feedthrough gain, i.e. $N(\rho) := D_o(\rho)$. In order to analyse how $N(\rho)$ 'approximates' the true generator

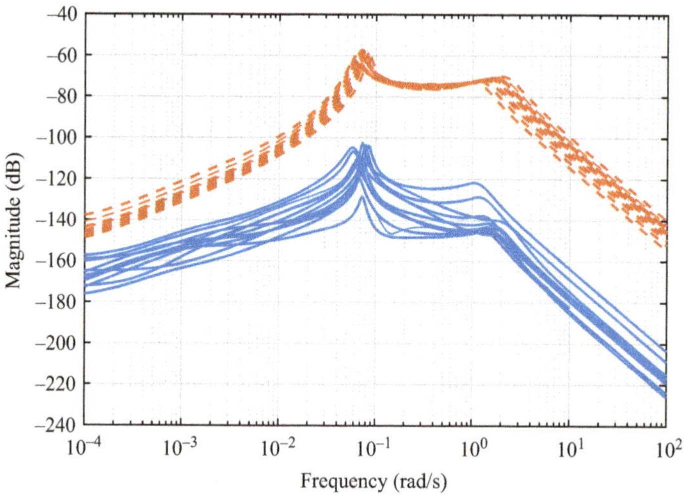

Figure 11.4 Bode plot of the systems $G(\rho, \mathscr{I})N(\rho, \mathscr{I})$ (solid blue) and $G(\rho, \mathscr{I})N(\rho)$ (dashed red) from the first input to the output q. Different lines correspond to different flight speed and height.

system the bode plots of $G(\rho, \mathscr{I})N(\rho, \mathscr{I})$ and $G(\rho, \mathscr{I})N(\rho)$ are computed for different values of ρ. The worst results are obtained from the first input of the kernel to the output q. This input–output combination is shown in Figure 11.4. It can be seen that $G(\rho, \mathscr{I})N(\rho)$ is very close to zero in this case as well, so $N(\rho)$ is a suitable approximation of the nullspace. It is important to emphasise that $N(\rho)$ cannot be determined without the dynamical kernel, because the augmented input matrix $[B(\rho)^T \ D(\rho)^T]^T$ has full column rank for all $\rho \in \Omega$ so it does not have a nullspace.

11.4.4 Fault signal tracking

The next component of the reconfigurable control architecture is the construction of the K_N controller that is responsible for tracking the fault signal $f = u_\ell - A_\ell(s)y_{K_B,\ell}$, where u_ℓ is the output of the ℓth (faulty) actuator and $y_{K_B,\ell}$ is the ℓth output of the baseline controller. Since the kernel is static, this trajectory tracking problem is equivalent to finding for each time t a suitable $u_N(t)$ such that the ℓth entry of $N(\rho(t))u_N(t)$ be equal to $f(t)$. As we have strict and asymmetric constraints for the actuator outputs (Table 11.1), this problem is solved by parameter-dependent quadratic programming:

$$\min_{u_N(t)}(r(t) - N(\rho(t))u_N(t))^T W(r(t) - N(\rho(t))u_N(t)) \tag{11.21}$$

$$\text{w.r.t.} \quad N(\rho(t))u_N(t) \geq L - y_{K_B}$$

$$N(\rho(t))u_N(t) \leq U - y_{K_B}$$

where the vector r of reference signals is defined as $r_i = 0$ if $i \neq \ell$ and $r_\ell = f$. Vectors L and U collect the lower and upper limits of the actuators and the diagonal matrix

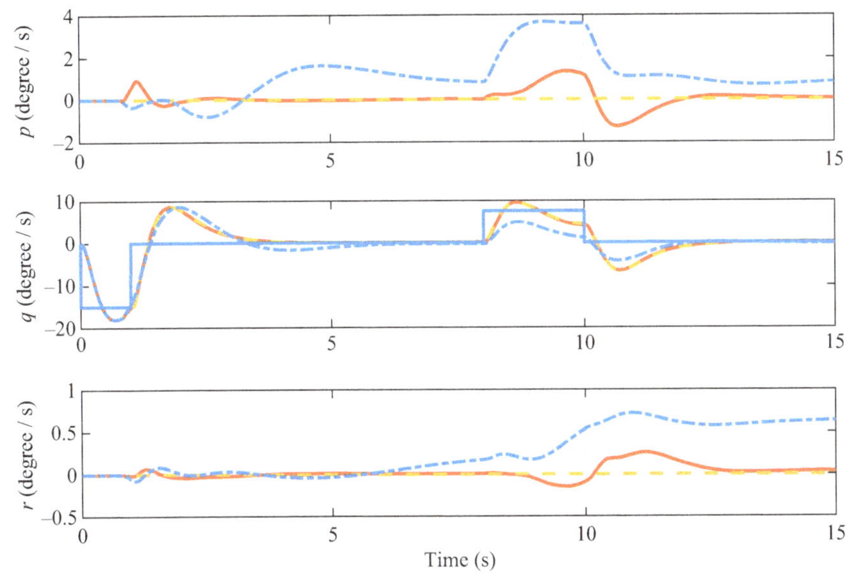

Figure 11.5 Simulation result of the angular rates (p, q, r) *in three different cases: without fault (yellow dashed), with fault and reconfiguration (red solid), and with fault (blue dashed-dot). The reference signal* q_{ref} *is plotted by solid blue line on the second subplot.*

W is used to weight the tracking errors [8], [13]. By defining r in the way above we prescribe 0 reference for the non-faulty inputs. This expresses our intention to keep the healthy inputs close to the value determined by the baseline controller. By varying the entries of W the contribution of each control input to the reconfiguration signal y_N can be tuned. In the following numerical simulations, $W = \text{diag}(w_1, \ldots, w_{n_u})$ is used with $w_\ell = 1{,}000$ and $w_i = 1$ for $i \neq \ell$, i.e. precise tracking is demanded for the fault signal, but no preferences are prescribed for the use of other control inputs.

11.4.5 Simulation results

To demonstrate the applicability of the proposed control reconfiguration method, the following fault scenario is presented: the pilot performs a pitch-rate doublet manoeuvre, by giving a q_{ref} command as shown in Figure 11.5. The manoeuvre starts at $t = 0$ s. It is assumed that at $t = 1$ s the right horizontal tail is jammed at its actual position. By assuming an ideal fault detection and isolation algorithm, the reconfiguration starts also at $t = 1$ s.

Simulation results are shown for the healthy, faulty and for the reconfigured cases in Figure 11.5.

In ideal (healthy) case, p is constantly zero, the fault causes roll rate excursions of 4 degree/s due to the asymmetrical deflections of the control surfaces. The proposed

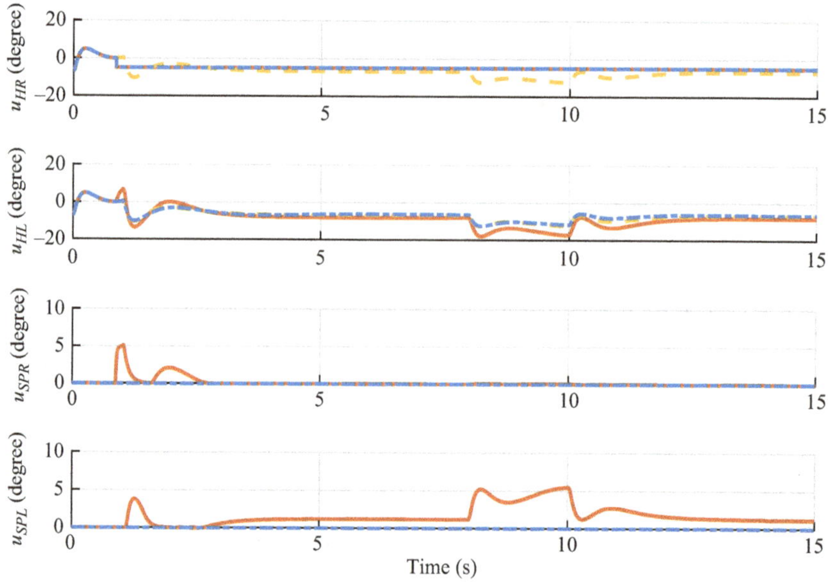

Figure 11.6 Simulation result of the inputs u_{HR}, u_{HL}, u_{SPR} and u_{SPL} in three different cases: without fault (yellow dashed), with fault and reconfiguration (red solid) and with fault (blue dashed-dot). The fault occurs at 1 s.

control allocation scheme is able to reduce it by at least a factor of 4 and the steady-state error is 0, showing the clear advantage of the proposed method. The nonzero transient error is due to the non-linear aircraft model. By testing the method on the LPV model only a negligibly small (less than 10^{-5}) error can be experienced due to the numerical inaccuracy of the simulation and the use of approximate actuator inverse.

Considering the pitch rate (q) variation, it can be seen that the healthy and reconfigured responses are very close to each other, while the uncompensated cases have more than 50% error and nonzero steady-state error. This shows that, although stability is not lost, the performance is significantly degraded.

The yaw rate (r) responses of the three cases are shown in the third subplot of Figure 11.5. Without fault r does not leave the longitudinal plane, while in the uncompensated case a small but non-negligible yaw rate response can be observed. When the reconfiguration scheme is applied the steady-state error is improved by an order of magnitude. (The nonzero error is again due to the non-linear characteristics of the simulation.)

Figure 11.6 shows the inputs u_{HR}, u_{HL}, u_{SPR} and u_{SPL} with and without reconfiguration. The contributions of the other inputs are small in all the simulation scenarios. It can be seen that the compensation of the faulty right horizontal tail required the active use of the left horizontal tail (u_{HL}) and the two spoilers (u_{SPR}, u_{SPL}). The latter two are not used in normal operations.

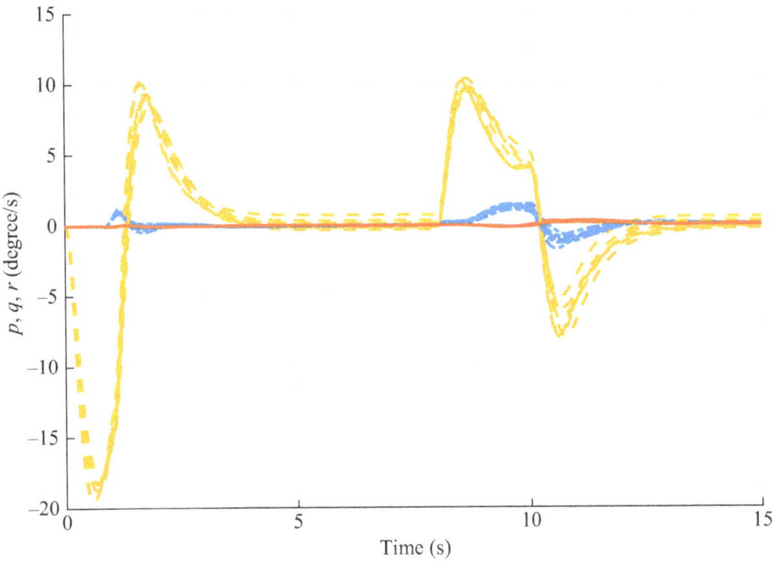

Figure 11.7 *Robustness analysis of the angular rates p (red solid), q (yellow dashed), r (blue dot-dashed) with different values of mass and inertia*

11.4.6 Robustness analysis

As our reconfigurable control is based on perfect cancellation of the plant dynamics, the performance of the baseline controller can be perfectly preserved only if the exact plant model is known. Since this is not the case in a real environment, the robustness of the method against the most critical parameter uncertainties has to be checked. For this, numerical simulations are performed with different values of inertia and mass parameters of the aircraft. The actual values of the parameters are selected randomly according to uniform distribution in a ±30% interval around their nominal values of: $I_{xx} = 950,000$, $I_{xz} = -52,700$, $I_{yy} = 6,400,000$, $I_{zz} = 7,100,000$, $mass = 8,944$, where the inertia is measured in (sl ft^2) and the mass is given in slug. This way 500 different scenarios are generated and tested. The results are summarised in Figure 11.7, where the angular rates are plotted for the seven most representative parameter values. It can be seen that the results are very similar in all cases, leading to a conclusion that the fault-tolerant controller is robust to the mass and inertia parameters.

11.5 Conclusion

This chapter has shown that an efficient control reconfiguration architecture can be developed for LPV systems by using the dynamic nullspace of the plant model. Though a complete design method is presented, there are several points where the proposed approach can be further improved: for example, finding systematic method for the stabilisation of the nullspace generator, extending the method for uncertain plant dynamics are good research directions for the future. Also, the reconfiguration architecture provides large freedom for the designer that can be exploited to achieve improvements in the design procedure. For example, there is no restriction for the synthesis method used to design the tracking controller: different control methods, e.g. model predictive controllers, constrained or robust synthesis methods can be applied to guarantee the satisfaction of hard input constraints and to improve the robustness properties of the closed-loop system. This freedom provides promising research directions for the future.

Acknowledgements

The authors would like to thank Peter Seiler (University of Minnesota) and Arnar Hjartarson (formerly MUSYN Inc.) for providing insight related to the B-1 simulator model, originally developed by David Schmidt. Also, the authors would like to thank the support of the GINOP-2.3.2-15-2016-00002 grant of the Ministry of National Economy of Hungary and by the European Commission through the H2020 project EPIC under grant no. 739592. The work was also supported by the János Bolyai Research Scholarship of the Hungarian Academy of Sciences.

References

[1] Goupil P, Marcos A. Industrial benchmarking and evaluation of ADDSAFE FDD designs. In: Fault Detection, Supervision and Safety of Technical Processes (8th SAFEPROCESS); 2012. p. 1131–1136.

[2] Lombaerts TJJ, Chu QP, Mulder JA, *et al.* Modular flight control reconfiguration design and simulation. Control Engineering Practice. 2011;19(6): 540–554.

[3] Hwang I, Kim S, Kim Y, *et al.* A survey of fault detection, isolation, and reconfiguration methods. IEEE Transactions on Control Systems Technology. 2010;18(3):636–653.

[4] Goupil P, Boada-Bauxell J, Marcos A, *et al.* AIRBUS efforts towards advanced real-time fault diagnosis and fault tolerant control. In: 19th IFAC World Congress, Cape Town, South Africa; 2014. p. 3471–3476.

[5] Ganguli S, Marcos A, Balas G. Reconfigurable LPV control design for Boeing 747-100/200 longitudinal axis. In: American Control Conference. vol. 5; 2002. p. 3612–3617.

[6] Alwi H, Edwards C, Tan C. Fault Detection and Fault-Tolerant Control Using Sliding Modes. London: Springer-Verlag; 2011.

[7] Johansen TA, Fossen TI. Control allocation – A survey. Automatica. 2013;49(5):1087–1103.

[8] Zaccarian L. Dynamic allocation for input redundant control systems. Automatica. 2009;45:1431–1438.

[9] Enns D. Control allocation approaches. In: AIAA Guidance, Navigation and Control Conference and Exhibit. American Institute of Aeronautics and Astronautics, Boston, MA; 1998. p. 98–108.

[10] Cristofaro A, Galeani S. Output invisible control allocation with steady-state input optimization for weakly redundant plants. In: Conference on Decision and Control (CDC); 2014. p. 4246–4253.

[11] Galeani S, Serrani A, Varanoa G, *et al.* On input allocation-based regulation for linear over-actuated systems. Automatica. 2015;52:346–354.

[12] Wu F, Dong K. Gain-scheduling control of LFT systems using parameter-dependent Lyapunov functions. Automatica. 2006;42:39–50.

[13] Vanek B, Peni T, Szabó Z, *et al.* Fault tolerant LPV control of the GTM UAV with dynamic control allocation. In: AIAA GNC; 2014.

[14] Varga A. The nullspace method-a unifying paradigm to fault detection. In: IEEE Conference on Decision and Control; 2009. p. 6964–6969.

[15] Varga A. On computing nullspace bases – A fault detection perspective. In: 17th IFAC World Congress; 2008. p. 6295–6300.

[16] Magni JF. Linear Fractional Representation Toolbox for Use With Matlab; 2006. Available from: http://w3.onera.fr/smac/?q=lfrt.

[17] D'Andrea R, Khatri S. Kalman decomposition of linear fractional transformation representations and minimality. In: American Control Conference (ACC); 1997. p. 3557–3561.

[18] Beck C, D'Andrea R. Noncommuting multidimensional realization theory: Minimality, reachability and observability. IEEE Transactions on Automatic Control. 2004;49(10):1815–1820.

[19] Szabó Z, Péni T, Bokor J. Null-space computation for qLPV Systems. In: 1st IFAC Workshop on Linear Parameter Varying Systems. vol. 48(26); 2015. p. 170–175.

[20] Silverman LM. Inversion of multivariable linear systems. IEEE Transactions on Automatic Control. 1969;AC-14(3):270–276.

[21] Wonham M. Linear Multivariable Control: A Geometric Approach. New York: Springer Verlag; 1985.

[22] Basile GB, Marro G. Controlled and Conditioned Invariants in Linear System Theory. Prentice Hall, Englewood Cliffs, NJ; 1992.

[23] Isidori A. Nonlinear Control Systems. London: Springer-Verlag; 1989.

[24] De Persis C, Isidori A. On the observability codistributions of a nonlinear system. Systems and Control Letters. 2000;40(5):297–304.

[25] Balas G, Bokor J, Szabo Z. Invariant subspaces for LPV systems and their applications. IEEE Transactions on Automatic Control. 2003;48(11): 2065–2069.

[26] Szabo Z, Bokor J, Balas G. Inversion of LPV systems and its application to fault detection. In: Proceedings of the 5th IFAC Symposium on fault detection supervision and safety for technical processes (SAFEPROCESS'03). Washington, DC, USA; 2003. p. 235–240.

[27] Bokor J, Balas G. Detection filter design for LPV systems—A geometric approach. Automatica. 2004;40(3):511–518.
[28] Bokor J, Szabo Z. Fault detection and isolation in nonlinear systems. Annual Reviews in Control. 2009;33(2):1–11.
[29] Aling H, Schumacher JM. A nine-fold canonical decomposition for linear systems. International Journal of Control. 1984;39(4):779–805.
[30] Alwi H, Edwards C, Marcos A. Actuator and sensor fault reconstruction using an LPV sliding mode observer. In: AIAA GNC; 2010. p. 1–24.
[31] Pfifer H, Seiler P. Robustness analysis of linear parameter varying systems using integral quadratic constraints. International Journal of Robust and Nonlinear Control. 2015;25:2843–2864.
[32] Hanema J, Lazar M, Tóth R. Tube-based LPV constant output reference tracking MPC with error bound. IFAC-PapersOnline. 2017;50(1).
[33] Besselmann T, Löfberg J, Morari M. Explicit MPC for LPV systems: Stability and optimality. IEEE Transactions on Automatic Control. 2012;57(9):2322–2332.
[34] Schmidt D. A Nonlinear Simulink Simulation of a Large Flexible Aircraft. MUSYN Inc. and NASA Dryden Flight Research Center; 2013.
[35] Georgiou TT. On the computation of the gap metric. Systems Control Letters. 1988;11:253–257.
[36] Vinnicombe G. Measuring Robustness of Feedback Systems. Department of Engineering, University of Cambridge, Cambridge; 1993.
[37] Boyd S, Ghaoui LE, Feron E, *et al.* Linear Matrix Inequalities in System and Control Theory. vol. 15 of Studies in Applied Mathematics. SIAM; 1994.
[38] Wu F. Control of Linear Parameter Varying Systems. University of California, Berkeley; 1995.

Chapter 12

Data-driven approaches to fault-tolerant control of industrial robotic systems

Yuchen Jiang[1,2] and Shen Yin[1]

Robotic systems in the modern industries are working in increasingly complex environment. The unexpected external disturbances and the abrupt change of working condition bring great challenges to the control of such systems. For some safety-critical applications, it is required to continuously stabilize the systems under faulty condition (including malfunctions and system failures). In this chapter, a data-driven fault-tolerant control (FTC) framework is introduced. As a semi-supervised machine learning technique that can learn from the rewards from external environment, reinforcement learning (RL) aims to maximize the long-term returns by, for instance, maintaining a value function. Based on the approximated value function, optimal control law can be derived. When using RL, although the convergence of the value function has been proved for some simple systems, such as the linear time invariant system, the internal stability of the close-loop system still remains unguaranteed. To deal with this problem, a novel framework for FTC system design based on Youla parameterization is introduced. It can be implemented in the modular and plug-and-play manner. It should be noted that in this work RL acts as a supervising module that calculates optimized Youla matrix Q_c in the design phase. Simulation results on a wheeled robot are provided to show the performance of the proposed approach in the continuously stabilizing framework. Some open questions and future work are summarized at the end of this chapter.

12.1 Background

Industrial robotic systems are the engineered systems constructed to increase production efficiency and lighten the workload by replacing or assisting human workers. The development of the robotic systems relies on the advances in mechatronics, computer science, manufacturing, material science and control science [1–4]. As robots are designed to release human efforts, the transformation of scientific and

[1]Department of Control Science and Engineering, Harbin Institute of Technology, Harbin, China
[2]Department of Electrical and Computer Engineering, Technical University of Munich, Munich, Germany

technological achievements are significant, which is expected to be smooth and efficient. To this end, the gap between the academic outputs and industrial applications should be filled.

Simplicity and low cost are two favorable aspects in the application of industrial robotic systems [5]. On one aspect, simple functional and structural design helps one to improve the generalization capacity, and as a result helps one to increase the stability and robustness [6,7]. Simple design also benefits the maintenance process by reducing the maintenance efforts and cutting down the downtime, so the benefit loss could be minimized. On another aspect, industrial enterprises seek for maximum profits. In many circumstances, factories prefer not to replace existing plants, equipment and devices unless there are strategic needs that the production line has to be updated. From the economics point of view, "cost" does not mean how much money are spent in purchasing new equipment but indicates the maximum expense to give up introducing new facilities. These acts will actually reduce the expected cost. Apparently, new and immature technologies are bundled with far more implicit risks than the traditional ones. As a result, it will lead to a situation where advanced technologies are available but are not widely adopted due to practical concerns. Each year, hundreds to thousands of novel approaches are proposed and their effectiveness is testified, but they are still far from practical application.

Today, the biggest proportion of the robotic controllers applied in industry is still the PID controllers and the model predictive controllers (MPC) [8–10]. It is universally known that the PID parameters are difficult to tune so skilled workers/experts and long configuration time are in need. Regarding MPC, it is still unclear from the theoretical point of view how to guarantee the close-loop stability vigorously. But other existing "advanced" controllers such as those with adaptiveness design, neural network modeling and fuzzy rules are far from popular than PID and MPC in the practical applications. Where have these research outputs gone? The outlets, according to the authors' knowledge, go to two directions: the high technology industries and disruptive industries. The robotic systems related high technology industries include astronautics, national defense, electronic information and high-tech service industry while examples of the disruptive industries include autonomous vehicles, intelligent factories and the smart grid [11].

12.2 Introduction and motivation

System monitoring, fault diagnosis and FTC have been extensively investigated during the past three decades [12–15]. Successful applications have been seen in the area of industrial process control, aircrafts, robotics, etc. [16–18]. Traditional system monitoring and fault diagnosis schemes are generally based on mechanism models derived from the first principles [19].

However, heavy reliance on such models leads to defects in robotic system design, mainly due to the following reasons: (i) model mismatch problem. The model parameters cannot be accurately measured or well identified in practice [20]. As a result, fulfilling the monitoring, control and behavior prediction tasks based on such model are problematic. (ii) Increased system scales and complexity resulting from correlated control loops and multi-hierarchical structures lead to extreme modeling difficulties [21–23]. (iii) Even if the mechanism models are available, the performance of the model-based schemes is nonoptimal due to inevitable discrepancy between models and real systems [24]. From information flow point of view, there is information loss during the modeling because the model cannot capture all characteristic dynamics. Such loss of information is irreparable in the subsequence design and implementation procedures. On the other aspect, the variety of system specifications leads to inconsistencies in the model-based design and implementation, which also influences the overall efficiency of maintenance and life cycle management [25]. Therefore, considering the rapid variation of external requirements and the internal reformation and optimization, the robotic industry needs to develop in a dynamic and adaptive framework with data-driven realizations where system scalability is considered during the design phase [26–28]. Robust control strategies provided more stability margin and control performance margin to the controlled object while introducing much conservatism. Optimal control approaches aim to achieve maximum performance indices that are sometimes infeasible or difficult to reach.

In recent years, artificial intelligent and machine learning have received extensive focus from multiple areas of research [29–32]. For automatic control tasks, intelligent learning aided control schemes have been reported in the international conferences and workshops [33–38]. It is known that RL is a semi-supervised machine learning approach that learns from the rewards obtained from the external environment, to approximate a state value function or a state-action value function, or straightforwardly, an optimal control law. With strong adaptivity and robustness, the design of intelligent learning aided controllers has great advantages in the robotic systems. However, although the convergence of the value function has been proved for some simple systems (e.g. the linear time-invariant system), the internal stability of the close-loop system still remains a huge problem.

Based on the above observations, this work proposes to design the intelligent learning aided controllers under the framework of Youla parameterization. The goal is to design optimal controllers that have fault-tolerance capability and increased robustness.

The rest of the chapter is organized as follows. Section 12.3 presents some important fundamental and preliminary knowledge. This part formulates the controlled systems and basic assumptions and summarizes the adopted system representation techniques and the adopted observers. In Section 12.4, a novel continuous stabilization framework is introduced based on Youla parameterization and RL-aided fault-tolerant controller design approach. It is emphasized that the reward function will be designed via the evaluation of the systems' overall stability degree. After that, in Section 12.5, a simple simulation example on the dynamic steering control system for a wheeled robot is provided to illustrate the performance of the approach. In the last concluding section, the major open questions on the novel framework are discussed for future reference.

12.3 Data-driven control framework based on Youla parameterization

12.3.1 System description and preliminaries

The controlled object is represented with the following discrete-time state-space model:

$$x(k+1) = Ax(k) + Bu(k) + E_d d(k) + E_f f(k) + w(k) \tag{12.1}$$

$$y(k) = Cx(k) + Du(k) + F_d d(k) + F_f f(k) + v(k) \tag{12.2}$$

where x is the state variable, y is the observed output, u is the control input signal, d characterizes the disturbance and f indicates the fault signal. w and v are the noise terms. The system matrices (A, B, C, D) are with compatible dimensions. Matrices E_d, F_d, E_f, F_f represent the directions of disturbances corresponding to the states and outputs and the directions of faults corresponding to the states and outputs, respectively.

Remark 12.1. *The system states are driven by the control input u, disturbance d and fault f. Considering that there are existing disturbance rejection approaches [39,40] and disturbance identification/estimation approaches [41,42], d can be regarded as known inputs. It is assumed that d and f are uncorrelated to x, i.e. they do not influence the close-loop stability. As a result, the configuration of the close-loop poles depends entirely on the design of u.*

Remark 12.2. ***Basic assumptions***

- *The system can be linearized around working point, and nonlinearities are not notable around the working point. For systems with strong nonlinearities that cannot be tackled within the framework of this paper, it is recommended to refer to "Parameterization of nonlinear observer-based fault detection systems" [43] where stable residual generators are parameterized by a cascade connection of the process kernel representation and a post-filter.*
- *The system states are observable or partially observable. The system states are controllable or partially controllable.*
- *External disturbances can be online compensated.*
- *The control input signal is unbounded. There is no special constraints to the system states.*
- *Large overshoot at the initialization stage can be accepted.*

Some preliminaries on coprime factorization and Bezout's identity are provided in this part. This knowledge is useful for the design of nominal controller in the subsequent sections. Readers are recommended to refer to [44,45] for more details.

Let $G(z)$ be the transfer function matrix of the open-loop system. Find stable systems $\hat{M}(z), \hat{N}(z), M(z)$ and $N(z)$ over $\mathcal{RH}_\infty$ such that

$$G(z) = \hat{M}^{-1}(z)\hat{N}(z) = N(z)M(z)^{-1} \tag{12.3}$$

$$\begin{bmatrix} X(z) & Y(z) \\ -\hat{N}(z) & \hat{M}(z) \end{bmatrix} \begin{bmatrix} M(z) & -\hat{Y}(z) \\ N(z) & \hat{X}(z) \end{bmatrix} = \begin{bmatrix} I & 0 \\ 0 & I \end{bmatrix} \tag{12.4}$$

then $\hat{M}, \hat{N}$ are called the left coprime transfer matrices, N, M are the right coprime transfer matrices, and (12.3) represents the left and right coprime factorization of $G(z)$. Equation (12.4) is called the double Bezout's identity, and it is a key to derive some compact results during the mathematical simplification processes. The following state-space realizations of the coprime transfer matrices provide a simple solution to the coprime factorization problem.

$$\hat{M}(z) = (A - LC, -L, C, I) \qquad \hat{N}(z) = (A - LC, B - LD, C, D) \tag{12.5}$$

$$\hat{X}(z) = (A + BF, L, C + DF, I) \qquad \hat{Y}(z) = (A + BF, -L, F, 0) \tag{12.6}$$

$$M(z) = (A + BF, B, F, I) \qquad N(z) = (A + BF, B, C + DF, D) \tag{12.7}$$

$$X(z) = (A - LC, -(B - LD), F, I) \qquad Y(z) = (A - LC, -L, F, 0) \tag{12.8}$$

where F and L should be selected so that $A + BF$ and $A - LC$ are stable.

Two types of observers, i.e. the full-dimensional state observer fault detection filter (FDF) and the reduced-order output observer diagnostic observer (DO), which will be used for residual generation with data-driven solutions. The meaning of the matrices and variables are in align with [44]. It is proved in Theorem 12.2 and in Appendix A that the residual signals generated by them are equivalent under certain condition.

FDF:

$$\hat{x}(k) = A\hat{x}(k) + Bu(k) + L_{FDF} \cdot r(k) \tag{12.9}$$

$$\hat{y}(k) = C\hat{x}(k) + Du(k) \tag{12.10}$$

$$r(k) = y(k) - \hat{y}(k) \tag{12.11}$$

DO:

$$z(k+1) = Gz(k) + Hu(k) + L_{DO} \cdot y(k) \tag{12.12}$$

$$r(k) = Vy(k) - Wz(k) - Qu(k) \tag{12.13}$$

subject to the Luenberger conditions I–III:

I. G is stable.

II. $TA - GT = LC, H = TB - LD$.

III. $VC = WT, Q = VD$.

Definition 12.1. *v_s is called a parity vector if $v_s\Gamma_s = 0$. $\Gamma_s = [C \quad CA \quad CA^2 \quad \cdots \quad CA^{s-1}]$. $P = \{v_s | v_s\Gamma_s = 0\}$ is called the parity space, and s is the length of parity space.*

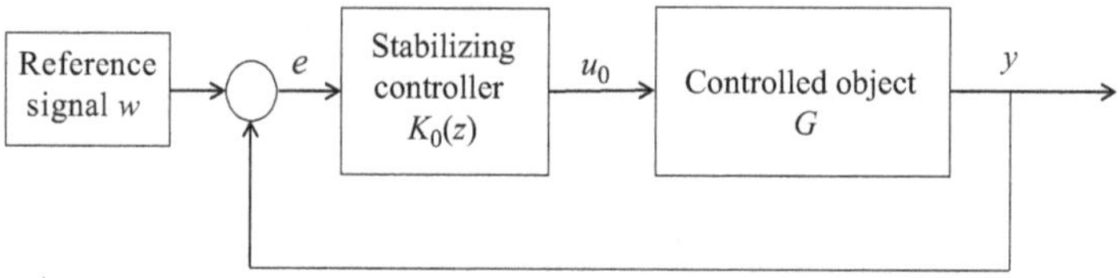

Figure 12.1 System structure

12.3.2 Youla parameterization of all stabilizing controllers

According to the dual principle in control theory, a system's observability and controllability are in a dual relationship. While the observers can be constructed based on the system model, it is also possible to build controllers directly from the system. Youla parameterization provides a realization approach to controller design based only on system information.

Theorem 12.1 (Youla parameterization). *If the system structure follows Figure 12.1 where the controlled object is defined as (12.1) and (12.2), then all linear controllers that internally stabilize the close-loop system can be parameterized as*

$$K(z) = [Y(z) - M(z)Q(z)][X(z) - N(z)Q(z)]^{-1} \tag{12.14}$$

$$= [\hat{X}(z) - Q(z)\hat{N}(z)]^{-1}[\hat{Y}(z) - Q(z)\hat{M}(z)] \tag{12.15}$$

According to Theorem 12.1, if the system model is available, it is straightforward to build stabilizing controllers with arbitrary Youla matrix Q. However, when the system model is hard to obtain, it is more favorable to use data-driven solutions that are based only on the input and output data.

The following inferences are all straightforward from Theorem 12.1 but give interpretations from different angles of view.

Inference 12.1 *For arbitrary $Q(z) \in \mathcal{RH}_\infty$, $K(z)$ in the form of (12.14) or (12.15) stabilizes the close-loop system described in Theorem 12.1.*

Inference 12.2 *All the controllers that stabilize the close-loop system described in Theorem 12.1 can be written in the form of (12.14) or (12.15). All the controllers that stabilize the close-loop system described in Theorem 12.1 can find a corresponding $Q(z)$.*

Inference 12.3 *If one stabilizing controller is known, then all stabilizing controllers can be obtained. Given two stabilizing controllers, they can transform to each other with a proper $Q(z)$.*

Based on Theorem 12.1, different forms of parameterization can be obtained (see Figure 12.2 and the box below). The readers are referred to [46] for more details. The first three forms are model dependent while the fourth is model independent.

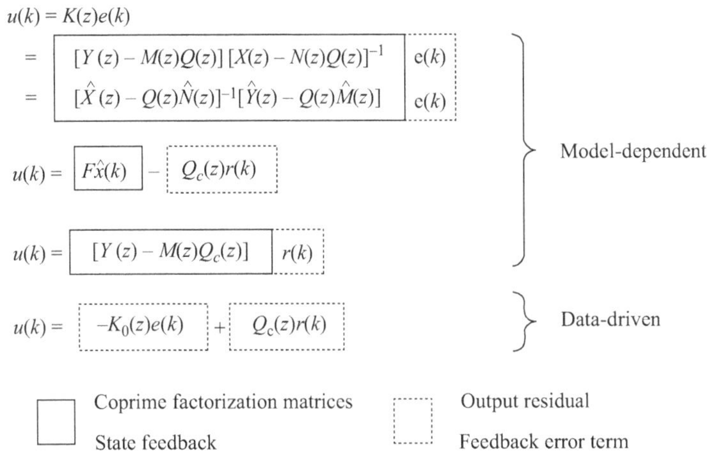

Figure 12.2 Different forms of parameterization

Parameterization of controllers: different forms

1. Youla parameterization: $X(z)$, $Y(z)$, $\hat{X}(z)$, $\hat{Y}(z)$, $M(z)$, $N(z)$, $\hat{M}(z)$, $\hat{N}(z)$ are related to the system's transfer function $G(z)$. The controller is parameterized with $Q(z)$.
2. State observer and residual generator-based control law: F is the stabilizing feedback gain, $\hat{x}(k)$ is generated by the state observer, $r(k)$ is generated by a residual generator. The control law is parameterized with $Q_c(z)$.
3. Residual generator-based control law: $Y(z)$ and $M(z)$ are related with the system's transfer function $G(z)$, $r(k)$ is generated by a residual generator. The control law is parameterized with $Q_c(z)$.
4. Residual generator-based control law: $K_{st}(z)$ is a stabilizing controller, $r(k)$ is generated by a residual generator (such as FDF, DO, or parity space-based residual generator). The control law is parameterized with $Q_c(z)$.

Next we will show in Theorem 12.2 and Inference 3.1 that the FDF and DO can be used to generate residual signals that guarantee the stability of the close-loop system. Theorem 12.3 reveals the condition of equivalence of the two residual generators when used for feedback control. The proofs of Theorems 12.2, 12.3 and Inference 3.1 are provided in Appendix A.

Theorem 12.2. *Consider the system structure following Figure 12.1 where the controlled object is defined as (12.1) and (12.2), if $K_{st}(z)$ is a stabilizing controller, and the residual $r(k)$ is generated by FDF as in (12.9), (12.10) and (12.11), then the control law $u(k) = -K_{st}(z)e(k) + Q_c(z)r(k)$ stabilizes the close-loop system.*

Theorem 12.3. *If the observer gains satisfy* $L_{DO} = T \cdot L_{FDF}$, *then the residuals generated by the FDF and DO have the following relationship:* $r_{DO}(k) = V \cdot r_{FDF}(k)$.

Inference 3.1 *Consider the system structure following Figure 12.1 where the controlled object is defined as (12.1) and (12.2), if* $K_{st}(z)$ *is a stabilizing controller, and the residual* $r(k)$ *is generated by DO as in (12.12) and (12.13), then the control law* $u(k) = -K_{st}(z)e(k) + Q_c(z)r(k)$ *stabilizes the close-loop system.*

The next theorem provides data-driven solutions to the DO's system matrices based on the parity vectors.

Theorem 12.4 (Parity space to MISO DO, [44,47]). *Given* $\alpha_s \in \Gamma_s^{\perp}$, *then the solution of the Luenberger equations II and III can be calculated by*

$$G = \begin{bmatrix} 0 & & & \\ 1 & \ddots & & \\ & \ddots & \ddots & \\ & & 1 & 0 \end{bmatrix}, L = -\begin{bmatrix} \alpha_{s,0} \\ \vdots \\ \vdots \\ \alpha_{s,s-1} \end{bmatrix}$$

$$\begin{bmatrix} H \\ Q \end{bmatrix} = \left(\alpha_s H_{u,s}\right)^T, W = \begin{bmatrix} 0 & \cdots & 0 & 1 \end{bmatrix}, V = \alpha_{s,s}$$

$$\text{with} \quad T = \begin{bmatrix} \alpha_{s,1} & \cdots & \cdots & \alpha_{s,s} \\ \vdots & & \iddots & \\ \vdots & \iddots & & \\ \alpha_{s,s} & & & \end{bmatrix} \begin{bmatrix} C \\ CA \\ \vdots \\ CA^{s-1} \end{bmatrix}.$$

In order to retain the correlation information in the output variables and reduce the online implementation complexity, approaches to MIMO DO design were introduced in [47]. The underlying idea is to construct η-independent MISO residual generators, which are driven by identical control input signals $u(k)$ and output signals $y(k)$.

$$z_i(k+1) = A_{z,i}z_i(k) + B_{z,i}u_i(k) + L_{z,i}y(k) \tag{12.16}$$

$$r_i(k) = g_i y(k) - c_{z,i}z_i(k) - d_{z,i}u(k) \qquad (i = 1, 2, \ldots, \eta) \tag{12.17}$$

subject to η groups of Luenberger conditions.

12.3.3 Plug-and-play control framework

In this part, we introduce a stabilizing control framework based on Youla parameterization. As shown in Figure 12.3, the nominal controlled object P_0 is a generalized system that includes the actuators and sensors connected to the actual object of control, and the input/output data of P_0 can be easily collected. The control demand u is composed of three parts, driven by the stabilizing controller K_{st}, the robustness controller $K_{rb}(Q_c)$ and the tracking controller K_{tr}, respectively. The stabilizing controller

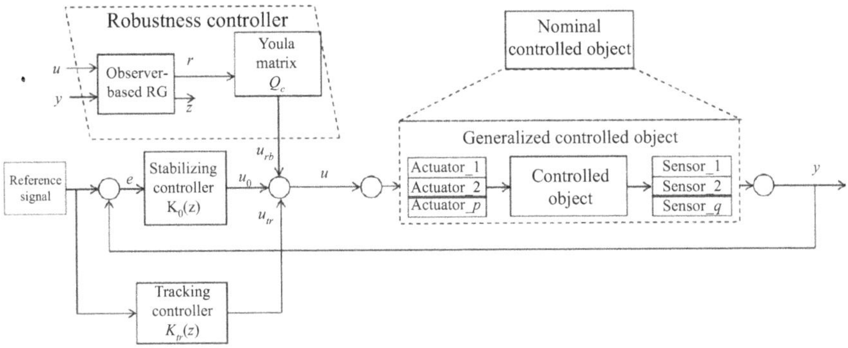

Figure 12.3 Plug-and-play control framework

and the robustness controller are feedback controllers while the tracking controller is a feedforward controller.

$$u = u_{st} + u_{rb} + u_{tr} \tag{12.18}$$

$$= K_{st}(z)e(k) + Q_c(z)r(k) + K_{tr}(z)w(k) \tag{12.19}$$

According to Inference 3.1, $K_{st}(z)e(k) + Q_c(z)r(k)$ can represent an arbitrary stabilizing controller of the close-loop system. Note that $K_{tr}(z)w(k)$ is feedforward and does not change the poles of the close-loop system, the system's stability will not be affected.

For some existing industrial systems, there are usually reliability concerns and economic concerns when updating the control system for better performance [48]. It is preferred not to make major modifications to the main stabilizing controllers [49,50]. In the proposed control framework, $K_{st}(z)$ can be regarded as the existing controller that stabilizes the close-loop system. The robustness controller plays the role of a plug-and-play module for performance improvement. The main advantage of such module is that during the configuration stage, the close-loop stability is continuously guaranteed, as indicated by the aforementioned theorems.

12.4 Reinforcement learning-aided approach to fault-tolerant controller design

Kaelbling *et al.* defined RL as "a way of programming agents by reward and punishment without needing to specify *how* the task is to be achieved" [51]. In using RL, we aim to develop learning-aided approach for the purpose of FTC system design with optimization. To this end, the selection of the reward function, value function evaluation (estimation) and controller update will be the major focuses in this part (Figure 12.4).

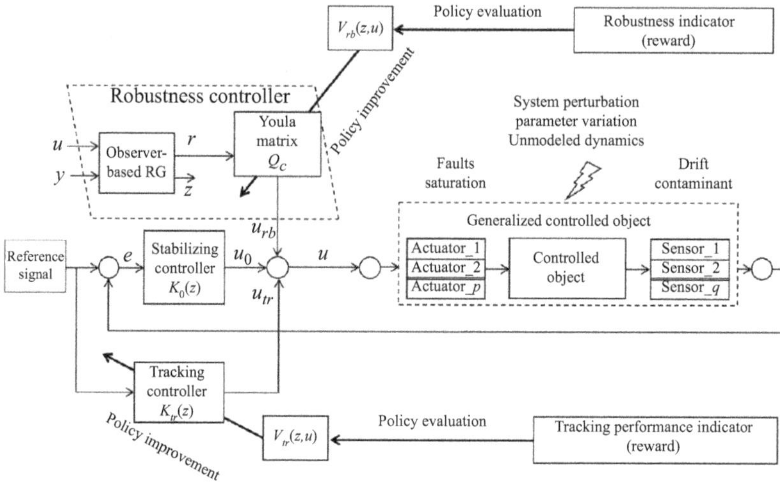

Figure 12.4 Fault-tolerant control with the aid of reinforcement learning

Considering that RL solutions are reached based more on the actual experiences and less on the model assumptions, it is important to guarantee the stability of the close-loop control system during the learning phase to avoid divergence of the states. It should be noted that in the Youla parameterization framework, on-policy RL control methods (such as SARSA) and the off-policy RL control methods (such as deep Q-learning) can be applied to optimize Q_c offline [52–56].

12.4.1 Applying RL to control system design

In order to apply RL to the control problems, the basic concepts in RL and control system design are summarized below, and the connections between them are established in Table 12.1, which shows the corresponding relation.

Reward is the instant evaluative feedback from the external environment (the controlled objective) to the agent (controller to be learned or optimized). *Return of a policy (control strategy)* is the accumulated reward over time following that policy (control strategy). *Partial return* is the accumulated reward over a limited period of time. *Value of a policy* is the expected return following that policy. *Policy* is the projection from the state space to the action space. In automatic control systems, it is defined as the control law of taking action $a \in A$ at state s.

12.4.2 Reward function design

A major difference that distinguishes RL from supervised learning schemes is that the RL designers are supposed to tell the agent “what to achieve” instead of “how to approach it” [57]. Toward completing the RL task, it is important to specify what the ultimate goal is and to feed the agents with to-the-point feedback, i.e. the *proper*

Table 12.1 Correspondence of reinforcement learning and control system design

Reinforcement learning	**Control system design**
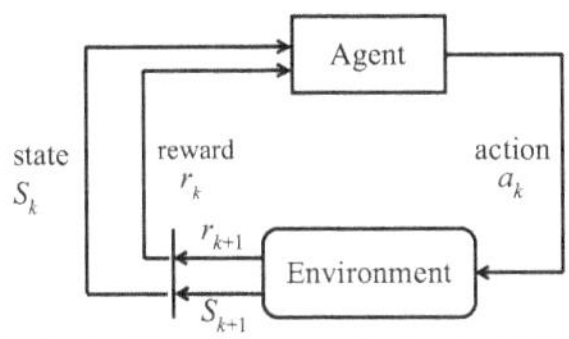	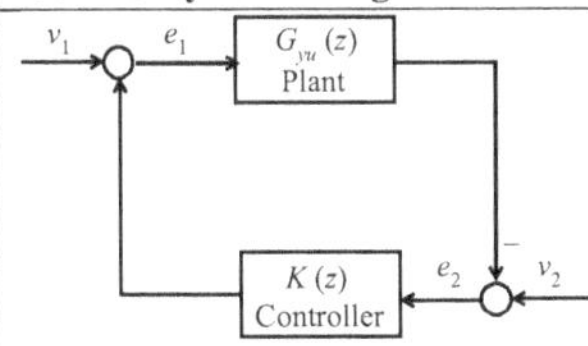
Agent	Controller
Environment	Close-loop system and external inputs
State/state-action pair	State
Policy	Control law
Action	Control demand/control input
Reward	(Evaluative index)
Value function	Value/cost function

instant rewards. Then the RL agents will gradually learn to maximize the rewards it receives.

Rewards must be provided to the agent in a way that in maximizing them the agent will also achieve the ultimate goals [57]. If the provided rewards misalign with the ultimate goal, the learning results will be faulty. For instance, in the task of designing a machine-vision-based object-picking robot arm, the agent should only be rewarded for successfully picking up the objects, not for achieving subgoals such as a successful object identification or successful trajectory planning. For the low-level controller design tasks, it is important to specify the control performance evaluation criteria *prior to* the learning process and provide equivalent online performance indicators as instant rewards.

In control theory, robustness indicates the systems' capability to maintain certain performances under perturbations. The robust controller design task aims to achieve close-loop robust stabilization and robust performance. Robust performance includes robustness against disturbances, robustness against faults, robust tracking performance, anti-saturation performance, etc. It should be noted that the robust performance problem can be equivalent to introducing virtual perturbations to the robust stabilization problem.

Most of the existing robust stabilization controller design approaches are based on explicit system models where the system parameters are assumed to be known as *a priori* knowledge. In the RL-based design framework, we will specify an evaluation criterion to quantize the degree of close-loop stability (how stable exactly) and feed it as reward to train the controller.

Recently, the concept of stability margin is highlighted in [58]. It can be seen as a metric to the system stability due to its internal connection with the system norm and singular value decomposition.

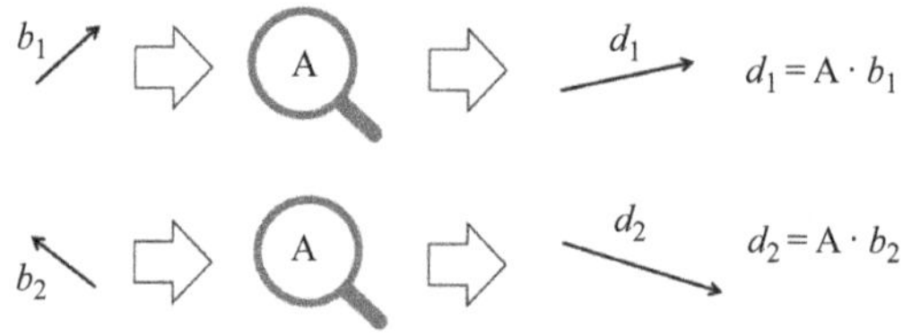

Figure 12.5 Interpretation of matrix's amplification effect

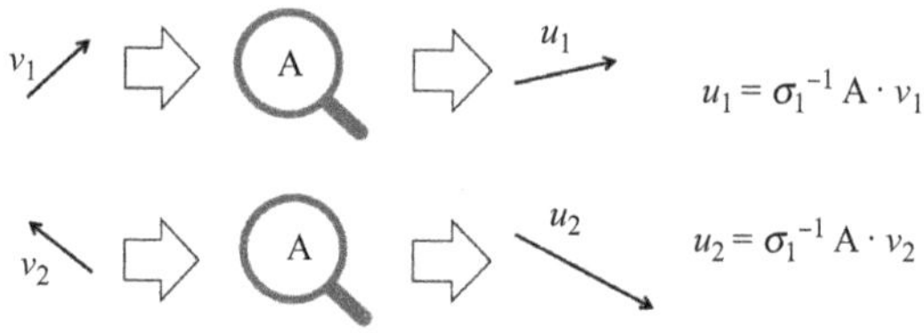

Figure 12.6 Interpretation of singular value decomposition

As shown in Figure 12.5, matrix A can be seen as an amplifier to the input vectors. The output vectors can be seen as modulated signals of the inputs with the multiplied matrix. Consider multiplying the matrix with an arbitrary vector. For different vectors, the "amplification" (also called "gain" in the control theory) of that vector is different and is related to the direction of the input vector. H_∞ norm is the upper bound of the amplification (maximum gain):

$$\|A\|_\infty = \max \frac{\|d\|}{\|b\|} \tag{12.20}$$

The graphical illustration also applies to transfer function matrices, where dynamics of the transfer functions also play a role in modulating the inputs. Signal norms and system norms characterize the dynamic systems while vector norms and matrix norms characterize the constant and static systems:

$$\|A(z)\|_\infty = \max \frac{\|d(z)\|}{\|b(z)\|} \tag{12.21}$$

The formulations of matrix norm in (12.20) and (12.21) are induced by vector norm and signal norm, respectively, which cannot be directly implemented in computer algorithms. A numerical feasible solution is achieved by singular vector decomposition (SVD). SVD can tell about exactly how much the matrix amplifies the input vector: if the input vector is in accordance with the direction of a left singular vector, the output direction must be aligned with that of the corresponding right singular vector, and the amplification (gain) is the corresponding singular value (see Figure 12.6). Thus, the maximum amplification gain can be determined with the largest singular value.

Based on the above interpretation of matrix norm and SVD, and according to the logics in Figure 12.7, the metric of close-loop stability can be designed using system's

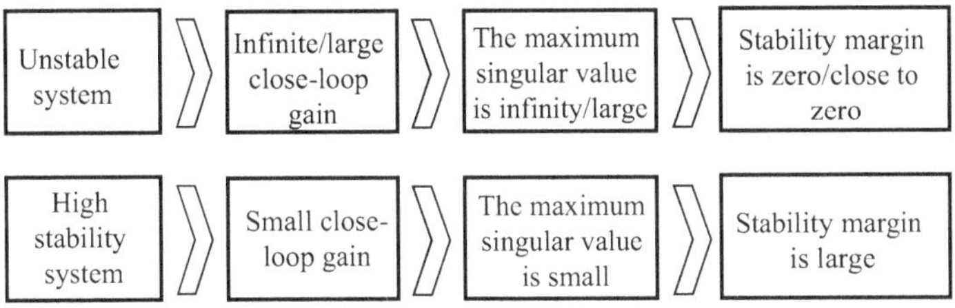

Figure 12.7 Evaluation of stability degree

H_∞ norm. Here, an algorithm of data-driven solution to the close-loop robustness indicator based on the stability margin is summarized.

Algorithm. Data-driven solution to the close-loop stability margin [58,59]

Step 1. Determine the length of stacked data and construct Hankel matrices.
Step 2. Calculate the stable image representation $\mathscr{I}_G$ using model-based approach or data-driven identification approach.
Step 3. Calculate the stable kernel representation $\mathscr{K}_K$ using model-based approach or data-driven identification approach.
Step 4. Calculate the close-loop robustness indicator based on (A.12).

The basic formulation of the close-loop stability metric is referred to Appendix A as well as the recommended literature therein. For practical application, real-time calculation is required, and recursive calculation of the stability degree should better be adopted [60].

12.4.3 RL-based solution to Youla parameterization matrix

In the control-loop level design of the robotic systems, both the state space and the action space are much larger than those in the decision-making level. The valid values of states and control signals are infinite. In these cases, the tabular solution methods cannot be used. Instead, function approximation approaches should be used for maintaining a value function.

In using function approximation, the central task of the RL agent is to learn from existing data and improving its generalization capability to deal with the untaught circumstances (unmet states). Generally, the number of undetermined parameters (the dimension of θ) is much smaller than the number of system states. The update of these parameters is subtle because changing one parameter changes the estimated value of many states. Furthermore, in the so-called continuing task (compared against the episodic task) [57], the learning process should be implemented in an incremental manner. When new training data are available, the policy can be updated without too much delay.

RL can be applied to learn the Youla transfer matrix (Figure 12.8). First of all, the Youla transfer matrix Q_c is parameterized with stable systems represented in the "zero-pole-gain (zpk)" form, where the parameters of gains, numerators and

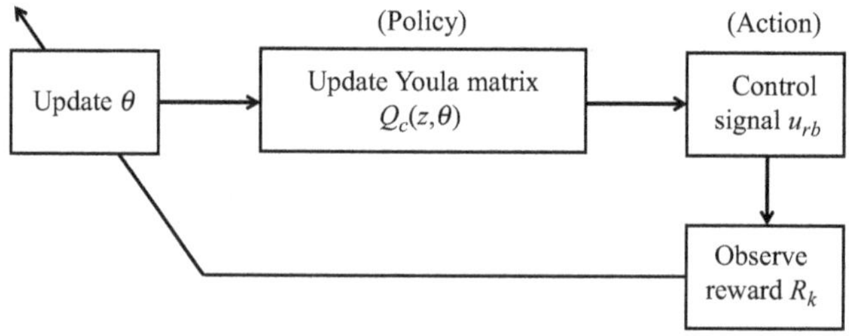

Figure 12.8 Update of the Youla parameterization matrix

denominators can be easily found and used to construct θ. z is the Z-transform operator:

$$Q_c(z,\theta) = k \cdot \frac{\prod_{j=1}^{j^{\max}} (z - n_j)}{\prod_{p=1}^{p^{\max}} (z - d_p)} \tag{12.22}$$

$$\theta = [k \;\; n_1 \;\; \cdots \;\; n_{j^{\max}} \;\; d_1 \;\; \cdots \;\; d_{p^{\max}}] \tag{12.23}$$

Use the close-loop stability metric at time k as the instant reward to the RL agent: $R_k = \delta_{cl}(z_k)$. The objective function $J(\theta)$ evaluates how much the estimated values deviate from the desired values:

$$J(\theta) = \sum_{z \in S} \mu(z) \left[U_k - \hat{q}(z,\theta)\right]^2 \tag{12.24}$$

where $\mu(z)$ is the weighing factor of the states, used for balancing the impact of observations. It can be defined as the percentage of time spent in state z. $\hat{q}(z,\theta)$ is the estimated state-action value function. Here θ plays the roles of both *actions* and *weights*. U_k is the expected return following a given control policy. In the continuing task setting, U_k is implemented with the differential form:

$$U_k = (R_{k+1} - \bar{R}) + (R_{k+2} - \bar{R}) + (R_{k+3} - \bar{R}) + \cdots \tag{12.25}$$

where $\bar{R}$ is the average-reward $\bar{R} = \lim_{h\to\infty}(1/h)\mathbb{E}[R_k | u^{\theta}_{0:k-1}] = \lim_{k\to\infty} \mathbb{E}[R_k | u^{\theta}_{0:k-1}]$ and $u^{\theta}_{0:k-1}$ indicates that $\{u_0, u_1, \ldots, u_{k-1}\}$ are driven by a fixed θ. By applying differential semi-gradient SARSA, we have the following update rule:

$$\theta_{k+1} = \theta_k + \alpha \left[R_{k+1} + \gamma \hat{q}(z_{k+1}, \theta_{k+1}) - \hat{q}(z_k, \theta_k)\right] \cdot \nabla \hat{q}(z_k, \theta_k) \tag{12.26}$$

where $\alpha > 0$ is the step-size parameter, $\gamma \in (0, 1)$ is the discount rate, and ∇ denotes the Hamilton operator:

$$\hat{q}(z,\theta) = \theta^T \cdot \xi(z) \tag{12.27}$$

Linear function approximation is applied to $\hat{q}(z,\theta)$. $\xi(\cdot)$ is called the feature vector or basis function of the linear model, which is used for construct features and can be defined in many different ways [57]. Following this, the update rule based on the differential semi-gradient SARSA can be derived as

$$\theta_{k+1} = \theta_k + \alpha \left[R_{k+1} + \gamma \theta_k^T \xi(z_{k+1}) - \theta_k^T \xi(z_k) \right] \xi(z_k) \tag{12.28}$$

12.5 Simulation study

12.5.1 Simulation setup

A dynamic steering control system for a wheeled robot is adopted in this study. The kinematics is described with the so-called one-track model, which is a simple but most popular form that is used in practical vehicles. The simulation model is in discrete-time, built on MATLAB®/Simulink® and ran on a PC with a 2.3 GHz Core i5 Processor. The sampling rate is 10 Hz, and the total simulation time is 100 s.

The system matrices of the linearized controlled object around the working point are as follows:

$$A = \begin{bmatrix} 0.6333 & -0.0672 \\ 2.0570 & 0.6082 \end{bmatrix} \qquad B = \begin{bmatrix} -0.0653 \\ 3.4462 \end{bmatrix} \tag{12.29}$$

$$C = \begin{bmatrix} -152.7568 & 1.2493 \\ 0 & 1 \end{bmatrix} \qquad D = \begin{bmatrix} 56 \\ 0 \end{bmatrix} \tag{12.30}$$

The system states are the slip angle (x_1) and yaw rate (x_2). The control input is the steering angle (u), and the outputs are the lateral acceleration sensor output (y_1) and the yaw rate sensor output (y_2). Both the states and the outputs are corrupted with normally (Gaussian) distributed white noise of small variances. The reference signal $w(k)$ is shown in Figure 12.9.

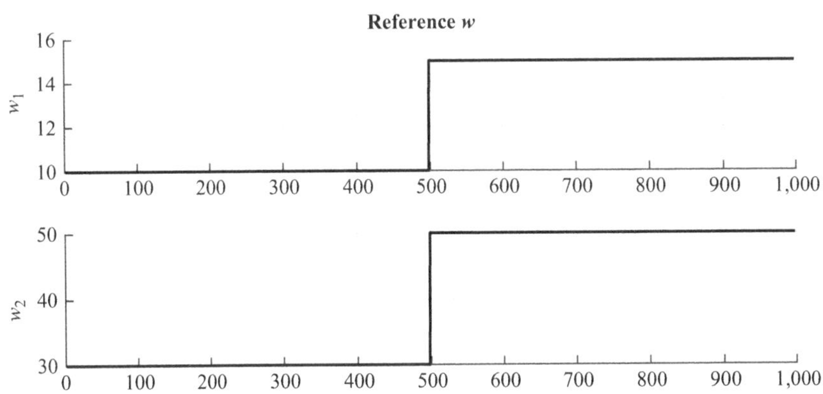

Figure 12.9 Reference signals

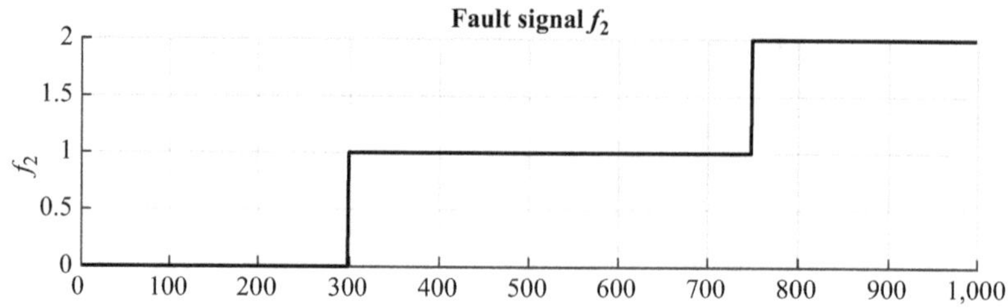

Figure 12.10 Introduced fault

Internal fault is introduced to the controlled object. Consistent with the system description, the fault signal is driven by the term $E_f f(k)$ in (12.1), where

$$E_f = \begin{bmatrix} 1 & 0 & 0 \\ 0 & 1 & 0 \end{bmatrix} \qquad f = \begin{bmatrix} 0 & f_2 & 0 \end{bmatrix}^T \tag{12.31}$$

As shown in Figure 12.10, f_2 is composed of two step signals, at the 30th and the 75th second, respectively.

An FDF was implemented to generate the residual signal, with

$$L_{FDF} = \begin{bmatrix} -0.0022 & -0.0135 \\ -0.0645 & 0.8250 \end{bmatrix} \tag{12.32}$$

It can be verified that L_{FDF} can stabilize the state tracking error dynamics. The tracking controller u_{tr} was implemented with a static feedforward gain $K_{tr} = [0.0007 \;\; 0.007]$. The stabilizing controller u_{st} was implemented with a stable dynamic system:

$$\mathscr{P}_{st} = \left[\begin{array}{c|c} A_{st} & B_{st} \\ \hline C_{st} & \mathbf{0} \end{array}\right]$$

where

$$A_{st} = \begin{bmatrix} -0.2916 & 0.0657 \\ 0.752 & 0.1027 \end{bmatrix}, B_{st} = \begin{bmatrix} 0.0135 & -0.1659 \\ -0.014 & 0.438 \end{bmatrix}, C_{st} = \begin{bmatrix} -0.1601 & 0.034 \end{bmatrix}.$$

12.5.2 Results and discussion

In order to illustrate the performance of the introduced framework and approaches, three configurations are designed for comparative study.

1. Fault-free condition: In the fault-free condition, system runs in normal operation status, and the system states in this condition are regarded as a reference to evaluate the fault-tolerance capability in the faulty condition.
2. Faulty condition without the robustness and/or tracking controller: Internal fault is introduced. The robustness/tracking controller is not plugged-in.
3. Faulty condition with the robustness and tracking controller: Internal fault is introduced. The robustness and tracking controllers are online.

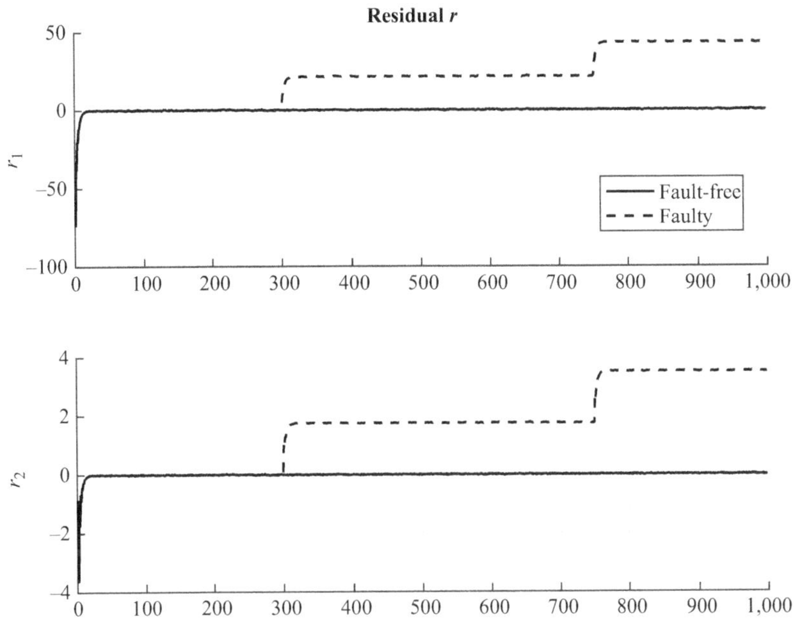

Figure 12.11 Residual signals

Figure 12.11 illustrates the residual signals generated by the FDF. It can be seen that the FDF successfully detected the happening of faults as both residuals deviate from zero after the 300th sample. The strength of fault is also reflected by the deviation degree off zero. The residuals doubled after the 750th sample compared with those in 300–750, which is consistent with the fault strength (see Figure 12.10).

Figure 12.12 compares the control performance of the state variables before and after the faults using different combination of controllers. In the fault-free scenario, the state variables were respectively stabilized at one value in each of the two stages as the reference values changed (x_1 was stabilized at approx. -1.1 and -1.8 rad, and x_2 was stabilized at approx. 4.9 and 8 rad/s). In the faulty scenario, both states were also stabilized, but at different equilibrium points. Furthermore, as indicated by the abrupt changes at the 300th and 750th samples, the states deviate more seriously if the fault strength increases.

Our goal of applying the FTC schemes is to weaken the deviation of the states and restore the normal states as in the fault-free condition. In Figure 12.12(a) and (b), the dotted lines show the controlled states with the stabilizing controller, and equipped without and with the robustness indicator, respectively. It can be seen that the controlled performance in Figure 12.12(b) is superior: both states were almost consistent with the ones in fault-free condition.

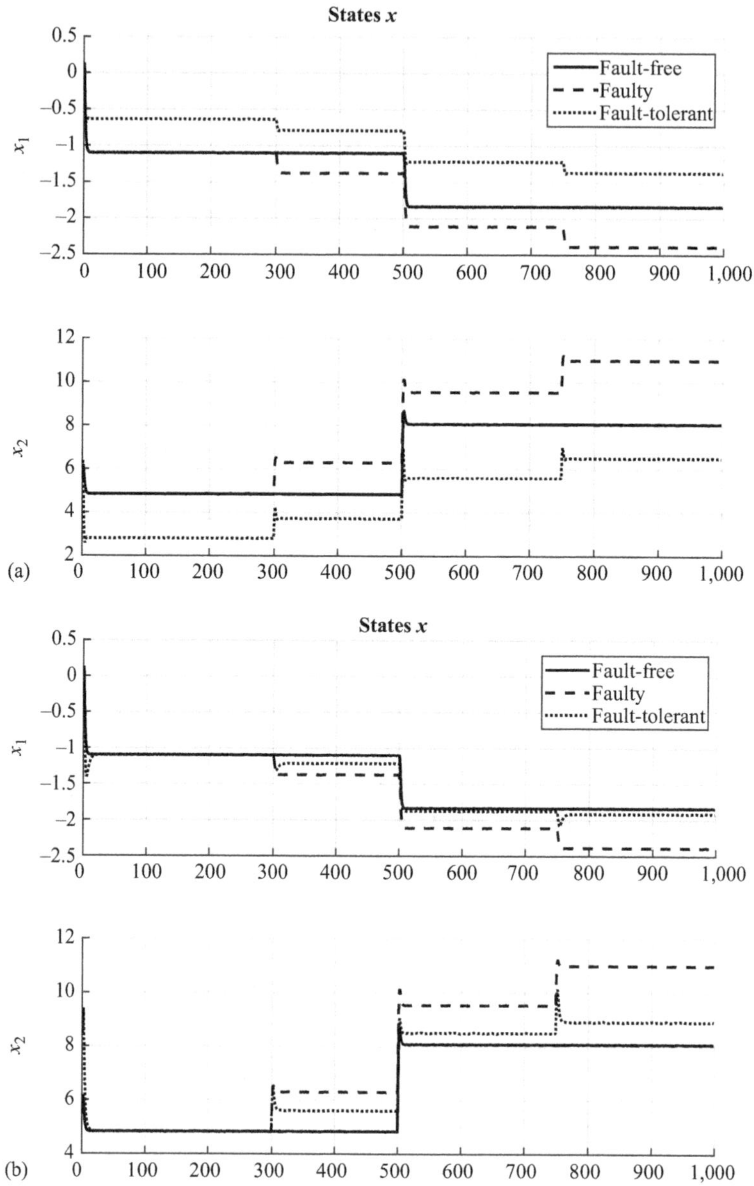

Figure 12.12 Comparison of state trajectories: (a) fault-tolerant without robustness controller and (b) fault-tolerant with robustness controller

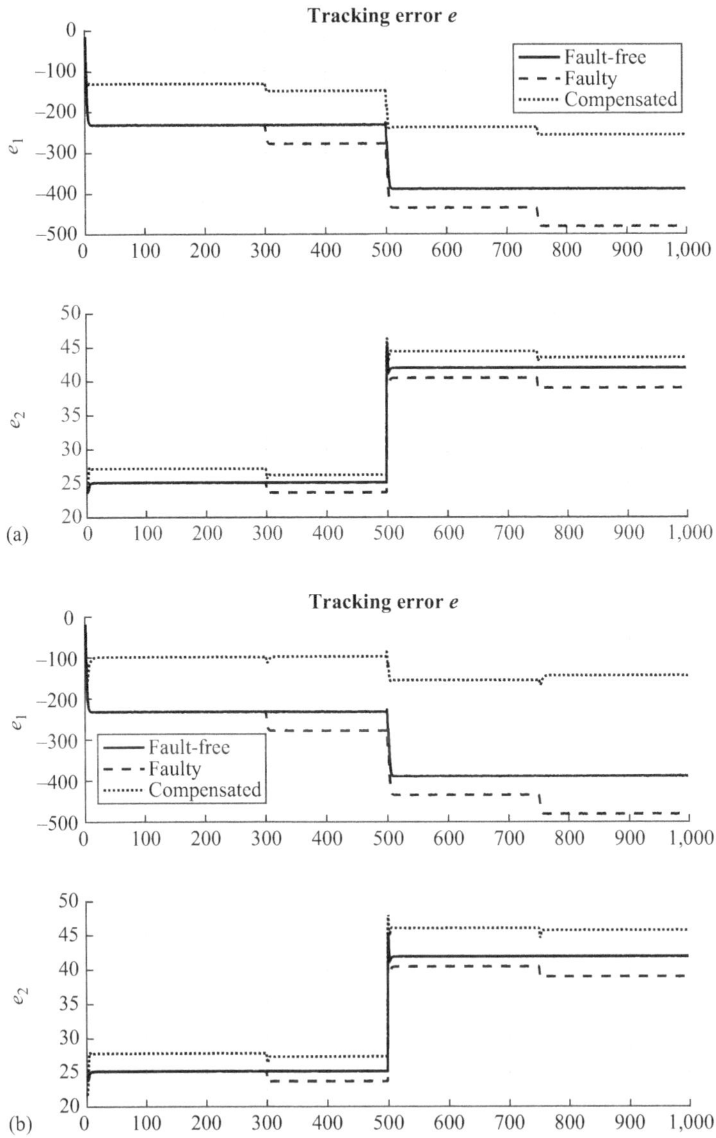

Figure 12.13 Comparison of tracking performance: (a) compensation with tracking controller and (b) compensation with tracking controller and robustness controller

Figure 12.13 illustrates the tracking performance of the proposed approaches. The tracking error of the first output variable was deteriorated after the fault happens, which circumstance is similar to the one happened to the state variables. The dotted line in Figure 12.13(a) is the results obtained with compensation of the tracking controller. The tracking error e_1 is noticeably suppressed. However, there is still serious performance degradation when the faults happened at the 300th and the 750th sample. To refine the fault-tolerant performance, a combination of tracking and robustness controllers was tested, and the corresponding results are shown in Figure 12.13(b). It can be observed that the tracking performance is less influenced by the faults.

12.6 Open questions about the framework and future work

This part summarizes the challenges in the design of RL-aided controllers under the data-driven continuously stabilizing FTC framework.

1. Conventional Youla parameterization is applicable to linear systems. Some robotic system control problems involve intrinsic or strong nonlinearities. For such systems, the audiences are recommended on the one hand to use nonlinear parameterization forms such as in [43], and on the other hand, design a neural network system to construct a nonlinear Youla gain (the learning approach to the Youla gain need to be redesigned as well).
2. In this work, RL acts as a supervising module that calculates optimized Youla matrix Q_c in the design phase. Further efforts are required for online learning-based FTC with the aid of pattern recognition techniques to gain awareness about the working condition and external environment. Given that normal working condition variations and malfunctioning/faults can be properly defined and dealt with, the adaptiveness would be greatly enhanced.
3. For applications to mobile robots, especially the autonomous wheeled robot, both aspects of algorithm realization and hardware implementation/deployment should meet the real-time requirement. To this end, (online) learning-aided controller block needs to be designed using fast approximation methods and dimension reduction techniques. Hardware acceleration can be achieved by optimized and integrated design, as well as dedicated system-on-chip design. Neural network engines can be applied to accelerate computations.
4. The available data from multiple sensors may have different sampling rates and are in different scales. They also present different characteristics regarding distortion degree, contamination, etc. How to synchronize the measurements and to achieve effective data association is a basic problem in digitalized and computer-based controller design tasks. Some potential solutions include the time stamp techniques and downsampling approaches.
5. The feedback controllers designed under the Youla parameterization framework only improve the system dynamics by reassigning the locations of the close-loop poles. The tracking controller applied in this work is conventional and not satisfactory enough. A separate learning aided feedforward controller can be designed to improve the tracking performance under faulty condition.

Appendix A

Proof of Theorem 12.2

According to the FDF as in (12.9)–(12.11),

$\hat{x}(k) = (zI - A + LC)^{-1}\left[(B - LD)u(k) + Ly(k)\right]$

$r(k) = y(k) - C\hat{x}(k) - Du(k)$

$= y(k) - C(zI - A + LC)^{-1}\left[(B - LD)u(k) + Ly(k)\right] - Du(k)$

$= \left[I - C(zI - A + LC)^{-1}L\right]y(k) - \left[C(zI - A + LC)^{-1}(B - LD) + D\right]u(k)$

$= \hat{M}(z)y(k) - \hat{N}(z)u(k)$ Let $K_{st}(z) = \hat{X}_0^{-1}(z)\hat{Y}_0(z)$.

Based on Theorem 12.1, the following control law stabilizes the close-loop system:

$u(k) = -K(z)e(k)$

$= -\left[\hat{X}_0(z) - Q(z)\hat{N}(z)\right]^{-1}\left[\hat{Y}_0(z) - Q(z)\hat{M}(z)\right]e(k)$

$\Rightarrow \left[\hat{X}_0(z) - Q(z)\hat{N}(z)\right]u(k) = -\left[\hat{Y}_0(z) - Q(z)\hat{M}(z)\right]e(k)$

$\Rightarrow \hat{X}_0(z)u(k) = Q(z)\hat{N}(z)u(k) - \hat{Y}_0(z)e(k) - Q(z)\hat{M}(z)e(k)$

$u(k) = \hat{X}_0^{-1}(z)\left[Q(z)\left(\hat{N}(z)u(k) - \hat{M}(z)e(k)\right) - \hat{Y}_0(z)e(k)\right]$

$= \hat{X}_0^{-1}(z)\left[Q(z)\left(\hat{M}(z)y(k) - r(k) - \hat{M}(z)(y(k) - w(k))\right) - \hat{Y}_0(z)e(k)\right]$

$= \hat{X}_0^{-1}(z)\left[Q(z)\left(-r(k) + \hat{M}(z)w(k)\right) - \hat{Y}_0(z)e(k)\right]$

$= -\hat{X}_0^{-1}(z)Q(z)r(k) - \hat{X}_0^{-1}(z)\hat{Y}_0(z)e(k) + \hat{X}_0^{-1}(z)\hat{M}(z)w(k)$

Let $Q_c(z) = -\hat{X}_0^{-1}(z)Q(z)$,

$u(k) = Q_c(z)r(k) - K_{st}(z)e(k) + \hat{X}_0^{-1}(z)\hat{M}(z)w(k)$

Since $w(k)$ has no influence on the close-loop stability, $u(k) = -K_{st}(z)e(k) + Q_c(z)r(k)$ also stabilizes the system. ■

Proof of Theorem 12.3

According to the DO as in (12.12) and (12.13),

$z(k) = (zI - G)^{-1}(Hu(k) + Ly(k))$

$r_{DO}(k) = Vy(k) - Wz(k) - Qu(k)$

$= \left[V - W(zI - G)^{-1}L\right]y(k) - \left[W(zI - G)^{-1}H + Q\right]u(k)$

$= \hat{M}_{DO}(z)y(k) - \hat{N}_{DO}(z)u(k)$

where

$\hat{M}_{DO}(z) = V - W(zI - G)^{-1}L$

$\hat{N}_{DO}(z) = W(zI - G)^{-1}H + Q.$

According to the FDF as in (12.9)–(12.11),

$\hat{x}(k) = (zI - A + LC)^{-1}\left[(B - LD)u(k) + Ly(k)\right]$

$r_{FDF}(k) = y(k) - C\hat{x}(k) - Du(k)$

$= y(k) - C(zI - A + LC)^{-1}\left[(B - LD)u(k) + Ly(k)\right] - Du(k)$

$= \left[I - C(zI - A + LC)^{-1}L\right]y(k) - \left[C(zI - A + LC)^{-1}(B - LD) + D\right]u(k)$

$= \hat{M}_{FDF}(z)y(k) - \hat{N}_{FDF}(z)u(k)$

where

$\hat{M}_{FDF}(z) = I - C(zI - A + LC)^{-1}L$

$\hat{N}_{FDF}(z) = C(zI - A + LC)^{-1}(B - LD) + D.$

Perform an equivalent transformation on FDF with a transformation matrix T: $\bar{x} = Tx$, $x = T^{-1}\bar{x}$, then

$$\bar{x}(k+1) = TAT^{-1}\bar{x}(k) + TBu(k) + TL_{FDF}r(k)$$
$$\hat{y}(k) = CT^{-1}\bar{x}(k) + Du(k)$$

$$\hat{M}_{FDF}(z) = I - CT^{-1}(zI - TAT^{-1} + TL_{FDF}CT^{-1})^{-1}TL_{FDF}$$
$$\Rightarrow V \cdot \hat{M}_{FDF}(z)$$
$$= V - VCT^{-1}(zI - TAT^{-1} + TL_{FDF}CT^{-1})^{-1}TL_{FDF}$$
$$= V - W(zI - TAT^{-1} + TL_{FDF}CT^{-1})^{-1}TL_{FDF}$$
$$\hat{M}_{DO}(z) = V - W(zI - G)^{-1}L_{DO}$$
$$= V - W(zI - (TA - L_{DO}C)T^{-1})^{-1}L_{DO}$$
$$= V - W(zI - TAT^{-1} + L_{DO}CT^{-1})^{-1}L_{DO}$$

Let $L_{DO} = TL_{FDF}$, then $V \cdot \hat{M}_{FDF}(z) = \hat{M}_{DO}(z)$.

Similarly,

$$\hat{N}_{FDF}(z) = CT^{-1}(zI - TAT^{-1} + TL_{FDF}CT^{-1})^{-1}(TB - TL_{FDF}D) + D$$
$$\Rightarrow V \cdot \hat{N}_{FDF}(z)$$
$$= W(zI - TAT^{-1} + TL_{FDF}CT^{-1})^{-1}(TB - TL_{FDF}D) + VD$$
$$\hat{N}_{DO}(z) = W(zI - G)^{-1}H + Q$$
$$= W(zI - (TA - L_{DO}C)T^{-1})^{-1}(TB - L_{DO}D) + VD$$
$$= W(zI - TAT^{-1} + L_{DO}CT^{-1})^{-1}(TB - L_{DO}D) + VD.$$

Since $L_{DO} = TL_{FDF}$, we have $V \cdot \hat{N}_{FDF}(z) = \hat{N}_{DO}(z)$.

Therefore,

$$r_{DO}(k) = \hat{M}_{DO}(z)y(k) - \hat{N}_{DO}(z)u(k)$$
$$= V\left(\hat{M}_{FDF}(z)y(k) - \hat{N}_{FDF}(z)u(k)\right)$$
$$= V \cdot r_{FDF}(k).$$ ■

Proof of Inference 3.1

According to the proof of Theorem 12.2, the following control law stabilizes the close-loop system: $u(k) = -K_{st}(z)e(k) + \bar{Q}_c(z)r_{FDF}(k)$. According to Theorem 12.3, $r_{DO}(k) = V \cdot r_{FDF}(k)$ if $L_{DO} = TL_{FDF}$. Then, $u(k) = -K_{st}(z)e(k) + \bar{Q}_c(z)V^{\dagger}r_{DO}(k)$. Let $Q_c(z) = \bar{Q}_c(z)V^{\dagger}$, then $u(k) = -K_{st}(z)e(k) + Q_c(z)r_{DO}(k)$ stabilizes the close-loop system. ■

Basics of the close loop stability metric

Consider the control diagram in Figure A.1,

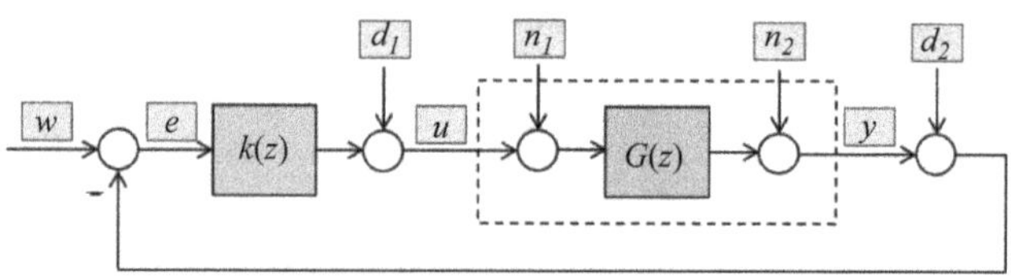

Figure A.1 Control diagram of the close-loop system

$$G_{ud_1} = (I + KG)^{-1} \qquad G_{ud_2} = -K(I + GK)^{-1} \tag{A.1}$$

$$G_{un_1} = -KG(I + KG)^{-1} \qquad G_{un_2} = -K(I + GK)^{-1} \tag{A.2}$$

$$G_{yd_1} = G(I + KG)^{-1} \qquad G_{yd_2} = -GK(I + GK)^{-1} \tag{A.3}$$

$$G_{yn_1} = G(I + KG)^{-1} \qquad G_{yn_2} = (I + GK)^{-1} \tag{A.4}$$

$$u = G_{ud_1}(z)d_1 + G_{ud_2}d_2(z) + G_{uv_1}(z)n_1 + G_{uv_2}(z)n_2 \tag{A.5}$$

$$y = G_{yd_1}(z)d_1 + G_{yd_2}d_2(z) + G_{yv_1}(z)n_1 + G_{yv_2}(z)n_2 \tag{A.6}$$

$$\begin{bmatrix} u \\ y \end{bmatrix} = \begin{bmatrix} G_{ud_1} & G_{ud_2} & G_{un_1} & G_{un_2} \\ G_{yd_1} & G_{yd_2} & G_{yn_1} & G_{yn_2} \end{bmatrix} \begin{bmatrix} d_1 \\ d_2 \\ n_1 \\ n_2 \end{bmatrix} := G_{cl}(z) \begin{bmatrix} d_1 \\ d_2 \\ n_1 \\ n_2 \end{bmatrix} \tag{A.7}$$

$$G(z) = N(z)M^{-1}(z) \qquad K(z) = \hat{V}^{-1}(z)\hat{U}(z) \tag{A.8}$$

$$\delta_{cl} = ||G_{cl}(z)||_{\infty}^{-1} \tag{A.9}$$

$$= \left\| \begin{bmatrix} G \\ I \end{bmatrix} (I + KG)^{-1} \begin{bmatrix} K & I \end{bmatrix} \right\|_{\infty}^{-1} \tag{A.10}$$

$$= \left\| \begin{bmatrix} N \\ M \end{bmatrix} (\hat{V}M + \hat{U}N)^{-1} \begin{bmatrix} \hat{U} & \hat{V} \end{bmatrix} \right\|_{\infty}^{-1} \tag{A.11}$$

$$= ||\mathscr{I}_G \mathscr{K}_K||_{\infty}^{-1} \tag{A.12}$$

The audience are recommended to refer to literature [58,59] for more details.

References

[1] Luo Z. Robotics, Automation, and Control in Industrial and Service Settings. Hershey, PA: IGI Global; 2015.

[2] Robla-Gomez S, Becerra VM, Llata JR, *et al.* Working together: a review on safe human-robot collaboration in industrial environments. IEEE Access. 2017;5:26754–26773.

[3] Pettersson M, Olvander J. Drive train optimization for industrial robots. IEEE Transactions on Robotics. 2009;25(6):1419–1424.

[4] Jatta F, Legnani G, Visioli A. Friction compensation in hybrid force velocity control of industrial manipulators. IEEE Transactions on Industrial Electronics. 2006;53(2):604–613.

[5] Kuljaca O, Swamy N, Lewis FL, *et al.* Design and implementation of industrial neural network controller using backstepping. IEEE Transactions on Industrial Electronics. 2003;50(1):193–201.

[6] Xie X, Lam J. Guaranteed cost control of periodic piecewise linear time-delay systems. Automatica. 2018;94:274–282.

[7] Xie X, Lam J, Li P. Robust time-weighted guaranteed cost control of uncertain periodic piecewise linear systems. Information Sciences. 2019;460–461:238–253.

[8] Xie W, Bonis I, Theodoropoulos C. Data-driven model reduction-based nonlinear MPC for large-scale distributed parameter systems. Journal of Process Control. 2015;35:50–58.

[9] Lauri D, Rossiter J, Sanchis J, *et al.* Data-driven latent-variable model-based predictive control for continuous processes. Journal of Process Control. 2010;20(10):1207–1219.

[10] Kouro S, Cortes P, Vargas R, *et al.* Model predictive control – a simple and powerful method to control power converters. IEEE Transactions on Industrial Electronics. 2009;56(6):1826–1838.

[11] Jiang Y, Yin S, Kaynak O. Data-driven monitoring and safety control of industrial cyber-physical systems: basics and beyond. IEEE Access. 2018;6(1):47374–47384.

[12] Gao Z, Liu X, Chen MZQ. Unknown input observer-based robust fault estimation for systems corrupted by partially decoupled disturbances. IEEE Transactions on Industrial Electronics. 2016;63(4):2537–2547.

[13] Gao Z. Fault estimation and fault-tolerant control for discrete-time dynamic systems. IEEE Transactions on Industrial Electronics. 2015;62(6):3874–3884.

[14] Hwang I, Kim S, Kim Y, *et al.* A survey of fault detection, isolation and reconfiguration methods. IEEE Transactions on Control Systems Technology. 2010;18(3):636–653.

[15] Chen H, Jiang B, Ding SX, *et al.* Probability-relevant incipient fault detection and diagnosis methodology with applications to electric drive systems. IEEE Transactions on Control Systems Technology. 2018:1–8. Available from: 10.1109/TCST.2018.2866976.

[16] Naik AS, Yin S, Ding SX, *et al.* Recursive identification algorithms to design fault detection systems. Journal of Process Control. 2010;20(8):957–965.

[17] Park J, Hur J. Detection of inter-turn and dynamic eccentricity faults using stator current frequency pattern in IPM-type BLDC motors. IEEE Transactions on Industrial Electronics. 2016;63(3):1771–1780.

[18] Yin S, Rodriguez J, Jiang Y. Real-time monitoring and control of industrial cyberphysical systems with integrated plant-wide monitoring and control framework. IEEE Industrial Electronics Magazine. 2019;13(4):38–47.

[19] Henao H, Capolino G, Fernandez-Cabanas M, *et al.* Trends in fault diagnosis for electrical machines: a review of diagnostic techniques. IEEE Industrial Electronics Magazine. 2017;8(2):31–42.

[20] Wang J, Qin SJ. Closed-loop subspace identification using the parity space. Automatica. 2006;42(2):315–320.

[21] Luo H, Zhao H, Yin S. Data-driven design of fog computing aided process monitoring system for large-scale industrial processes. IEEE Transactions on Industrial Informatics. 2018;14(10):4631–4641.

[22] Stankovic M, Stankovic S, Stipannovic D. Consensus-based decentralized real-time identification of large-scale systems. Automatica. 2015;60:219–226.
[23] Jiang Y, Yin S. Recent advances in key-performance-indicator oriented prognosis and diagnosis with a MATLAB toolbox: DB-KIT. IEEE Transactions on Industrial Informatics. 2018. Available from: 10.1109/TII.2018.2875067.
[24] Chen H, Jiang B, Lu N. A newly robust fault detection and diagnosis method for high-speed trains. IEEE Transactions on Intelligent Transportation Systems. 2018:1–11. Available from: https://doi.org/10.1109/TITS.2018.2865410.
[25] Muradore R, Fiorini P. A PLS-based statistical approach for fault detection and isolation of robotic manipulators. IEEE Transactions on Industrial Electronics. 2012;59(8):3167–3175.
[26] Ferdowsi M, Benigni A, Lowen A, *et al.* A scalable data-driven monitoring approach for distribution systems. IEEE Transactions on Instrumentation and Measurement. 2015;64(5):1292–1305.
[27] Hirzinger G, Bals J, Otter M, *et al.* The DLR-KUKA success story: robotics research improves industrial robots. IEEE Robotics & Automation Magazine. 2005;12(3):16–23.
[28] Garcia E, Jimenez MA, Santos PGD, *et al.* The evolution of robotics research. IEEE Robotics & Automation Magazine. 2007;14(1):90–103.
[29] Sant I, Pedret C, Vilanova R, *et al.* Advanced decision control system for effluent violations removal in wastewater treatment plants. Control Engineering Practice. 2017;49(1):60–75.
[30] Ding J, Modares H, Chai T, *et al.* Data-based multiobjective plant-wide performance optimization of industrial processes under dynamic environments. IEEE Transactions on Industrial Informatics. 2016;12(2):454–465.
[31] Vijaykumar S, Dsouza A, Schaal S. Incremental online learning in high dimensions. Neural Computing. 2005;17(12):2602–2634.
[32] Klanke S, Vijaykumar S, Schaal S. A library for locally weighted projection regression. The Journal of Machine Learning Research. 2008;9:623–626.
[33] Kohn W, Zabinsky Z, Nerode A. A micro-grid distributed intelligent control and management system. IEEE Transactions on Smart Grid. 2015;6(6):2964–2974.
[34] Zhang B, Du H, Lam J, Zhang N, Li W. A novel observer design for simultaneous estimation of vehicle steering angle and sideslip angle. IEEE Transactions on Industrial Electronics. 2016;63(7):4357–4366.
[35] Palanisamy M, Modares H, Lewis FL, *et al.* Continuous-time Q-learning for infinite-horizon discounted cost linear quadratic regulator problems. IEEE Transactions on Cybernetics. 2015;45(2):165–176.
[36] Hwang KS, Lin JL, Yeh KH. Learning to adjust and refine gait patterns for a biped robot. IEEE Transactions on Systems, Man, and Cybernetics: Systems. 2015;45(12):1481–1490.
[37] Lee JY, Park JB, Choi YH. Integral reinforcement learning for continuous-time input-affine nonlinear systems with simultaneous invariant explorations. IEEE Transactions on Neural Networks and Learning Systems. 2015;26(5):916–932.

[38] Luo B, Liu D, Huang T, *et al.* Model-free optimal tracking control via critic-only Q-learning. IEEE Transactions on Neural Networks and Learning Systems. 2016;27(10):2134–2144.
[39] Sira-Ramirez H, Linares-Flores J, Garcia-Rodriguez C, *et al.* On the control of the permanent magnet synchronous motor: an active disturbance rejection control approach. IEEE Transactions on Control Systems Technology. 2014;22(5):2056–2063.
[40] Castaneda LA, Luviano-Juarez A, Chairez I. Robust trajectory tracking of a delta robot through adaptive active disturbance rejection control. IEEE Transactions on Control Systems Technology. 2015;23(4):1387–1398.
[41] Zhang P, Ding SX. Disturbance decoupling in fault detection of linear periodic systems. Automatica. 2007;43(8):1410–1417.
[42] Huang B, Kadali R. Dynamic Modeling, Predictive Control and Performance Monitoring: A Data-driven Subspace Approach. Springer-Verlag, London; 2008.
[43] Yang Y, Ding S, Li L. Parameterization of nonlinear observer-based fault detection systems. IEEE Transactions on Automatic Control. 2016;61(11):3687–3692.
[44] Ding SX. Data-Driven Design of Fault Diagnosis and Fault-Tolerant Control Systems. Springer-Verlag, London; 2014.
[45] Luo H, Yin S, Liu T, *et al.* A data-driven realization of the control-performance-oriented process monitoring system. IEEE Transactions on Industrial Electronics. 2019. Available from: DOI: 10.1109/TIE.2019.2892705.
[46] Ding S, Yang G, Zhang P, *et al.* Feedback control structures, embedded residual signals, and feedback control schemes with an integrated residual access. IEEE Transactions on Control Systems Technology. 2010;18(2):352–367.
[47] Jiang Y, Yin S. Design approach to MIMO diagnostic observer and its application to fault detection. IEEE Annual Conference of Industrial Electronics Society. 2018.
[48] Stefano R, Marcello F, Giancarlo F. Plug-and-play decentralized model predictive control for linear systems. IEEE Transaction on Automatic Control. 2013;58(10):2608–2614.
[49] Bendtsen J, Trangbaek K, Stoustrup J. Plug-and-play control—modifying control systems online. IEEE Transaction on Control Systems Technology. 2013;21(1):79–93.
[50] Stoustrup J. Plug-and-play control: control technology towards new challenges. European Journal of Control. 2009;15(3):311–330.
[51] Kaelbling LP, Littman ML, Moore AW. Reinforcement learning: a survey. Journal of Artificial Intelligence Research. 1996;4:237–285.
[52] Lillicrap TP, Hunt JJ, Pritzel A, *et al.* Continuous control with deep reinforcement learning. 33rd International Conference on Machine Learning. 2016.
[53] Rottger M, Liehr A. Control task for reinforcement learning with known optimal solution for discrete and continuous actions. Journal of Intelligent Learning Systems and Applications. 2009;1:28–41.

[54] Williams RJ. Simple statistical gradient-following algorithms for connectionist reinforcement learning. Machine Learning. 1992;8(3–4):229–256.

[55] Duan Y, Chen X, Houthooft R, *et al.* Benchmarking deep reinforcement learning for continuous control. 33rd International Conference on Machine Learning. 2016;48.

[56] Huang B, Wu H, Huang T. Off-policy reinforcement learning for H_∞ control design. IEEE Transactions on Cybernetics. 2015;45(1):65–76.

[57] Sutton RS, Barto AG. Reinforcement Learning: An Introduction. The MIT Press, Cambridge, Massachusetts, London, England; 2018.

[58] Koenings T, Krueger M, Luo H, *et al.* A data-driven computation method for the gap metric and the optimal stability margin. IEEE Transactions on Automatic Control. 2018;63(3):805–810.

[59] Liu T, Luo H, Li K, *et al.* A data-driven method for SKR identification and application to stability margin estimation. 44th Annual Conference of the IEEE Industrial Electronics Society. 2018;p. 6223–6228.

[60] Gao Z, Ding SX, Cecati C. Real-time fault diagnosis and fault-tolerant control. IEEE Transactions on Industrial Electronics. 2015;62(6):3752–3756.

Chapter 13
Conclusions

Andrea Monteriù[1], Alessandro Freddi[1], and Sauro Longhi[1]

The aim of this book was to provide possible solutions to the problem of fault diagnosis and fault-tolerant control (FTC) of robotic and autonomous systems. A total of 44 international authors contributed to the book in the form of 12 chapters, describing both theoretical findings and challenging applications.

In detail, Chapters 1–5 covered the topics of fault diagnosis and FTC applied to unmanned vehicles, with a main focus on aerial and marine applications, which represent the most challenging ones from a diagnosis and control point of view. In Chapter 1, an active FTC strategy was proposed to accommodate actuator faults and model uncertainties for an unmanned quadrotor helicopter. The simulation results showed that the proposed scheme could effectively estimate and accommodate various kinds of actuator faults. Chapter 2 provided a tutorial solution to solve the tracking control problem for a birotor unmanned aerial vehicle (UAV): such a birotor configuration could be employed in those cases where a quadrotor UAV has experienced a failure. Simulations showed that the birotor could reach any position in the Cartesian space, and thus it could follow a safe emergency trajectory. Chapter 3 introduced the control method of the quad tilt-rotor UAV in the transition procedure under rotor-tilt axle stuck fault. Numerical results showed the stability of the vehicle in the presence of fault. Chapter 4 presented an unknown input observer approach for fault and icing diagnosis and an accommodation procedure in UAVs equipped with a redundant suite of effectors and actuators. The fault diagnosis algorithm was validated in simulation on the model of an Aerosonde UAV. In Chapter 5, a fault-failure-tolerant control scheme for the wave adaptive modular vessel catamaran was proposed. Simulation results showed that the system was able to complete a complex track even under severe actuator faults and failures, such as lock-in-place for the azimuth angle or loss of efficiency of the thrust driver.

Chapters 6–8 moved the attention from unmanned vehicles to full-fledged mobile robots with manipulation capabilities. In Chapter 6, a fault-tolerant model-based control approach for the head subsystem of a TIAGo humanoid robot was presented. Simulation results showed that although the problem is not completely answered, the proposed methodology presented a first line of action toward a final fulfilling solution,

[1]Department of Information Engineering, Università Politecnica delle Marche, Ancona, Italy

upon which improvements can be made. Chapter 7 presented a distributed framework for detecting and isolating faults in multi-manipulator robotic systems: in this case, the theoretical findings have been corroborated with experimental results on a setup composed of three heterogeneous robots. Chapter 8 presented a nonlinear optimal control method for aerial robotic manipulators with flexible joints: this controller achieves the solution of the optimal control problem for the aerial robotic system under model uncertainties and external perturbations and has been validated in simulation.

Chapters 9 and 10 described two typical scenarios of how fault detection and isolation and FTC can cope with safety and autonomy of modern transportation systems, thus giving an idea on the future of autonomous transportation systems. Chapter 9 proposed a nonlinear active FTC system applied to an aircraft model: faults were decoupled from aerodynamic disturbances, such as vertical wind gusts, and simulations on a high-fidelity aircraft model showed reliability of the active fault-tolerant scheme. Chapter 10 proposed a reconfigurable trajectory tracking control design method for autonomous in-wheel electric vehicles. The efficiency of the proposed method was demonstrated in a real-data CarSim simulation, showing significant energy savings.

Chapters 11 and 12 proposed two common theoretical approaches to fault diagnosis and fault-tolerance in modern robotic and autonomous systems. Chapter 11 showed that efficient control reconfiguration architecture can be developed for LPV systems by using the dynamic nullspace of the plant model, and it was tested on an aircraft model as example. Finally, Chapter 12 faced the problem of FTC on a different perspective, namely, that of data-driven approaches. The chapter proposed a way to design intelligent learning-aided controllers under the framework of Youla parameterization, so that robust and fault-tolerant controllers can be designed and implemented in a modular and plug-and-play manner. The approach was applied to a simulated wheeled robot as example.

The techniques and applications considered in the book cannot clearly cover all the problems related to faults in the fields of robotics and autonomous systems; nevertheless, some general considerations emerge by analyzing the results and conclusions of each chapter.

First, model-based techniques were shown to be very effective for solving several fault-related problems in robots and autonomous systems, such as fault estimation, trajectory tracking in the presence of faults and uncertainty or retaining stability in the presence of actuation failure. Moreover, these approaches can be adapted to different systems as long as they share similar models. However, the complexity to tune observer and controller parameters makes it difficult to quickly deploy such techniques to real systems: this is an important aspect that should be taken into account in the near future, in order to facilitate the application of fault detection and diagnosis (FDD) and FTC techniques to everyday applications.

Second, as the complexity of the system to be diagnosed increases, the number of possible faults increases as well and the definition of FTC strategies becomes even more difficult. This is especially true when the problem is moved from a single robot to a fleet of robots, where diagnosis and tolerance are to be considered not only at agent level but also at fleet level. The development of FDD and FTC techniques that take into account single agents as well as fleets of robots is a very interesting (and open) field of research to be considered in the near future.

Third, the integration of data-driven approaches with model-based ones is a very promising way to cope with model mismatches and uncertainties: "learning from data" is a necessary step to adapt FDD and FTC techniques and increase both safety and autonomy of robotic systems. How to integrate such techniques, however, is still an open problem to solve.

Index

www.ingramcontent.com/pod-product-compliance
Lightning Source LLC
LaVergne TN
LVHW011847200926
848484LV00002B/409